FISHERIES MANAGEMENT

FISHERIES MANAGEMENT

EDITED BY

ROBERT T. LACKEY

AND

LARRY A. NIELSEN

Department of Fisheries and
Wildlife Sciences,
Virginia Polytechnic Institute and
State University, Blacksburg,
Virginia 24061

A HALSTED PRESS BOOK

JOHN WILEY & SONS

NEW YORK – TORONTO

© 1980 by
Blackwell Scientific Publications

All rights reserved. No part of this
publication may be reproduced, stored in
a retrieval system, or transmitted,
in any form or by any means,
electronic, mechanical, photocopying,
recording or otherwise without the
prior permission of the copyright owner.

First published 1980

Published in the U.S.A. and Canada
by Halsted Press,
a Division of John Wiley & Sons, Inc.,
New York

Library of Congress
Cataloging in Publication Data

Fisheries management.

  Includes index.
  1. Fishery management.   I. Lackey, Robert T.
II. Nielsen, Larry A.
SH328.F56   333.95′6   80–20028
ISBN 0–470–27056–X

Printed in the
United States of America

Contents

v

PRINCIPLES OF MANAGEMENT

MANAGEMENT OF FISHERIES

Preface

The objective of this book is to present a broad discussion of fisheries management for use by students as a textbook and by professionals as a reference. In this sense, we felt that the knowledge and experience represented by the 16 authors, all experts in their topics, far surpassed the possible inconvenience of an edited text. The diligence and cooperation of the authors have shown this to be the case from the viewpoint of the editors, and we are confident that readers will agree.

Since the time of preparation of the text, two authors have changed their professional affiliations. Robert T. Lackey is with the National Water Resources Analysis Group, U.S. Fish and Wildlife Service, Kearneysville, West Virginia, and Michael K. Orbach has moved to the Center for Coastal and Marine Studies, University of California at Santa Cruz. Dora R. May Passino, of the Great Lakes Fishery Laboratory, U.S. Fish and Wildlife Service, Ann Arbor, Michigan, and Adam A. Sokoloski both prepared their chapters on personal time and at private expense.

Several authors wish to express special thanks to those individuals which contributed directly to the preparation of their chapters. Charles E. Warren and William J. Liss acknowledge the careful review of Chapters 2 and 3 by Terry Finger, Hiram Li, Becky McClurken, Dale McCullough, and Marion Ormsby. John Cairns, Jr. acknowledges the help of Darla Donald in the editing and preparation of Chapter 10 and Betty Higginbotham and Angela Miller for the final typed copy. Peter A. Larkin recognizes the contributions of Carl Walters and Helen Hahn to the preparation of Chapter 11. Vincent F. Gallucci wishes to thank M. Griben, T. Quinn, II, and C. Rawson, who enthusiastically criticized his contribution to Chapter 5, and P. Smith, who make several useful comments. Ms. C. Anderson cheerfully endured the retypings that result when the authors are from different universities. Chapter 5 was partially supported by the Washington Sea Grant Program and the Oregon State University Sea Grant Program. Chapter 14, by J. L. McHugh, is Contribution 268 from the Marine Sciences Research Center, State University of New York, Stony Brook.

The editors wish to thank T. Thorn, who prepared the index. They especially thank Dr. Gerald H. Cross, Department Head of Fisheries and Wildlife Sciences at Virginia Tech, and Mr. Robert Campbell, editor for Blackwell Scientific Publications, for their encouragement throughout the project.

L.A.N.

Blacksburg, Virginia   R.T.L.

CHARACTERISTICS OF FISHERIES

Chapter 1
Introduction

LARRY A. NIELSEN AND ROBERT T. LACKEY

1.1 PERSPECTIVE

If the average citizen were asked to picture a fisheries scientist, he probably would imagine a biologist who spent his time collecting fish, identifying them, and describing their life histories. This caricature, however, is a fairly accurate impression of an ichthyologist, not a fisheries scientist; the study of fish is only part of the broader study of fisheries. In most universities, the study of fisheries is conducted within a College of Agriculture or Natural Resources rather than in a biology or life sciences department. Programs in agricultural sciences character-istically address such topics as economics, farm and ranch management, processing of animal and plant products, and techniques for efficient harvest as well as biologically-oriented topics such as genetics, nutrition, and disease. Similarly, a fisheries manager not only deals with fish and their biological characteristics, but also must consider such things as who will use the resource, how much it will cost to manage effectively, and how environmental changes will affect the resource.

A fishery may be defined in this broad sense as a system composed of three interacting components: the aquatic biota, the aquatic habitat, and the human users of these renewable natural resources. Under this definition, we find a diversity of systems ranging from whale fisheries in the open ocean to brook trout fisheries in isolated alpine lakes, from subsistence fisheries for seal in northern Canada to fisheries for aquarium species in the tropics. Each component of a fishery influences how that fishery performs, and understanding the entire system and all its parts is essential to the successful management of a fishery.

1.2 AQUATIC BIOTA

Most fisheries are classified in terms of the product harvested, that is, the target species, for human use (Rounsefell 1975). Thus we speak of fishing for salmon, trawling for cod, or spearing bullfrogs. The types of organisms which are used directly or indirectly by humans cross virtually all taxonomic categories.

Whereas the human consumption of aquatic organisms in North America is limited to a relatively small number of fishes and shellfishes such as flatfish, salmon, tuna, shrimp, and oysters, diets in many cultures include a large variety of aquatic organisms. In Japan, where annual consumption of fisheries products approaches 40 kg per person, the weekly menu for a typical family may include several types of finfish, shrimp, squid, sea urchin, and kelp.

Beyond their uses as harvestable products, however, aquatic organisms affect in many ways the nature of a fishery. The type and number of organisms which make up the aquatic food web partially will determine how many target organisms are produced and can be harvested. Natural competitors, predators, and parasites may reduce the abundance of target organisms or alter the size and quality of the desired products. The increasing emphasis on maintaining rare species provides another dimension in which aquatic organisms affect decisions regarding fisheries management. Approximately 100 aquatic species are considered endangered or threatened, and that number promises to grow in the future.

Fig. 1.1. Aquatic resources are typically categorized by target species: *top left* Northern pike, a popular recreational fish; *top right* loggerhead turtle, a species threatened by extinction; *bottom left* blue crab, a valuable crustacean in coastal fisheries; *bottom right* oyster, a mollusc harvested by commercial and recreational fishermen. Top left photograph by courtesy of U.S. Forest Service, remainder by courtesy of Fish and Wildlife Service.

The so-called finfishes are vertebrates in classes Agnatha, Chondrichthyes, and Osteichthyes and are the most common target species of traditional fisheries. Finfishes range from species such as menhaden, which are of low value and are used for industrial products, to highly valued species such as tuna and salmon. Recreational uses of the aquatic biota are centered around the finfishes, which accounts for the restricted perspective from which many Americans view fisheries. Angling in freshwater is aimed primarily at a few species of predatory fish, principally sunfishes, bass, perch, trout, and catfish. The diversity of finfishes in marine habitats is much higher, and the variety of the angling catch also is greater.

Certain species of mammals are the targets of highly-publicized fisheries. Catches of whales, dolphins, seals, and sea otters were much greater before 1900 than at present, both because of great declines in abundance and because substitutes for the products of such fisheries are now available. Emotional attachments to mammalian species and competition among nations for their capture, however, foster continued interest out of proportion with the current commercial value of these fisheries.

Crustacea constitutes perhaps the most important group of fisheries organisms. The larger forms, including shrimps, lobsters, crayfish, and crabs, are heavily exploited for human consumption. While the total catch by weight is much lower than that of finfishes, the commercial value of all crustaceans landed in the United States is approximately equal to the value of all landings of finfishes. In addition, the smaller forms of crustaceans are members of most aquatic food chains leading to harvestable products. The 'krill' which are strained by baleen whales are crustaceans, as are the cladocerans and copepods which are eaten by most finfishes.

Mollusca forms a commercially important group which illustrates the breadth of uses from aquatic resources. Clams, oysters, squid, and octopus are examples of molluscs which are highly valued for human consumption. The value of commercial landings of molluscs in the United States is about one-third the value of crustaceans. Prior to extensive use of plastics, the shells of molluscs were the raw material for the production of buttons and other decorative ornaments. Bivalue molluscs also are the source of natural pearls, and cultivation of molluscs for pearl production is an important industry. The collection and marketing of shells remains an important recreational activity and tourist industry in coastal resorts.

Representatives of most other phyla, including aquatic plants, have some commercial and recreational importance, directly or indirectly. Many fisheries are of little importance on a world-wide scale, but may be important in localized areas where density of an organism is high enough to permit harvest and where a local market exists. Many types of aquatic organisms which presently are not utilized are likely to become targets of fisheries as demands for food and other products become greater with increasing world population. The great diversity of the aquatic biota is a valuable resource of itself, and exploitation of that diversity will increase as knowledge of the distribution, abundance, and usefulness of various organisms increases.

1.3   AQUATIC HABITAT

The aquatic habitat includes the non-living aspects of the aquatic ecosystem. The study of aquatic habitats is the subject of disciplines called physical limnology (freshwater) and oceanography (marine), and understanding gained from these fields is essential to the proper management of fisheries resources. While it is tempting to assume that our experience with terrestrial habitats is sufficient to guide management of aquatic habitats, we must realize that the aquatic habitat is foreign to land animals and adaptations to aquatic habitats usually are quite different than adaptations observed on land.

Aquatic habitats may be classified by a variety of criteria including salinity, size, depth, origin, and rates of water flow. On the basis of salinity, waters are described as freshwater if the concentration of dissolved ions is lower than 0.5 parts per thousand, saltwater if the concentration is higher than 35 parts per thousand, and estuarine or brackish in between those concentrations. Whereas the salinity of most marine systems is very similar, fresh and brackish waters vary greatly in concentration of ions, depending on location, watershed use, and season.

Freshwater systems can be divided into standing waters, including lakes, ponds, and reservoirs, and flowing waters, including streams and rivers. Natural lakes are formed in many ways, but most in the northern hemisphere were formed by glacial activity within the last 20,000 years. Glacial lakes contain primarily coldwater or coolwater fish species and, because the lakes are young in an evolutionary sense, the same species occur across broad geographic regions. Some large lakes, such as Lake Baikal in Siberia and Lake Victoria in Africa, are over one million years old and contain many species which are found only in these lakes.

Standing waters are created by man when watersheds are dammed; the impounded waters can be defined as ponds if smaller than two hectares and reservoirs if larger than two hectares. More than four million ponds exist in the United States, mostly in agricultural areas where they have been constructed to aid in soil and water conservation. Although ponds are common parts of the rural landscape, their management for fisheries resources is frequently restricted by small size, private ownership, poor water quality, and other conflicting uses. In Asia, aquaculture is practiced in small ponds which are stocked with various species of carp and are fertilized with wastes from other domestic animals.

Reservoirs occur when relatively large streams or rivers are dammed for controlling floods, producing power, irrigating farmland, or establishing domestic water supplies. Reservoirs generally are constructed to serve several purposes, and the production of fish is usually of secondary importance. Nevertheless, almost one-third of the total recreational fishing in North America occurs in reservoirs, mostly on those larger than 200 hectares. Large reservoirs in Europe generally are managed for the production of commercial crops. The habitat in reservoirs is intermediate between a lake and a stream, which provides both great opportunity and difficulty for management.

Flowing waters are divided loosely into streams or rivers on the basis of size and location. Rivers are larger flowing waters which typically discharge into the marine environment. Throughout history, rivers have been an important resource for man, providing transportation, sewage disposal, and water supply for riverside communities. A quick glance at a map will reveal that most major cities lie along rivers and that most rivers are densely settled with cities of all sizes. Fisheries in rivers frequently suffer because industrial and other uses render the water unsuitable for fish life. Aside from residential species, most rivers contain migratory fish populations which use the freshwater habitat for spawning and for rearing of young fish and spend their adult life in the ocean.

Streams generally are small and drain the upper sections of watersheds. Because streams carry little water and may contain water for only part of the year, they are not used for the same variety of purposes as rivers. Streams provide variety to the landscape and contain ideal habitats for trout and other species important for recreational fishing. More than any other aquatic system, a stream is intimately associated with the surrounding watershed; water level and quality is extremely variable, depending on weather and land use patterns.

Estuaries are the boundaries between freshwater and marine systems and are extremely important from a fisheries standpoint. All migratory species, including salmons, eels, and striped bass, pass through estuaries on their journey between fresh and salt water. Because estuaries receive all the nutrients carried by river water and because they are shallow and the water is well-mixed, the production of aquatic organisms is very great. This high productivity supports valuable fisheries for many aquatic invertebrates and is ideal for the rapid growth of young fish which will later migrate to marine habitats. Estuaries also are important for human commerce, and major cities exist on the shores of most estuaries. The impact of dense human population often conflicts with the ability of an estuary to remain a suitable habitat for fisheries.

Marine systems may be divided arbitrarily into coastal habitats, where water is shallower than 200 meters, and open ocean habitats. Coastal habitats occur adjacent to continental land masses and encompass traditional fisheries areas such as the Grand Banks near Newfoundland, the North Sea, and the Barents Sea. Coastal habitats are similar to estuaries because they are relatively shallow,

Fig. 1.2. The aquatic habitat is diverse.

FACING PAGE *top* open oceans are important for many purposes, only one of which is fishing; *bottom* estuaries are the 'nurseries' of major fish populations;

THIS PAGE *top* rivers and streams are important for many reasons, not the least of which is as a corridor for migratory fishes; *bottom* lakes, ponds, and reservoirs support many different types of fishing activities.

Photographs courtesy of U.S. NOAA (facing page top), Fish and Wildlife Service (facing page bottom), National Park Service (this page top), and Soil Conservation Service (this page bottom).

have high inputs of nutrients from the adjacent continents, and are very productive. They also are vulnerable to disturbance by the intensity of human activity along the coast.

Despite the fact that oceans cover three-quarters of the earth's surface, the productivity of fisheries resources from the open oceans is rather low. The open ocean sometimes is described as an aquatic desert in which low concentration of essential nutrients is the limiting factor. The great depth of the oceans prohibits the circulation of nutrients to the upper waters where photosynthesis can occur. The difficulty of locating and catching fish in the open ocean also makes most fisheries impractical. Large species such as whales and tuna can be harvested profitably, but only by very modern fleets utilizing advanced technology.

1.4   HUMAN USES

The third component of a fishery consists of the uses and users of the resources. More than any other topic in fisheries management, consideration of the human component is generally ignored. Most fisheries scientists and managers are oriented toward the biological sciences and find study of human aspects difficult or irrelevant. This attitude impedes the management of natural resources since economic, political, or social realities may overrule the wisest plans for biological or habitat management.

Commercial uses of fisheries can generally be categorized as either large industrial or small localized operations. The large industrial fisheries are characteristic of the more developed nations which apply modern technology and large capital expenditures to the capture or culture of fisheries organisms. Such fisheries utilize highly-sophisticated networks including airplanes and radar to locate fish and a diversity of ships which capture fish, store and process the raw products, and transport finished products directly to markets. Within lesser developed countries, fisheries are pursued on a smaller scale. Catches are intended for use by the fisherman and his family group (subsistence fisheries) or for a local market. Techniques used in subsistence fisheries rely heavily on manual labor, and methods may have changed little over thousands of years.

Use of aquatic systems for public recreation tends to occur only in industrialized nations. Likewise, concern for aquatic organisms and environments is restricted to areas with very high standards of living. In Europe and especially in North America, the indirect economic benefits associated with recreational fishing are substantial and lead to competition between recreational and commercial uses of particular fisheries.

1.5   SPECIAL ATTRIBUTES OF FISHERIES

The many-faceted nature of fisheries and fisheries management creates an exciting environment for the professional manager. Opportunities for progress in management of biota, habitat, and users exist if the fisheries professional is willing to augment his own capabilities with the tools and approaches of many disciplines. The opportunities of any diverse management system, however, are accompanied by a set of factors which work against effective management. Fisheries management is no exception.

Fisheries are parts of complex and unpredictable ecosystems. Floods, droughts, temperature cycles, and fluctuations in the abundance of organisms are examples of factors which are uncontrollable and often occur with little or no warning. When scores of such factors operate within a single ecosystem, the result may resemble a kaleidoscope—ever-changing and never predictable. Even for the few fisheries systems which have been well studied, the predictive capability of fisheries managers is disappointingly poor. In such cases, we speak of *random* or *stochastic* processes, which is just another way of stating that our understanding is insufficient to allow prediction.

Aquatic systems also are highly *dynamic*. In this sense, dynamic means that many things are occurring which may not be detectable when the system is observed at a single instant. The number of fish in a pond may be about the same in April of each year, but during the year, some fish have hatched, grown, died, and been caught. Each of those dynamic processes may be affected by the environment, the abundance of fish, and each other. These processes also tend to reduce the influence of changes to a system, whether such changes are undesirable, as in the case of a spill of pollutant, or desirable, as when a fisheries manager tries to improve fishing. Thus, dissecting the system into its component parts is essential to understanding how the system might respond to different conditions.

Added to the difficulties of a complex and dynamic ecosystem is the inescapable fact that aquatic ecosystems are *foreign* to man. Men are terrestrial creatures who can directly observe aquatic systems in superficial ways only. The wealth of literature, art, and mythology which relates to the ocean and its creatures is indicative of the aura of mystery that surrounds the aquatic habitat. Because men cannot experience the habitat as a fish, whale, or clam would, fisheries scientists must rely on indirect observations. We lower nets, bottles, thermometers, and cameras into the water and attempt to explain what is happening and why with the help of statistical analysis. While statistics is a helpful tool for analyzing data, statistics cannot decide whether the data are *accurate*, that is, representative of the true situation. In terrestrial ecosystems, the accuracy of data is evaluated by comparing sampling data to direct human observations. In aquatic systems, however, such comparisons usually are impossible and the accuracy of fisheries data must be regarded with caution.

When we consider the human component of fisheries management, we confront the problem of defining and measuring the benefits which man receives from management. Clearly, the users of fisheries systems (that is, the fishermen) receive benefits, but how should these benefits be measured? Should we measure the numbers or pounds of fish caught, dollars in the pocket, or amount of recreational use, and how should we compare these? Beyond this is the larger question of defining all the potential beneficiaries of the fishery. Fishermen are only a portion of society; benefits and costs accrue to people who may never catch, eat, or see a fish, but may be for or against recreational and commercial fishing on philosophical grounds. In a world where the availability of natural resources is decreasing and the number of people is increasing, conflict among users of fisheries is inevitable. While it may be discomforting to think of comparing the existence of an endangered species with the production of hydroelectric power, control of floods, or creation of recreational fishing, such comparisons must be made.

The distillation of all these attributes reveals one last characteristic of fisheries management: decisions must be made in the presence of *uncertainty*. There never really is enough information to make a totally confident decision. Also, the quality of information will vary from subject to subject; predictions of how many fish will be in a lake may be very precise while the question of how many fishermen will pursue them may have only a vague or approximate answer. It is an unsettling reality for fisheries managers, most of whom have been trained as biological scientists, to make decisions based on imperfect information. In the broad sense, however, public fisheries are perfect miniatures of society as a whole and are impacted by all the factors which affect all public activities.

The obvious corollary to the above characteristics is that success in fisheries management in large part depends on the personal experience, intuition, and luck of the manager. Indeed, the question of whether fisheries management is an art or a science is a favorite subject for debate by fisheries professionals (professors usually argue that management is science, managers usually argue for art!). While no one can debate the fact that experience improves judgment, the problem, in the words of an old axiom, is that 'experience is a great thing, but it always comes just after you need it.' The scientific principles provided by the study of the natural and the social sciences are the tools with which the creative manager can effectively manage a fishery.

The use of multiple criteria for evaluating success in fisheries management programs has created many possibilities for innovative programs. The basic ideas of population dynamics have flourished to provide better understanding of how organisms respond to fishing. Availability of computers and sophisticated hand-calculators invites every manager to consider complex questions and frees him from many tedious tasks formerly associated with the job. Most nations of the world have extended their areas of coastal jurisdiction for fisheries purposes to 200 miles; this change alone has drastically altered the nature of fisheries management. The environmental movement has brought fisheries management under the scrutiny of a bigger public with different priorities.

Whether the enlarged scope for fisheries management will encourage or discourage future fisheries programs cannot be predicted. When placed within a systems context, a subsystem such as fisheries may be an inconsequential or even detrimental component in the overall goal of improving the quality of life for people. The probability of this is small, but the probability is large that fisheries as we now know them will be changed greatly in the future. Having meaningful inputs into that future is the goal of fisheries management.

1.6   ORGANIZATION OF BOOK

This book is divided into three sections: *Characteristics of Fisheries, Principles of Fisheries Management*, and *Management of Fisheries.*

The *first* section of this book introduces the components of a fishery. The study of fisheries is a synthetic science, calling upon many disciplines to provide the necessary background for management. Chapters 2–6 describe the principles of aquatic ecology, the life histories of important fisheries organisms, the mechanics of describing fish populations, and the social-political-economic realm in which users of fisheries function.

The *second* section of the book addresses the principles which apply to the management of fisheries. Chapters 7–10 describe the systems approach, and the subsystems concerning economic, environmental, and policy questions. Chapter 11 addresses management objectives, providing a synthesis of the previous ten chapters.

The *third* section of the book casts the principles and features of previous sections in the context of management for major habitats. Variation in the characteristics of particular habitats and the way fisheries in these habitats are used creates unique management situations and strategies for each habitat type. Aquaculture is included in this section because the culture of fish occurs in a highly-controlled habitat which is uniquely different than the natural habitats of fisheries management.

1.7   REFERENCES

Bennett G.W. (1971) *Management of Lakes and Ponds*, 2nd edn. New York: Van Nostrand Reinhold.

Benson N.G. (1970) *A Century of Fisheries in North America*. Bethesda, Maryland: Spec. Publ. No. 7, Am. Fish. Soc.

Calhoun A. (1966) *Inland Fisheries Management*. Sacramento: Calif. Dept. Fish Game.

Carlander K.D. (1969) *Handbook of Freshwater Fishery Biology, Vol. I*. Ames, Iowa: Iowa State Univ. Press.

Carlander K.D. (1977) *Handbook of Freshwater Fishery Biology, Vol. II*. Ames, Iowa: Iowa State Univ. Press.

Cushing D.H. (1968) *Fisheries Biology*. Madison: Univ. Wisconsin Press.

Everhart W.H., A.W.Eipper, and W.D.Youngs (1975) *Principles of Fishery Science*. Ithaca, New York: Cornell Univ. Press.

Frey D.G. (ed.) (1966) *Limnology in North America*. Madison: Univ. Wisconsin Press.

Gerking S.D. (ed.) (1967) *The Biological Basis of Freshwater Fish Production*. Oxford: Blackwell Scientific Publications.

Gulland J.A. (1974) *The Management of Marine Fisheries*. Seattle: Univ. of Washington Press.

Hynes H.B.N. (1970) *The Ecology of Running Waters*. Toronto: Univ. of Toronto Press.

Lackey R.T. (1974) *Introductory Fisheries Science*. Blacksburg, Virginia: Sea Grant, Virginia Polytechnic Institute and State University.

Lagler K.F. (1956) *Freshwater Fishery Biology*. Dubuque, Iowa: Wm.C.Brown.

Reid G.K. (1962) *Ecology of Inland Waters and Estuaries*. New York: Van Nostrand Reinhold.

Rounsefell G.A. (1975) *Ecology, Utilization, and Management of Marine Fisheries*. St Louis: C.V.Mosby.

Royce W.F. (1972) *Introduction to the Fishery Sciences*. New York: Academic Press.

Warren C.E. (1971) *Biology and Water Pollution Control*. Philadelphia: W.B.Saunders.

Chapter 2
Adaptation to Aquatic Environments

CHARLES E. WARREN AND WILLIAM J. LISS

2.1 THE STUDY OF ADAPTATION

The individual spring chinook salmon is adapted to move from its feeding areas in the Pacific Ocean into the Columbia River in January or February. Then it moves up-river into the Snake and Salmon Rivers, finally to spend the summer in some headwater stream in the mountains of Idaho. There in the fall it will spawn and die. Its progeny emerge from the gravel and eventually descend to the sea to complete the cycle. The individual salmon is adapted in many ways to complete its own life history. Indeed, its life history pattern as a whole is adapted to the kinds of places, or habitats, and environmental conditions occurring where the salmon is in the different periods of its life: the individual salmon is adapted to persist. But it is also adapted to reproduce, to have a function or play a role in the persistence of its population. Thus, through the life histories and reproduction of its individual organisms, the spring chinook population is adapted to persist. And in this, the salmon population must depend upon the stream communities, the estuarine communities, and the marine communities to which it is adapted and within which it plays various roles, as predator, as prey, and in other known and in unknown ways.

Knowledge of the adaptation of fish and other aquatic organisms is the foundation of fisheries science and management, a foundation we must continue to learn and extend. When we log the forests or irrigate the meadows or dam the rivers, we change the habitat and environment to which the spring chinook salmon was adapted. It has some life history and evolutionary capacity to adapt to such change, but we should know something of its limits. And when we troll in the sea and gill-net in the rivers, we are adding another mortality factor to the environment of the salmon, a factor to which the population must adapt if it is to persist. Successful management depends on our knowledge of this. Still, we seek knowledge of adaptation not simply because it is useful but because we desire to understand living organisms.

Although each kind of living organism like the spring chinook salmon is unique in its life history pattern of adaptation, all kinds of living organisms face rather common problems of staying in suitable habitat and persisting under

15

changing environmental conditions. Knowledge of adaptation cannot be simply a vast aggregation of facts, which alone are not even interesting. Rather we want to understand the principles according to which any sort of living organism adapts to habitat and environment. It is this kind of knowledge that gives meaning to all the facts that otherwise would be isolated. We intend what follows to be more of the nature of principles of adaptation than of isolated facts.

2.2   ORGANISMIC AND ENVIRONMENTAL SYSTEMS

2.2.1   The spaciotemporal extension of organismic systems

In response to the question 'What is an adaptive trait?', Dobzhansky (1956) noted that the adaptive value of any characteristic of an organism can be seen only in the context of its entire life history pattern. He was saying that, to explain and understand the characteristics of an organism, we should think of the organism not simply as any particular state but rather as the entire sequence of states from birth to death. The adaptive value of the 'chloride' secreting cells appearing in the gills of young salmon in freshwater can be understood only when it is known the fish will soon enter an estuary and the sea, where these specialized cells will function in osmoregulation. It is the entire life history pattern that is adapted for persistence and reproduction. The entire life of an organism is an event, in the sense of Whitehead (1927) an enduring object extending through time as well as space.

But we are also concerned with the adaptation of populations and even biological communities. We can define a population as a group of interbreeding individuals that have relatively little reproductive contact with individuals from other such groups of the same species. As with individual organisms, the population can be thought of as extending from its origin to its end, from colonization to extinction. With this concept, population growth, persistence, and evolution can be explained. A biological community is most usefully conceived of as extending from colonization through all the successional stages including the climax. For in this way, we have a biological community whose development can be explained. These three kinds of systems, individual organism, population, and biological community, each at a different level of organization, can be thought of as being *organismic systems*. Different though they be, they share certain abstract organismic properties important to their adaptation. Now, what about the environments to which these organismic systems must adapt?

2.2.2   The relativity and coextension of environmental systems

We cannot conceive of any natural organismic system as persisting except in some environment that provides it with space, time, information, energy, materials, and otherwise suitable conditions. Neither can we conceive of an environment except in relation to some system. This is so for the individual spring chinook salmon, it is so for its population, and it is so even for the different biological communities in which the salmon lives. Because we have defined an organismic system to extend from some original state to some terminal state, so also must the environment of that organismic system extend in space and time, as we have endeavored to show in Fig. 2.1. The environmental system of the

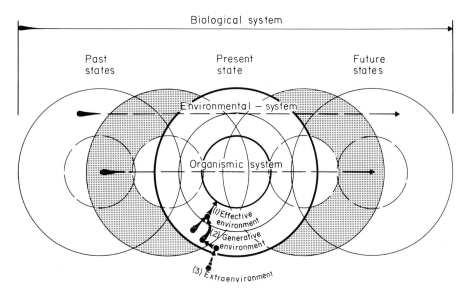

Fig. 2.1. A biological system viewed throughout its existence as being composed of an organismic system together with its spatially and temporally coextensive environmental system. In this view, an organismic system is not any one of its states but rather the continuum of states from its origin to its end. The environmental system is composed of an effective environment and a generative environment, which are shown only for one state but which extend as parts of the environmental system. After Warren, Allen and Haefner (1979). Reprinted from *Behavioral Science*, volume 24, No. 5, 1979, by permission of James Grier Miller, M.D., Ph.D., Editor.

individual salmon extends with it and affects the salmon throughout its life, from the gravel of its redd, through its stream and marine existence, until it returns again to spawn and die. In principle, the environmental systems of populations and communities are similarly correlative and coextensive, although they are more difficult to visualize, because populations and communities themselves are more difficult to visualize than are individual organisms.

Now what is this environmental system to which each organismic system must adapt? Environmental system is not an easy concept to grasp, because it is an abstract concept, a theoretical concept. That is we cannot directly and fully observe, measure, or otherwise evaluate an environment. We can know that some factor is in the environment of, say, an individual organism only by showing that the organism is affected by it. Thus we can show dissolved oxygen and food organisms to be factors in the environment of the salmon. All those factors or conditions that directly impinge with effect on the organism we can call the *effective environment*. All those sequences of factors causally leading to factors in the effective environment we can call the *generative environment*, as illustrated in Fig. 2.1 for one state of the environmental system. Mayflies that are eaten by juvenile salmon are in their effective environment; but the algae the mayflies eat and the nitrate the algae depend upon are in the generative environment of the salmon. Together, the effective and the generative phases of the environment form the *environmental system* that coextends with the individual organism and to which it must adapt throughout its life history. For conceptual and observational reasons, it is not desirable that we allow the generative environment to extend indefinitely to include 'everything.' So, at some vague boundary, we may wish to relegate all else of possible importance to an extra-environment (Fig. 2.1). Populations and communities have effective environments, generative environments, and thus coextensive environmental systems that are different from those of individual organisms but that can abstractly or theoretically be conceived of in much the same way.

2.2.3   The codetermination of organismic and environmental systems
If we are to learn about adaptation in the most general, interesting, and useful way, we must first grasp what is most fundamental to any organismic system. And then we must attempt to see how the fundamental nature of an organismic system is determined by and yet determines its environmental system. An organismic system such as a chinook salmon extends together with its environmental system from its origin to its end. At any stage in its life history, the individual salmon has a certain capacity—let us call it *realized capacity*—to do certain things: to feed, to grow, to migrate, to develop further, and so on. But just what it does depends not only on this realized capacity but also on the state of the environmental system at that time. At different times in its life, the salmon has the realized capacity to do different things, but not everything. As a parr or a smolt or a feeder in the sea, the salmon does not have the realized capacity to reproduce: yet, if it lives, it will develop this capacity, the environment permitting. Thus, even the parr in the stream has the *potential capacity* to reproduce, and from this potential capacity the realized capacity to reproduce can develop. Finally, environmental conditions being suitable in the stream, spawning performance will occur when four years later the salmon returns as an adult.

Figure 2.2 is an attempt to make these ideas very general for any sort of organismic system. From some original potential capacity, an organismic system develops through a sequence of realized capacities, according to the prevailing environmental system. Both environment and potential capacity determine the realized capacity—what the organismic system is capable of—at any time. And what the system actually does—what its *performance* is—at any time is determined by its realized capacity and the state of the environment at that time.

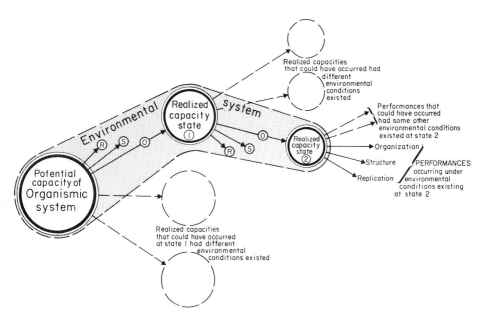

Fig. 2.2. A diagrammatic view of an organismic system showing that from some potential capacity the organismic system passes through a particular sequence of realized capacities jointly determined by its potential capacity and the actually prevailing environmental system. With different prevailing environmental systems, different sequences of realized capacities develop. For any realized capacity such as state 2, different sets of performances will occur in connection with different sets of environmental conditions. After Warren, Allen and Haefner (1979). Reprinted from *Behavioral Science*, volume 24, No. 5, 1979, by permission of James Grier Miller, M.D., Ph.D., Editor.

Thus, the environment is involved in determining what the realized capacity and performance of an organismic system will be. But the potential and realized capacities determine what could possibly be suitable environments for an organismic system, for they determine what is necessary for the system to persist and replicate. This is so for the population of spring chinook and for its freshwater and marine communities as well as for the individual chinook. Potential capacity, realized capacity, performance, and environmental relations are so fundamental to our understanding of life we can speak generally about them in terms of organismic systems.

2.2.4   Organismic incorporation of subsystems together with their environments

At any stage of its development, an organismic system *incorporates* not only its subsystems but their level-specific environments as well. This is not difficult to see, but its consequences are profound: biological adaptation, indeed life itself, would otherwise not be possible. The individual chinook salmon *incorporates* its organ systems, which in turn incorporate the organs, and so on down to cells and beyond. In so doing, the individual achieves its capacities and performances by

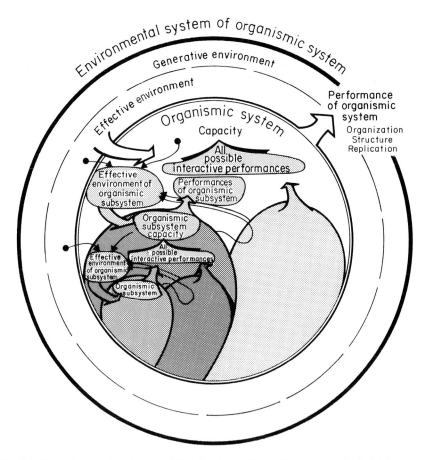

Fig. 2.3. Organismic subsystems and their level-specific environments, all of which are incorporated by an organismic system having its own particular environment. Any performance of the organismic system or any one of its subsystems is determined by its particular capacity and environment. After Warren, Allen and Haefner (1979). Reprinted from *Behavioral Science*, volume 24, No. 5, 1979, by permission of James Grier Miller, M.D., Ph.D., Editor.

'bringing together' in an organized way the capacities and performances of its subsystems. But it protects and controls these subsystems—these organ systems and so on—by incorporating or embodying them together with their *level-specific environments*. Thus the environment of the heart of the adult salmon in the sea is in no sense the sea. Rather, bathed in osmotically-regulated fluid in the pericardial sac and affected by other organs, the heart of the salmon has its own particular environment, through which the salmon as a whole controls the performance of the heart.

Now this commonplace example is not trivial. Particular though it may be, it is an example of a fundamental principle of *biological organization*: the incorporation of the capacities and the environments of organismic systems (Fig. 2.3). Thus, if we were to take the lowest level subsystem in Fig. 2.3 to be the chinook parr in the headwater stream, then the next higher level subsystem could be the population of parr, and the entire organismic system could be the stream community including all the species of stream organisms. This, then, would be an example of a *hierarchically organized* system.

2.3   ADAPTATION TO AN ENVIRONMENT

2.3.1   An environment as a system

We have now only to make very clear certain implications of the concept of environmental system before we can introduce a general and invariant definition of biological adaptation. At any time, the effective environment of an individual organism—or any other organismic system—consists of a set of environmental factor values, shown as $E1$ in Fig. 2.4a. Among these factors will be both physical and biological ones: temperature, oxygen, food organisms, parasites, and so on, all changing in level. Any set of environmental factor levels such as $E1$ is a *performance* of the generative environment that produced it. Thus it is a performance of an *environmental system*. But that environmental system can and will produce other sets of factor levels $E2$, $E3$, and so on, as in Fig. 2.4a. For an environmental system to have such performances, it must have the *capacity* for those performances (Fig. 2.4c). Thus, like any other system, an environmental system has at any time a capacity that consists of all its possible performances.

2.3.2   Adaptation as concordance of organismic and
          environmental systems

Let us return to the individual chinook salmon so as to first make visualizable our general concept of the nature of biological adaptation. At any place and time in its life history pattern, the individual salmon, if it is to survive to reproduce, must perform in a way adapted to the effective environment. Thus, if the environment were as $E1$ in Fig. 2.4a, then to be adapted the salmon would need to perform as

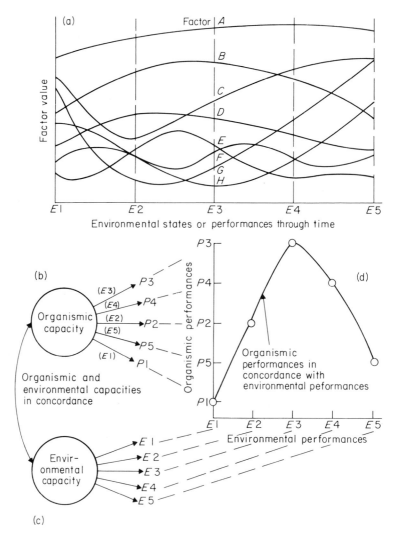

Fig. 2.4. (a) shows that any state or performance of any experimental or natural environmental system consists of a set of factor levels. (b) shows that, based on some capacity, a performance of an organismic system is determined by a state of its environmental system. (c) shows environmental states or performances to be based on some environmental capacity. (d) shows a graphical relationship between organismic and environmental performances. In adapted organismic systems, there must be present and probable future concordance of their capacities as well as their performances with those of their environmental systems.

$P1$ in Fig. 2.4b. But the environment will soon change, perhaps to $E2$ and the performance of the salmon will need to be $P2$ if it is to be adapted. Over a period of time during which the capacity of the individual salmon does not change much, the effective environment may pass through states of performances $E1$, $E2$, $E3$, $E4$, and $E5$, to which the salmon must respond with performances $P1$, $P2$, $P3$, $P4$, and $P5$, respectively, as shown in Fig. 2.4d. This set of performances of the salmon at least partially represents its capacity during that time. And the set of performances of the environmental system ($E1$, $E2$, etc.) similarly represents its capacity (Fig. 2.4c). Now, in a very important sense, for the salmon to be adapted its capacity must be adapted to the capacity of its environmental system, just as its performance must be adapted to the performance of its environmental system. For if individual chinook salmon were to have only the capacity to adapt to a few performances of the environmental system and not to many other performances within the capacity of the environmental system, no salmon could live to reproduce. The capacities as well as the performances of a salmon must generally be adapted to the capacities as well as the performances of its environmental system. This is so for any organismic system. How can we get a suitably universal statement of the principle?

The word *concordance* is suited to capture just the meanings we need: *harmonious relations* and a *set of rules* maintaining such relations. Adapted organisms, like the salmon, and other adapted organismic systems appear to be in harmonious relations with their environmental systems. And they behave as though they were obeying a set of rules established to maintain these relations. Indeed, such rules would be *biological laws of adaptation*, if they were stated in a general and invariant way. The capacities as well as the performances of the individual chinook salmon are thus in concordance with those of its environmental system, as are those of its population and community with their environmental systems. More generally, for an organismic system to be adapted, there must be *present and probable future concordance* of its *capacity* as well as its *performances* with those of its environmental system (Fig. 2.4b,c,d). In short, it is this that most fundamentally is biological adaptation, and it is this that we seek to explain and understand in biology.

Before we go on to consider adaptation of individual organisms, there is one distinction that will prove helpful. An individual organism like the chinook salmon is adapted to persist as a whole throughout its life history: let us call this *persistence adaptation*. Such persistence adaptation together with some other adaptations favor reproduction of the individual salmon. Now reproduction does not favor the individual salmon's persistence: it is not persistence adaptation. Rather, reproduction very importantly favors the persistence of the salmon's population. The individual salmon is thus adapted to favor the persistence of its population: let us call this *functional adaptation*. Any organismic system, as a whole, is adapted to persist; but it is also adapted to function or play a role in favoring the persistence of the more incorporating system of which it is a part.

2.4   ADAPTATION OF LIFE HISTORY PATTERNS

2.4.1   Organismic and environmental time and pattern

Although our immediate concern in this section will be the adaptation of life history patterns—by which we mean the individual organism through time—life history patterns can be understood only in the context of the populations and species of which they are parts. This is because the functional adaptation of a life history pattern is the roles that it plays in determining the distribution, abundance, and persistence of its population and species. The environments to which the individual, the population, and the species are adapted can be thought of as multidimensional patterns extending in time and space. The units of time and space that concern us are relative to the species of interest, especially its life history pattern. Thus it is not any sort of absolute time or space that is important to our understanding of species adaptation. Rather, it is time in units of life history duration, say mean generation length (T), the mean length of time between births of parents and births of their offspring. And it is space in units of mean or maximum movement (M) of organisms or reproductive propagules of the species. It is the species-specific scale of time and space of a rotifer, a mayfly, the spring chinook salmon, or any other species.

A fundamental problem of each kind of life is to keep itself in places of a kind suitable for its persistence and reproduction. To consider this problem in its most general form, we do not need to consider environments, their factors, and their favorability as such. Rather, we can simply define *habitat* as a kind of place suitable for the reproduction and persistence of organisms of a kind.

This problem has to do with patterns of *habitat availability* in time and space relative to the generation length and movement of individuals of particular species. To make this more visualizable, we can separate habitat availability into *habitat extension* in time and space and *habitat separation* in time and space. How long in relative time and how far in relative space do particular habitats extend? And how much time elapses between the existence of habitats at particular locations and by what distances are existing habitats separated in space? Perhaps we can make the problem faced by each species even more visualizable by imagining a large panel of lights of different sizes, each light representing a habitat existing at a location only when it is turned on. Habitat extension in time is represented by how long particular lights are lit; and extension in space can be represented by the sizes of particular lights. Habitat separation in time can be thought of as the periods between when particular lights are lit; and habitat separation in space can be the distance between lights that are on. Any imaginary species that could reproduce and persist over the habitat system represented by this panel of winking light must have mean generation length and mean and maximum distances of movement that could keep its individuals and thus populations where lights are on.

In an already greatly simplified form, we attempted in Fig. 2.4a to illustrate how a few environmental factors might vary through time or space. Now let us suppose that these and all other environmental dimensions (or factors) can, in some sense, be collapsed into one, which we will call *environmental quality* for reproduction and persistence. This may help us to understand possible patterns of habitat and environmental conditions in time and space, because we can now represent them in three dimensions, as we have done in Fig. 2.5.

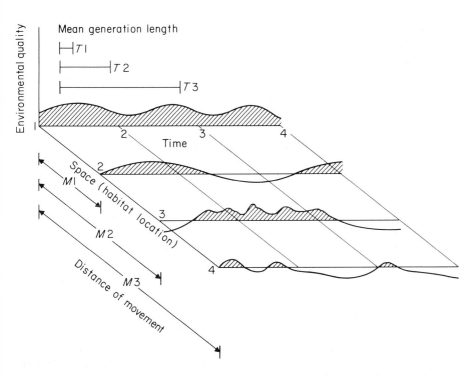

Fig. 2.5. A very idealized or simplified representation of how environmental patterns in space and time might relate to different mean generation lengths (*T*) and distances of movement (*M*) of life history patterns. Environmental quality, say for reproduction and persistence of individuals and populations, is a collapsing into one dimension of all relevant factors. At different locations in space, different patterns of environmental quality occur through time. Environmental quality is suitable at a given location only for the shaded portions of the pattern. For a population to persist, mean generation length and distance of movement must allow reproduction and persistence of individuals at one or more habitat locations.

At location 1 in space, there is a habitat that persists throughout the time represented in Fig. 2.5, a place at which environmental conditions are always suitable for reproduction and persistence of populations 1, 2, and 3, whose individuals have mean generation lengths $T1$, $T2$, and $T3$, and maximum distances of movement of $M1$, $M2$, and $M3$. These populations would probably adapt in different ways to the changing environmental quality at location 1. The changes in environmental quality at location 1 occur over several generation lengths ($T1$) of population 1. Thus the adaptive characteristics of individuals in population 1 would be likely to change from one generation to the next as a result of natural selection. Such evolutionary change is an important means of adapting to an environmental change on this relative scale of time. But, for population 3, an entire cycle of environmental quality occurs in about the mean generation length ($T3$). Thus each individual, to survive and reproduce, must have the capacity to adapt physiologically or behaviorally to this entire pattern of environmental change.

Now Fig. 2.5 does not make explicit whether or not the habitats are distributed continuously in space from location 1 through location 4. If they are, then adaptation to habitat availability will not be a big problem for any of the three populations as environmental quality changes through time. But if the habitats are discontinuous in space and occur only precisely at the four locations, then adaptation to habitat availability and environmental conditions is a different sort of problem for each of the populations, because of the differences in the mean generation lengths (T) and movements (M) of their individuals. Population 3, with the relatively long movements ($M3$) of its individuals, can essentially use habitats 1 through 4 as though they were continuous, and thus take advantage of favorable environmental quality and avoid unfavorable environmental conditions. But population 1, with the shorter movements ($M1$) of its individuals, may not be able to do this.

2.4.2   Classes of life history patterns

A life history pattern can be classified in any of an indefinite number of ways, each according to some life history characteristic. The way we choose depends on our explanatory objective and our point of view. A 'well-rounded' or full classification would include at least the major 'dimensions' of a life history pattern: life form, trophic or nutritional type, size, length of life, reproductive pattern, and developmental pattern. But for particular explanations, such as theories, we very often treat organisms as though they had only one life history 'dimension,' such as reproductive class. This may permit mathematical or other simple representations of some theoretical interest. But the assumptions making it possible usually severely limit the meaning of the theoretical representation.

Plants, lacking the mobility of most animals, often evolve ecotypes or races specially adapted to very local conditions, sometimes not very many meters apart

(Bradshaw 1972). Plants, as distinguished from animals, in this and other ways form life history classes. Among plants, *life form classes* can be distinguished (Raunkiaer 1934, 1937), based, for instance, on whether their perennial growth bud is above or below the soil surface, the latter providing a degree of protection against winter conditions. A life form sort of classification was developed for birds (Salt 1953), and one could be of value for aquatic life. *Trophic classification*, as a basis for explaining nutrition and energy and material transfer, has attracted much attention in recent years.

Size is profoundly involved in the internal and environmental problems a species of organisms faces in persisting and in its possible life history and evolutionary modes of adapting. Based on these and other considerations, MacArthur and Wilson (1967) proposed that life history patterns may, for some theoretical purposes, be usefully classified as being either *r-selected* or *K-selected*. Among other characteristics, *r*-selected species tend to have relatively smaller size, shorter lives, higher rates of increase, and to devote more of their total energy to reproductive products than do *K*-selected species (Table 2.1). Of special theoretical interest is the idea that the per capita rate of increase of *r*-selected species is much more sensitive to increases in their own density and to the density of their competitors than is that of *K*-selected species (Fig. 2.6). Theoretically, *r*-selected species would rapidly colonize such ephemeral habitats as those

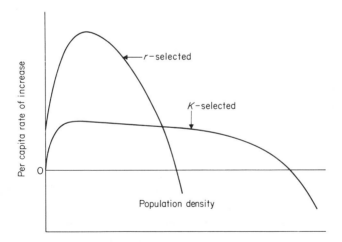

Fig. 2.6. Curves illustrating two general forms of dependence of per capita rate of increase on population density. So-called *r*-selected species have relatively high rates of increase at low population densities but are very sensitive to increases in population density. The rate of increase of *K*-selected species is supposed to be relatively less influenced by increases in population density but not to reach such high values. Thus, it is argued that *K*-selected species are less sensitive to intraspecific and interspecific competition than *r*-selected species. After Pianka (1972). Reprinted from *American Naturalist* by permission of The University of Chicago Press. © 1972 by The University of Chicago Press.

Table 2.1. Possible characterization of *r*-selected and *K*-selected species and their environments. After Pianka (1970). Reprinted from *American Naturalist* by permission of The University of Chicago Press. © 1970 by The University of Chicago Press.

	r-selection	*K*-selection
Climate	Variable and/or unpredictable; uncertain	Fairly constant and/or predictable; more certain
Mortality	Often catastrophic, non-directed, density independent	More directed, density dependent
Survivorship	Often Type III	Usually Types I and II
Population size	Variable in time, non-equilibrium; usually well below carrying capacity of environment; unsaturated communities or portions thereof; ecologic vacuums; recolonization each year	Fairly constant in time, equilibrium; at or near carrying capacity of the environment; saturated communities; no recolonization necessary
Intra- and interspecific competition	Variable, often lax	Usually keen
Selection favors	1 Rapid development 2 High maximal rate of increase, r_{max} 3 Early reproduction 4 Small body size 5 Single reproduction	1 Slower development 2 Greater competitive ability 3 Delayed reproduction 4 Larger body size 5 Repeated reproductions
Length of life	Short, usually less than 1 year	Longer, usually more than 1 year

represented in Fig. 2.5 as occurring widely separated in time at habitat location 4, which they would reach from similar habitats sporadically occurring at other locations. *K*-selected species would, on the other hand, be favored in more persistent habitats having more stable environmental conditions, perhaps something like we represent at habitat location 1 in Fig. 2.5.

The *reproductive class* of a life history pattern—oversimply, on a scale from asexual to sexual—is of profound importance in determining individual, population, and species adaptive capacities and performances. So also would be *developmental class*, taking into account such matters as germination requirements, adaptative developmental plasticity, resting or resistant stages, and flowering or other reproductive conditions. Adaptative developmental plasticity would probably be a primary mode of adaptation of population 2 with generation length $T2$ in the variable environments represented at habitat locations 2 and 3 in Fig. 2.5. Resistant stages could allow a population to persist at habitat locations 2 and 3 even through periods when these locations are superficially not habitats, because of unsuitable conditions for development and reproduction.

Finally, on the basis of any of their requirements, organisms can be classified as being *generalists* or *specialists*. In Fig. 2.7, we illustrate that organisms may be generalists or specialists for one, the other, or both of two environmental factors.

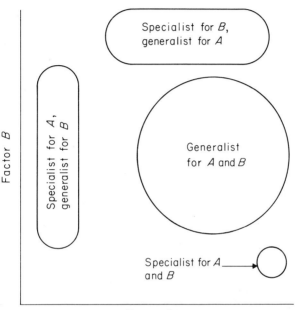

Fig. 2.7. Idealized representations of the ranges of values of environmental factors over which organisms of different species could persist. If some third environmental factor were represented and its ranges plotted along a third axis, the ranges of persistence of organisms for factors A and B would probably change, even to the extent of changing the category of a species from specialist to generalist.

Relatively few species of organisms can be either generalists or specialists with respect to all their requirements. The classes generalist and specialist have been of some interest in predation, competition, and evolutionary theory.

2.4.3   Adaptation of aquatic life history patterns
For the individual aquatic organism to persist and reproduce, it must locate and maintain itself in its particular kind of habitat. The spring chinook salmon

migrates some 1600 kilometers to reproduce in a headwater stream in Idaho, its reproductive habitat. There the emergent juveniles live awhile, before beginning their trip to the sea, where for perhaps three years they feed and grow. The individual salmon is adapted in many ways to locate and maintain itself in these habitats. Its persistence and finally its reproduction function to favor the persistence of its population.

Once located in a habitat, to persist and reproduce, an individual organism must be adapted to environmental conditions there. There must be trophic adaptation to meet energy and material requirements. The young salmon in the stream feeds predominantly on insects; later in the sea it will feed first on plankton, then on herring-like fish. Thus there is present and probable future concordance of the trophic characteristics of the salmon and its environments.

Aquatic organisms must also be concordant with physical conditions in their habitats. For some conditions, like temperature, most aquatic organisms are *conformers*: their body temperatures nearly follow those of their medium. For other conditions, aquatic organisms are *regulators*: regulation of ionic composition of bodily fluids is universal among aquatic organisms. But not all organisms regulate their osmotic pressure: fish osmoregulate; most crustaceans in the sea do not. Suitable levels of one physical factor in the environment of an organism are in part determined by prevailing levels of other factors.

The temperature, salinity, and oxygen requirements of fish and other individual organisms are in part determined by their history in relation to these environmental factors, especially by recent levels of exposure (Brett 1952; Fry 1964; Shepard 1955). Fish recently exposed to high temperatures can tolerate higher temperatures; those recently exposed to low oxygen concentrations can tolerate lower concentrations. We say the fish are *acclimated* to high temperatures or to low oxygen concentrations. This is a special case of short-term *developmental adaptation*. Learned behavior is another form of developmental adaptation, a very important one. Much of the concordance of the capacities and performances of life history patterns with those of their environments is achieved developmentally. An individual propagule, dispersed widely from the parent stock, may come to be located in a habitat very different from that of the parents. But developmental adaptation leading to different realized capacities and performances make it possible for the organism to persist and reproduce. Thus a propagule of arrowleaf has the capacity to develop a terrestrial form in that habitat, a shallow-water form in such a pond, and a deepwater form where that is adaptive (Fig. 2.8). Each of these growth forms is a performance of the arrowleaf concordant with an environmental system performance, a terrestrial, a shallow-water, or a deep-water habitat. Thus a single genotype has the potential capacity to produce different growth forms while the environmental system has the potential capacity to produce different habitats. There is concordance of arrowleaf and environmental capacities as well as performances.

Fig. 2.8. Adaptive developmental responses of the arrowleaf, *Sagittaria sagittifolia*, to a terrestrial habitat, a shallow water habitat, and a deeper water habitat. The different growth forms developing are adapted to the environment in which they occur. The aquatic forms, lacking a cuticle, can absorb nutrients directly from the water, which supports them. The terrestrial forms are self-supporting, require a protective cuticle, and must get their mineral nutrients from the soil. After Wallace and Srb (1961).

Functional adaptation—the role the individual organism plays in insuring the persistence of its population and species—has received perhaps the most attention in recent ecological thought. Of the energy and material resources available to an organism, how much should be expended for its persistence and how much for reproduction to favor the persistence of its population? And, of the resources available for offspring, should many small young (*r*-selected, see Table 2.1) or fewer large young (*K*-selected) be produced? Should these be produced in a single group or in several groups, very early in life or later? And should an organism be sexual, asexual, or alternating in its pattern of reproduction? Which of these patterns or combinations of patterns of reproductive functional adaptation have evolved in different species has depended on initial genetic capacities and long sequences of effective environments. Theories of 'life history strategy' (Stearns 1976) are attempts to generalize about the relations between classes of life history patterns and classes of habitats or environments. Because reproductive pattern is the part of the life history functionally relating life histories to populations, theories of life history strategy are nearly entirely in terms of reproduction. In order to make functional adaptation of life history patterns most understandable, we will consider them in their population and species contexts.

2.5   ADAPTATION OF SPECIES POPULATIONS

2.5.1   The nature of population adaptation

It is said that years ago individual spring chinook salmon entering the Columbia River were on the average even larger than now. A gill net fishery, such as exists in the Columbia, can select against larger fish. It is also known that an intensive early or late fishery can alter the timing of salmon runs. It is as though salmon populations were altering the size of their individuals and the timing of their migrations so as to avoid the gill net fisheries. Indeed, even though biologists explain such evolutionary adaptation by natural selection, its functional consequence is to tend to reduce mortality occasioned by a particular selective factor: the gill net fishery. But the size of the giant spring chinook may have been an adaptation for the long migration to headwater streams in Idaho; so also may the timing of the run. Even though the spring chinook could adapt to the fishing, because of this it may become less well adapted in other ways, which may not become apparent until other environmental conditions become exceptionally severe. At such a time, relieving of fishing pressure may not be enough to allow the population to recover, for its adaptive capacity may have been too permanently altered. We need to know far more about the problem of adaptation than we do, if we are to more adequately manage our fisheries.

In considering the nature of population adaptation, we must be clear about the habitat availability and the environment to which a population must adapt. An environment is not simply a suitable kind of place but rather is a system of factors ultimately impinging with effect upon a population during its existence in time and space. But a population is a diffuse sort of thing, and this makes conception of its environment much more difficult than conception of the environment of an individual organism. Figure 2.9 may help us to distinguish between the environments of individual organisms and the environment of their population and thus make clearer the concept of population environment. Each individual organism is influenced over some range of each of its environmental factors. For two environmental factors, two-dimensional projections of the range of each factor influencing each of a group of individuals are shown in Fig. 2.9. The range of each environmental factor that influences the group—the 'population'—is greater than that of any individual and is enclosed by the outer boundary of individual environmental conditions. The environment of a population is generally much more extensive and variable in time and space than that of any individual.

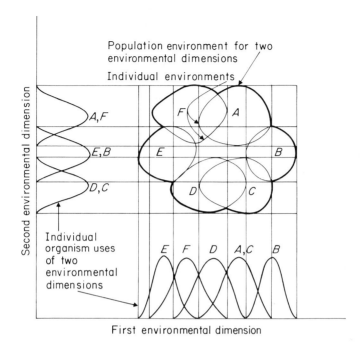

Fig. 2.9. An idealized or simplified illustration of how the limits of individual use of two environmental dimensions or factors determine the limits of population use of these dimensions, or determine the environment of the population with respect to these dimensions. The population environment differs from those of the individuals in that the individuals are parts of the environments of one another but cannot logically be in the environment of the population because they are parts of the population. (Suggested by Figure 6.7, Pianka 1974.)

The population, then, must have the capacity and exhibit the performances necessary to cope with this more extensive and variable environment. This capacity resides, first, in all the different life history patterns that can develop in a given generation. This can be called the capacity for *life history adaptation* of a population. Second, the population has the capacity from one generation to the next, through natural selection, to change the different life history patterns that can develop. This can be called the capacity for *evolutionary adaptation* of a population.

Life history adaptive capacity and evolutionary adaptive capacity ought to be considered together, as we must suppose they occur in nature. Perhaps we can extend Levin's (1968) concept of a 'fitness set' to make this visualizable. In Fig. 2.10, coordinate axes for individual adaptive capacity for two different sets of environmental conditions are shown. Each small circle represents the coordinate values of the possible life history patterns of an individual organism, in terms of

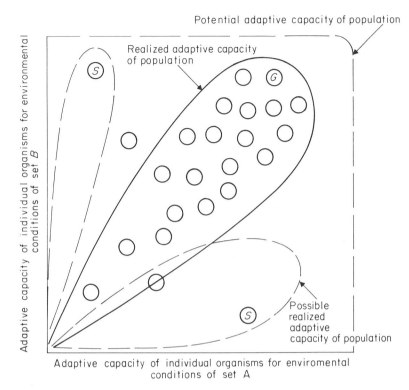

Fig. 2.10. An idealized or simplified representation of the adaptive capacity of a population. The adaptive capacity of a population includes the life history adaptive capacity of its individuals as well as the evolutionary adaptive capacity to change these. Because the circles represent the heritable adaptive capacities of individual organisms with respect to two different sets of environmental conditions, the circles enclosed by the solid ellipsoidal line altogether represent the total realized adaptive capacity of the population with respect to these two sets of factors. Evolution could change the realized adaptive capacity of a population to include any part of its potential adaptive capacity. Specialist (*S*) and generalist (*G*) individuals are located in this space.

individual adaptive capacity for the two different sets of environmental conditions. Remember that our conception of an individual organism is its life history pattern, and that its zygote, in some sense, entails its potential capacity to develop all possible life history patterns according to prevailing environments. Thus each small circle represents a heritable set of adaptive life history patterns, such as any individual organism can be thought to possess. Enclosed within the ellipsoidal line are many such individual potential capacities to develop life history patterns. Which life history patterns develop and which ones are most adaptive will depend upon which set of environmental factors is most prevalent.

In any generation, the population has this *life history adaptive capacity*. But from one generation to the next, there can come a change in the proportions of different individual life history capacities and patterns, according to prevailing environmental conditions. At any time, the capacity for this change is the *realized evolutionary capacity* of a population, which together with the life history adaptive capacity forms the *realized adaptive capacity* of the population. Over very long periods of evolutionary time, this could conceivably come to lie anywhere between the coordinate axes: the *potential adaptive capacity* of the population (Figs. 2.2, 2.10). All of this will depend upon the combination of sets of prevailing environmental conditions. The times, places, and ways of fishing are very important parts of the prevailing environmental conditions of any exploited fish population. And, short of extinction, the most profound effect that any fishery can have upon a fish population is to alter its potential and realized adaptive capacities.

Now populations like other organismic systems exhibit not only persistence adaptation but also *functional adaptation*. Most discussion of the functional adaptation of populations has pertained to their roles in biological communities. But populations are also parts of another sort of biological system, the biological species as a system of more or less loosely coupled populations (Mayr 1963). Biological species have adaptive capacities and performances based on those of their individual life history patterns and their populations. Indeed the adaptive capacities of species, because of the greater extension of species in space and time, must be greater than those of individual organisms and even populations.

2.5.2 Adaptation of aquatic populations
Species of aquatic organisms from distinctive phylogenetic backgrounds have adapted in a rich variety of ways, to diverse habitats. Because we have paid some attention to different classes of habitats and environments, different classes of life history patterns, and evolutionary change in these, we can now generalize about some of the major patterns of adaptation of aquatic organisms.

Small organisms tend to have the shortest generation lengths and the highest intrinsic rates of increase. One can quite generally view short generation length and high capacities for increase, especially when these are coupled with high capacities for dispersal, as adaptations favoring a species' taking advantage of opportunities of habitat availability occurring briefly and uncertainly in time and space. A series of habitat locations having environmental patterns similar to that illustrated at location 4 in Fig. 2.5 would be like this. And, of the three populations illustrated, only population 1 with generation length $T1$ and movement $M1$ could adapt to such a pattern of habitat availability and environmental conditions. Species of rotifers that live in rain puddles take advantage of such uncertain and temporary occurrences of habitats, which they reach by sexually produced propagules. On developing, these individuals can reproduce parthenogenetically. As a group, rotifers would be considered to be *r*-selected species (Table 2.1), though in this they vary in degree.

Williams (1975) proposed several models of considerable value in ordering reproductive life history strategies. The *'aphid-rotifer model'* he characterized as a class of organisms sexually producing widely dispersed propagules, which develop into parthenogenetically or asexually reproducing forms. The sexually-produced propagules ensure genetic diversity from which forms suited to diverse but uncertain environmental conditions are likely to be selected. Should only a single propagule reach a habitat, parthenogenetic reproduction would still favor colonization. But should several propagules reach a habitat, competition and selection between clones would favor the persistence of the best adapted. The life history of *Daphnia* places it in this class. *Daphnia* are well-adapted to lakes having seasonally favorable conditions such as illustrated at location 2 in Fig. 2.5. During the favorable season, a population of parthenogenetically reproducing *Daphnia*, because of their relatively short generation length, can evolve so as to stay well adapted to changing conditions. But, with the onset of severe water conditions, there occurs sexual production of resistant eggs, from which forms most suitable to environmental conditions the following spring can be selected.

A great many species of marine plants and animals have life history patterns falling into a class typified by Williams' *'strawberry-coral model.'* These organisms extend their populations through an environmental gradient in space by asexual reproduction until environmental limits are reached, as when the 'runners' of strawberry plants reach unsuitable soil. But such limits can be transcended to other suitable habitats by sexually produced propagules such as the seed of the strawberry. Williams suggests that many sponges, sessile polychaetes, bryozoans, and protochordates start life as widely-dispersed, sexually-produced planktonic young, relatively few of which colonize suitable surfaces and then reproduce asexually to occupy the available habitat.

Of course, the production of a multitude of widely dispersed young so common in marine organisms as a means of colonizing need not be accompanied by asexual reproduction. The oysters and mussels of the intertidal zone are in this class. Williams (1975), with his *'elm-oyster'* model, suggests that, for organisms an adult of which occupies more space than thousands of the juveniles, selection from the greater variability of sexually produced young is far more important than the redundancy of cloning. Another way of classifying intertidal organisms would be as organisms synchronizing their breeding times at optimal points in cycles, such as the tidal cycles that are shorter than their life history patterns. Of course, intertidal and other organisms must also synchronize their activities to seasonal and other cycles.

The life history capacities and performances, or patterns, of individual organisms of particular species have evolved to be concordant with the capacities and performances of their environmental systems. For this to have occurred, life history capacities and patterns must be heritable—must have a genetic basis. Such important life history performances as age at first reproduction, growth and thus size, and number of broods of young produced by a female are assumed to be heritable and adapted to the environment in which a species has evolved. These individual capacities and performances contribute to the adaptive capacities and performances of species populations. Theories of life history strategy deal with this, and some of the examples that have been developed are of special interest to fisheries science and management.

For a population to be relatively constant in number, on the average, one of the offspring produced by an individual over its reproductive life must survive to reproductive maturity. If environmental conditions can be expected to be favorable for reproduction only one in four years, then an average individual in a population would need to reproduce each year for at least four years to favor persistence of its population. For environments that are quite variable and only infrequently favor successful reproduction and survival of progeny to maturity, Murphy (1968) has postulated that life history characteristics such as long life, multiple reproductions (iteroparity), and late maturity will be favored.

Among herring-like fish, those living in the most variable environments mature at older ages, have the longest reproductive life spans, and thus have the most broods per individual (Murphy 1968), as shown in Table 2.2. The Peruvian anchovy reproducing under the stable winter conditions of the Peruvian Current, matures at 1 year of age and has a reproductive life span of but 2 years. At the other extreme, herring of a stock reproducing in the variable Atlantic Polar

Table 2.2. Reproductive characteristics of several populations of herring-like fishes in relation to environmental variability, which is assumed to increase with increase in variance of spawning success. After Murphy (1968). Reprinted from *American Naturalist* by permission of The University of Chicago Press. © 1968 by The University of Chicago Press.

Population	Age at first maturity in years	Reproductive span in years	Variation in spawning success (highest/lowest)
Herring (Atlanta-Scandian)	5–6	18	25×
Herring (North Sea)	3–5	10	9×
Pacific sardine	2–3	10	10×
Herring (Baltic)	2–3	4	3×
Anchovy (Peru)	1	2	2×

Front region mature at 5 or 6 years and then have a reproductive span of perhaps 18 years. Intensive fisheries are known to greatly reduce the average length of life of fish in exploited stocks. The Pacific sardine matures at an age of 2 or 3 years and has the capacity for a reproductive span of perhaps 10 years (Table 2.2), presumably adapting this species to the variable environment along the Pacific coast of the United States. After many years of an intensive fishery, which greatly reduced the reproductive life span of the average individual to a few years or less, the sardine population and fishery collapsed (see Chapter 14). On the basis of simulation modeling, Murphy (1967) has argued that if the fishery had not reduced the reproductive life span, this might not have occurred. This is an example of a fishery altering the adaptive capacity of a fish population to the extent that the population is no longer economically important.

Distinctive populations of Atlantic salmon ascend many rivers along the Northern Atlantic coast of North America to spawn. These populations differ in their life history characteristics in ways that adapt them to their particular freshwater and marine environmental systems (Schaffer and Elson 1975). The mean age and size at first spawning in these Atlantic salmon tend to be higher in long, fast rivers that are more difficult to ascend. Evidentally, larger and, thus, older fish are better able to survive the ascent of these rivers to spawn. Commercial fishing, on the other hand, tends to select for earlier reproductive age by harvesting the older, larger fish. Thus, the fishery selects against life history traits that evolved to adapt populations to their particular environments. In adapting to fisheries by changing their life history characteristics, populations may become less well adapted to other aspects of their environments. By making problems of this sort apparent, life history theory of adaptation can be of much importance to fisheries science and management.

2.6   ON THE NATURE OF BIOLOGICAL COMMUNITIES

Individual organisms and populations of particular species do not, in nature, occur alone. Rather, they occur in combination with other species of organisms. The nature of these combinations, which are called *biological communities,* is of much interest to biologists. Communities are formed by systems of populations and these are the context within which populations and life history patterns are to be understood. Beyond this, communities themselves are objects of biological interest, to be explained and understood in their own right.

Other species provide habitat and food and may be predators, parasites or competitors of any species population. That is, other populations are part of the environment of any particular population and with it they form a *system of populations.* The populations in such a system tend to be adapted one to another, both to persist as individual populations and to play their respective roles in favoring the persistence of the system as a whole. Thus these populations exhibit both persistence and functional adaptation.

Within biological communities, not all species populations are in close or direct interaction with all other species populations. In a lake community, for example, the plankton and fish populations of the pelagic habitat are in close interaction one with another. So also are the populations of microflora, microfauna, and macrofauna of the benthic habitat, which form a complex system of populations. The populations of the littoral zone of a lake form another such system of populations, not in close interaction with those of the pelagic and benthic habitats, but then not independent of them either. A lake community is composed of at least three major systems of populations, between which nutrient and other relations exist.

The *structure* of such a community can be thought of as the species composition, the relative distribution and abundance of the species, and the superficial appearance the arrangement of species gives the community. Through time, the structure of communities tends to change—there is *community development* or succession. And a community persists throughout such development. Structure, development, and persistence are community level performances. Biological communities have capacities for such performances just as other organismic systems have capacities for the performances they exhibit. These capacities of communities result from the ways in which communities incorporate the capacities and environments of systems of populations, particular populations, and life history patterns of individual organisms. That is, these capacities result from community *organization*. Their capacities and performances adapt aquatic communities as wholes to develop and persist in the environments that lakes, streams, estuaries, and the sea provide for them.

2.7   REFERENCES

Bradshaw A.D. (1972) Some of the evolutionary consequences of being a plant. *Evol. Biol.* **5**, 25–47.

Brett J.R. (1952) Temperature tolerance in young Pacific salmon, Genus *Oncorhynchus. J. Fish. Res. Board Can.* **9**, 265–282.

Dobzhansky T. (1956) What is an adaptive trait? *Am. Nat.* **90**, 337–347.

Fry F.E.J. (1964) Animals in aquatic environments: fishes. *In* D.B.Dill, E.F.Adolph, and G.G.Wilber (eds.), *Adaptation to the Environment.* (Handbook of Physiology, Section 4). pp.715–728. Washington: American Physiological Society.

Levins R. (1968) *Evolution in Changing Environments.* Princeton: Princeton University Press.

MacArthur R.H. and E.O.Wilson (1967) *The Theory of Island Biogeography.* Princeton: Princeton University Press.

Maxwell N. (1974) The rationality of scientific discovery. *Philosophy Sci.* **41**, 123–153; 247–295.

Mayr E. (1963) *Animal Species and Evolution.* Cambridge: Belknap Press.

Murphy G.I. (1967) Vital statistics of the Pacific sardine (*Sardinops caerulea*) and the population consequences. *Ecology,* **48**, 731–736.

Murphy G.I. (1968) Pattern in life history and the environment *Am. Nat.* **102**, 391–403.

Pianka E.R. (1970) On r- and K-selection. *Am. Nat.* **104**, 592–597.

Pianka E.R. (1972) r and K selection or b and d selection? *Am. Nat.* **106**, 581–588.

Pianka E.R. (1974) *Evolutionary Ecology.* New York: Harper and Row.

Raunkiaer C. (1934) *The Life Forms of Plants and Statistical Plant Geography.* Oxford: Clarendon.

Raunkiaer C. (1937) *Plant Life Forms.* Oxford: Clarendon.

Salt G.W. (1953) An ecologic analysis of three California avifaunas. *Condor,* **55**, 258–273.

Shepard M.P. (1955) Resistance and tolerance of young speckled trout (*Salvelinus fontinalis*) to oxygen lack, with special reference to low oxygen acclimation. *J. Fish. Res. Board Can.* **12**, 387–446.

Stearns S.C. (1976) Life-history tactics: A review of ideas. *Q. Rev. Biol.* **51**, 3–47.

Wallace B. and A.M.Srb (1961) *Adaptation.* Englewood Cliffs: Prentice-Hall.

Warren C.E., M.Allen, and J.W.Haefner (1979) Conceptual frame-works and the philosophical foundations of general living systems theory. *Behavl. Sci.* **24**, 296–310.

Whitehead A.N. (1927) *Process and Reality.* New York: Macmillan.

Williams G.C. (1975) *Sex and Evolution.* Princeton: Princeton Univ. Press.

Chapter 3
Ecology of Aquatic Systems

WILLIAM J. LISS AND CHARLES E. WARREN

3.1 POPULATION DISTRIBUTION AND ABUNDANCE

3.1.1 Populations as objects of study

It is often very difficult to perceive the boundaries of populations in space and time. Yet we believe that populations exist and, as a result of their organization, have the capacity to exhibit performances that are different from the performances of the individuals that compose them. As a result of the reproduction of the individuals composing them, populations have the capacity to persist through many individual generations. In addition, populations develop, that is, their structure and organization change through time. Population evolution is associated with changes in the genetic structure of populations through time that tend to maintain concordance of the population and its environment. This sort of population adaptation was dealt with in the last chapter. The distribution and abundance of populations also vary through time and from place to place. Explanation of these population characteristics continues to be a major task of ecology.

3.1.2 Accounts of population abundance

Populations do not increase without limit, and population decrease to zero seems to be relatively infrequent. Thus it appears that population abundance is somehow controlled or regulated. The mechanisms of natural regulation of population abundance have been a controversial topic in ecology since the early twentieth century, when Howard and Fiske (1911) proposed that two kinds of 'agents' or 'factors' were responsible for controlling the abundance of populations. Smith (1935) later called these *density-dependent factors* and *density-independent factors*.

If per capita (per individual, percentage, or relative; see Chapter 5) death rate (d) increases and per capita birth rate (b) declines as population density (N) increases (Fig. 3.1a), then birth rate and death rate are said to be density-dependent. Population density tends toward the steady-state N_∞ and density-dependence is thus thought to be stabilizing. Density-dependent changes in b and d result in part from reduction in availability of resources (food, habitat) per

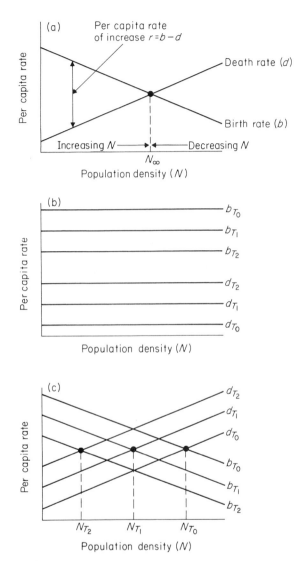

Fig. 3.1. (a) Density-dependent per capita birth rate (b) and death rate (d) are illustrated. The difference between b and d is the per capita rate of increase r. When $b > d$, N increases, but when $b < d$, N decreases. When $b = d$, $r = 0$ and the population has reached a steady state, N_∞. (b) Density-independent effect of a physical environmental factor, temperature, on per capita birth rate and death rate. T_0 is the temperature that is optimal for survival and reproduction. Temperatures T_1 and T_2 are progressively less favorable for survival and reproduction. (c) Interaction of density-dependent and density-independent effects in determining population abundance. N_{T_0} is the steady-state density at temperature T_0, N_{T_1} the steady-state density at T_1 and N_{T_2} the steady-state density at T_2. After Enright (1976).

individual in the population that is occasioned by increases in population density. The effect of physical environmental factors such as weather on death rates and birth rates may not depend upon population density (Fig. 3.1b). Unfavorable weather, such as very low or very high temperatures, may result in the death of a fixed percentage of the population, regardless of population density, and thus affect b and d in a density-independent manner.

Population abundance is influenced by density-dependent and density-independent factors acting together. Enright (1976) has developed a way of thinking about how population abundance is determined by both of these kinds of interactions (Fig. 3.1c). Temperature can be used as an example of a physical environmental factor that may cause density-independent effects. At a particular temperature, say T_0, birth rate and death rate change in a density-dependent manner, d increasing and b declining as N increases. As temperature is reduced from T_0 to T_2, conditions become less favorable for reproduction and survival (Fig. 3.1b), and consequently the density-dependent relationship between b and N is shifted downward while the density-dependent relationship between d and N is shifted upward (Fig. 3.1c). Such an account demonstrates how physical environmental factors that act in a density-independent manner interact with density-dependent factors to determine population abundance.

Changes in population biomass as well as changes in population number are density-dependent. In Fig. 3.2a, the density-dependent relationship between individual growth rate and population biomass of juvenile sockeye salmon from several of the lakes in the Babine-Nilkitkwa lake system is shown (Brocksen, Davis, and Warren 1970). Sockeye production is the total amount of tissue elaborated by the sockeye population and is determined as the product of growth rate and biomass. When this product is taken, a dome-shaped production curve is generated. How can the relationship between sockeye growth rate, biomass, and production be explained? The growth rate of individual sockeye was shown to be a function of the density of the food organisms, zooplankton (Fig. 3.2b). This growth curve approaches an asymptote where growth rate would be no longer limited by zooplankton density. Zooplankton density is a negative function of sockeye biomass (Fig. 3.2c). Thus, increased sockeye biomass resulted in a decline in the density of the zooplankton (Fig. 3.2c), and decreased zooplankton density resulted in decreased sockeye growth rate (Fig. 3.2b). The dome-shaped production curve is thus a result of reductions in zooplankton density and, consequently, sockeye growth rate that result from increased sockeye biomass.

Explanations such as this that couple the abundance of a population to the abundance of its prey organisms help to make clear the reasons for changes in population abundance. But explanation of the dynamics of a particular population should include more than just its food resource, because other factors such as predator and competitor populations and physical environmental factors

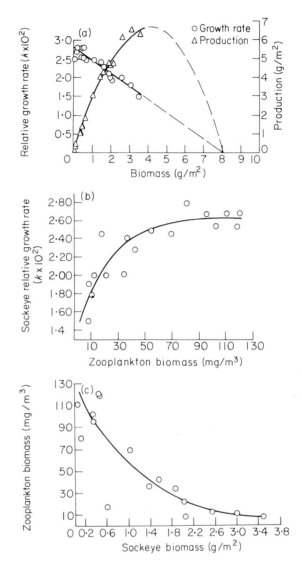

Fig. 3.2. Relationships between juvenile sockeye salmon relative growth rate, production, and biomass (a), sockeye relative growth rate and the biomass of the zooplankton upon which they fed (b), and sockeye biomass and zooplankton biomass (c) in the Babine-Nilkitkwa lake system, British Columbia, during four-month periods from June through September 1956 and 1957. After Warren (1971).

are also important in determining population distribution and abundance. We can think of a population as being embedded in a system of interacting populations that in large part determines its distribution and abundance. This helps to develop more adequate explanations. And it also draws our attention to a higher level of organization, the system formed by these interacting populations.

3.2 ORGANIZATION AND PERFORMANCES OF POPULATIONS IN SYSTEMS

3.2.1 Organizing principles of systems of populations

(*a*) *Predation and the Dynamics of Predation Systems*. A major objective of ecology is explanation of the structure and organization of systems of populations. The *structure* of a system of populations can be defined to include not only the species populations present in the system, but also their abundances and their distributions in time and space. We take the *organization* of a system of populations to be the inferred interrelations among the populations, which we must, at least in part, write onto the system to make it understandable. The organization of a system of populations underlies and determines its structure. Interrelations between species populations such as the processes of predation, competition, commensalism, and mutualism are an important part of system organization and are thus used to account for or explain the structure of systems of populations.

Stated simply, predation is the process of consumption of individuals (or parts of individuals) of one population (the prey) by individuals of another population (the predator). We can represent a sequence of predator-prey interactions as a 'food chain':

$$E > C \rightharpoondown H \rightharpoondown P \rightharpoondown R < I.$$

This system of populations is composed of a carnivore population, C, a herbivore population, H, a plant population, P, and plant resource concentration or intensity, R. Level of fishing effort, E, and plant resource input rate, I, constitute the environment of the system of populations. Complex food webs, rather than simple food chains, are the rule in natural systems. However, a simple predation system allows us to begin to understand how the biomass of each population in a system is related to the biomass of its predator and prey populations, and how such interrelationships between populations in a system and environmental factors such as I and E determine system structure.

The steady-state biomasses of the populations composing the system and, thus, system structure will be represented with interrelated systems of isoclines on phase planes (Fig. 3.3). The position and form of the isoclines can be deduced

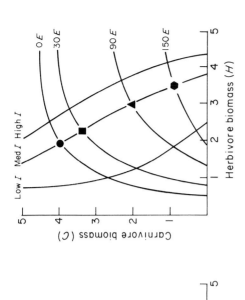

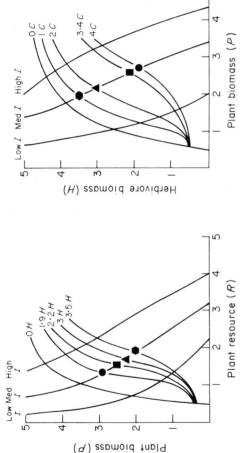

Fig. 3.3. Phase planes and isocline systems representing the interrelationships between populations in a predation system. Predator biomass is plotted on the *y*-axis of each phase plane and prey biomass is plotted on the *x*-axis. On each phase plane, the descending lines identified by different rates of plant resource input, *I*, are prey isoclines. Each prey isocline is defined as a set of biomasses of predator and prey where the rate of change of prey biomass with time is zero. The ascending lines on each phase plane are predator isoclines. Each predator isocline is defined as a set of biomasses of predator and prey where the rate of change of predator biomass with time is zero. Each intersection of a predator and prey isocline is a steady-state point where the rate of change of both predator and prey biomass with time is zero. At a particular level of *I* and *E*, a single steady-state point exists on each phase plane. the set of these points defining the steady-state biomasses of *C*, *H*, *P*, and *R*. The points that define the steady-state biomasses of *C*, *H*, *P*, *R* at Med *I* and 0*E* (circles), 30*E* (squares), 90*E* (triangles) and 150*E* (hexagons) are shown.

from response functions that represent the recruitment, production, loss to predation, nonpredatory losses such as disease, and yield to exploitation of the interacting populations (Booty 1976; Warren and Liss 1977). On each phase plane, prey biomass is plotted on the x-axis and predator biomass on the y-axis. The descending lines on each phase plane, identified by different rates of plant resource input I, are called prey isoclines. The ascending lines on each phase plane are called predator isoclines. Predator isoclines are identified by herbivore biomasses on the plant-resource phase plane, by carnivore biomasses on the herbivore-plant phase plane, and by levels of fishing effort on the carnivore-herbivore phase plane. On each phase plane, each intersection of a predator and prey isocline is a steady-state point. At any such steady-state point, neither the biomass of the predator nor the biomass of the prey will change so long as environmental conditions (I, E) remain constant. For the predation system described here, at a particular rate of resource input, I, and level of fishing effort, E, there exists a single steady-state point on each phase plane, the set of these points defining the mutual steady-state biomasses of C, H, P and R—the structure of the population system.

For example, at Med I and $0E$ in Fig. 3.3, carnivore, herbivore, and plant populations and plant resource attain steady-state biomasses of $4C$, $1.9H$, $2.8P$, and $1.4R$. These biomasses are indicated by the solid circles on the phase planes and together represent the structure of the system at Med I and $0E$. This steady-state point occurs at the intersection of the prey isocline identified by Med I and the predator isocline identified by $0E$ on the $C–H$ phase plane. On the $H–P$ phase plane, the point occurs at the intersection of the prey isocline again identified by Med I and the predator isocline identified by steady-state carnivore biomass $4C$. And, finally, on the $P–R$ phase plane, the steady-state point occurs at the intersection of the Med I prey isocline and the predator isocline identified by steady-state herbivore biomass $1.9H$.

It is apparent that the system of populations has many possible steady-state structures. That is, it has a capacity to exhibit different structures. Furthermore, the particular structure that is manifested depends upon the way the system is organized—the interrelationships between the populations—and the environment of the system.

What is the pattern of change that results when either I or E shift in magnitude? Let us begin with the steady-state structure generated by Med I and $0E$ (circles). Progressively higher levels of fishing effort ($0E$ to $150E$ on the $C–H$ phase plane) at this resource input rate result in reduction in the steady-state biomass of the carnivore and increase in herbivore steady-state biomass. In effect, increasing E shifts the steady-state point downward along the prey isocline identified by Med I and results in a reduction of the biomass of the predator, C. This allows the biomass of its prey, H, to increase. Furthermore, this increase in herbivore biomass leads to a reduction in steady-state plant biomass on the $H–P$

phase plane. And, finally, on the $P–R$ phase plane, the reduction in plant biomass resulting from increased herbivore biomass leads to an increase in the steady-state level of R. Thus, changes in level of fishing effort can lead directly or indirectly to changes in the biomasses of all populations in this simple system. The magnitudes of such indirect effects in natural systems of populations will vary greatly and will, in part, depend upon such factors as the number of prey populations upon which a particular predator forages.

Now, what sorts of changes will result from changing the plant resource input rate, I? Whereas changes in fishing effort shifted the steady-state point up or down the prey isoclines, changes in I result in shifts in the position of the prey isocline on each phase plane. Thus, at a given level of fishing effort such as $0E$, progressively higher rates of resource input (Low I to High I) result in higher steady-state biomasses of C, H, P, and R (Fig. 3.3).

The kinds of effects on system structure that differences in system environmental conditions can have, shown theoretically in Fig. 3.3, were empirically demonstrated for aquatic systems in laboratory stream channels (Warren 1971). Six stream communities composed of benthic algae, herbivorous insects, and juvenile cutthroat trout were exposed to three levels of light and two current velocities at each light level (Fig. 3.4). Plant biomass was affected most by light intensity but its effect was modified somewhat by current velocity. In general, over the light intensities and current velocities examined, increased light input resulted in increased plant biomass. And those streams that had the highest plant biomasses also maintained the highest biomasses of insects and trout.

For the predation system considered earlier (Fig. 3.3), a particular steady-state system structure potentially or theoretically exists for each particular combination of environmental factors I and E. As long as environmental conditions remain constant, the biomasses of the populations tend toward the steady-state points that define the steady-state structure for those conditions. In natural systems, however, the magnitudes of environmental factors such as plant resource input, I, vary seasonally and, at certain times of the year, fishing may cease. Thus, environmental factors may never be constant for long enough periods of time to permit the systems to reach steady-states. Changes in environmental conditions can therefore be thought of as shifting the location of steady-state points on phase planes. The biomasses of the populations change in response to this and are thus in constant pursuit of ever-shifting steady-state points.

(*b*) *Ecology of Production and Yield. Production* is the total amount of tissue elaborated by a population. It is determined as the product of relative growth rate and population biomass (see Fig. 3.2). *Yield* is the total amount of tissue harvested by a fishery and is one of the fates of fish production. Dome-shaped curves relating the steady-state production and yield of the exploited carnivore to its biomass can be derived from the relationship between steady-state carnivore

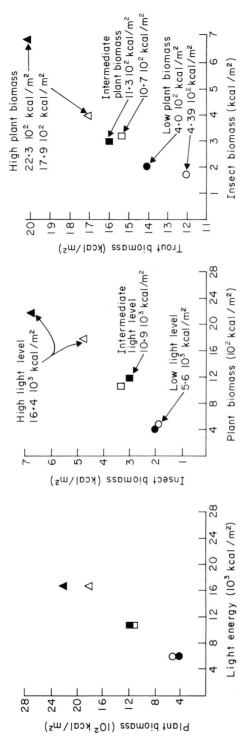

Fig. 3.4. The structure of aquatic communities in laboratory stream channels at different light levels and current velocities of 30 cm/sec (solid symbols) and 9 cm/sec (open symbols) in a spring 1968 experiment. After Warren (1971).

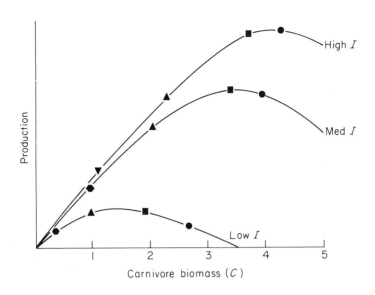

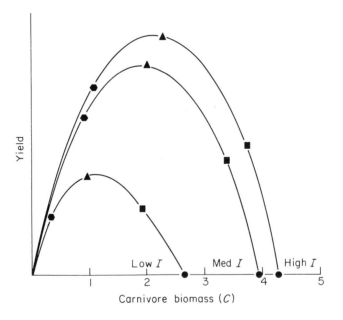

Fig. 3.5. Steady-state carnivore production and yield curves each derived from a prey isocline on the *C–H* phase plane (Fig. 3.3). Rate of light input *I* identifies the prey isocline from which each production and yield curve was derived. On each curve, the steady-state production, yield, and biomass of the carnivore population at levels of fishing effort 0*E* (circles), 30*E* (squares), 90*E* (triangles), and 150*E* (hexagons) are shown.

and herbivore biomasses defined by the prey isoclines on the *C–H* phase plane
(Fig. 3.3). Each production and yield curve is derived from a particular prey
isocline. Increases in plant resource input rate shift the prey isocline to the
right on the *C–H* phase plane and bring about increases in the magnitudes
of the carnivore production and yield curves (Fig. 3.5). Thus, families of
production and yield curves theoretically exist, each curve being causally related
to a particular plant resource input rate.

At a given rate of plant resource input, progressively higher levels of fishing
effort lead to lower steady-state carnivore biomasses and higher steady-state
herbivore biomasses (Fig. 3.3). Thus, increasing fishing effort shifts the steady-
state point downward along the prey isocline on the *C–H* phase plane (Fig. 3.3)
and to the left, toward the origin, along the production and yield curves (Fig.
3.5). Silliman (1968) was able to demonstrate the existence of families of yield
curves in exploited guppy populations (Fig. 3.6). These guppy populations were
fed different food rations and were exploited at different rates. The different food
levels are analogous to different capacities of the systems to produce food for
guppies. In natural systems, these different capacities are often ultimately related
to differences in the levels of environmental factors such as plant resource input,
I. In the exploited guppy populations, a steady-state yield curve was generated at
each food level.

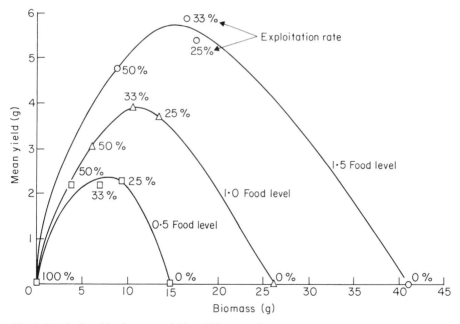

Fig. 3.6. Relationships between yield and biomass of guppy populations fed different
relative amounts of food and exploited at different rates. After Silliman (1968).

(*c*) *Interspecific Competition and Resource Utilization.* Interspecific competition occurs when different species populations utilize common resources that are in short supply. Competition, like predation, is a kind of interrelationship between populations. Thus competition is one of the ways in which populations are organized into systems of populations, and therefore it is a determinant of the distribution and abundance of organisms and of system structure. Pianka (1975) has correctly pointed out that competition should not necessarily be thought of as an all-or-none phenomenon. Rather it should be viewed as a process that operates more or less continually in the maintenance of system structure and population distribution, a process the intensity of which depends upon the relative densities of the competitors and the availability of resources.

Gause (1934) was one of the first to demonstrate, in very simple laboratory experiments with *Paramecium*, the effects of competitors on each others abundance and persistence. He showed that populations of *P. aurelia* and *P. caudatum* would grow and persist when held in separate cultures, but when they were placed together *P. caudatum* was driven to extinction. When *P. caudatum* was with *P. bursaria*, each population grew slower and reached lower steady-state densities than when they were grown in separate cultures. Similar results could be obtained theoretically if we were to add to the predation system considered earlier (Fig. 3.3) a species carnivore C^* as a competitor of C for the food resource herbivore H. Then, increasing demand would be placed upon the food resource and at each level of fishing effort E, the density of carnivore C would be less than its density would have been in the absence of carnivore C^*. Furthermore, the presence of C^* would result in a reduction in the magnitude of the production and yield curves of carnivore C (Fig. 3.7).

Interspecific competition is thought to promote differences in the use of resources by competing populations. Such *resource partitioning* (Schoener 1974) should tend to reduce the intensity of competition. A population's resources include habitat, food, and other factors that are required for its persistence. Resource utilization is usually evaluated by arranging some dimension of a resource such as substrate type, water depth, water velocity, prey type, or prey size along an axis and then determining the range and magnitude of utilization of the resource dimension by the population. Figure 3.8a shows the resource utilization pattern of seven hypothetical populations at a particular level of resource availability and one combination of densities of competing populations. Each resource utilization curve is a partial representation of the *niche* of a population along one of its dimensions. The populations partition the resource by utilizing different ranges of the resource continuum. The size of the range of resources utilized by a particular population is its *niche breadth*. Although populations utilize different resource ranges, there is some *niche overlap* in resource use between adjacent populations.

Roughgarden (1974) was able to demonstrate organization of this kind by

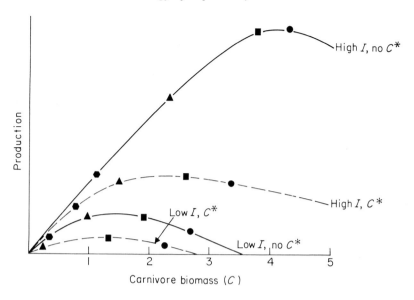

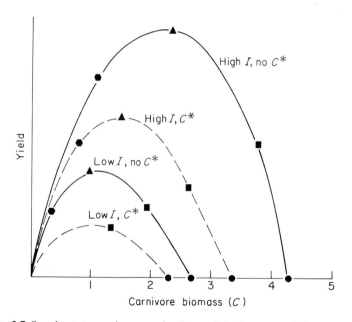

Fig. 3.7. Steady-state carnivore production and yield curves at different rates of light input are shown when a competing carnivore C^* is absent from the predation system shown in Fig. 3.3, and when it is present. The solid curves are identical to those shown in Fig. 3.6. On each curve, the steady-state production, yield, and biomass of the carnivore population at levels of fishing effort $0E$ (circles), $30E$ (squares), $90E$ (triangles), and $150E$ (hexagons) are shown.

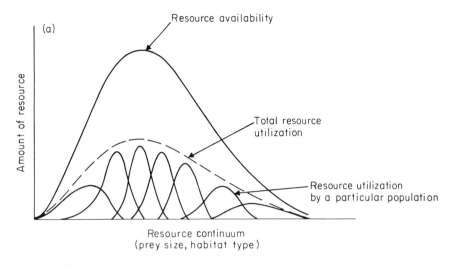

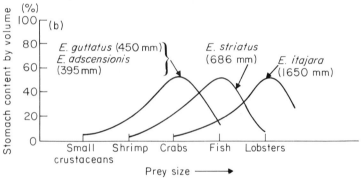

Fig. 3.8. (a) Curves showing the range and magnitude of utilization of a resource such as prey size or habitat type by each of seven hypothetical species populations at a particular level of availability of resources. Total resource utilization is the sum of the component species population utilization curves. After Pianka (1975). (b) Curves showing the utilization of prey resources by benthic groupers, genus *Epinephalus*. Data from Randall (1967), after Roughgarden (1974).

examining the patterns of food resource utilization of four coexisting species of benthic groupers, genus *Epinephalus*, on a coral reef (Fig. 3.8b). The various species of groupers differ morphologically in body size and, consequently, feed upon prey of different sizes and thus of different types.

There are two major ways that ecologists study the effects of interspecific competition on population distribution and abundance and the structure of systems of populations. One way involves adding, removing, or altering the density of one competitor and observing changes in resource utilization and density of the other competitor (see Colwell and Fuentes 1975, and Connell 1975, for detailed review). Werner and Hall (1976), by experimental manipulation,

Table 3.1. (a) Mean weight of individual fish of each *Lepomis* species in October at the end of the experimental period. Values are given in grams of dry weight± standard error. Weight of individual fish of each species at the beginning of the experiment was approximately 0.1 g (N > 100). (b) Percent contribution of each prey category to the diets of *Lepomis* species on the basis of dry weight and computed for the entire experimental period. After Werner and Hall (1976), copyright 1976 by the American Association for the Advancement of Science.

(a)

Fish	Species together	Species alone
L. macrochirus	1.29 ± 0.02	3.6 ± 0.15
L. gibbosus	1.21 ± 0.03	1.38 ± 0.04
L. cyanellus	1.34 ± 0.03	1.74 ± 0.04

(b)

Prey	*L. macrochirus*	*L. gibbosus*	*L. cyanellus*
	Species alone		
Vegetation dwellers	61	41	43
Benthic in- and epifauna	10	12	23
Open water zooplankton	8	1	1
Other	21	47	33
	Species together		
Vegetation dwellers	15	5	40
Benthic in- and epifauna	15	34	12
Open water zooplankton	33	6	4
Other	37	55	44

were able to demonstrate changes in resource utilization patterns in three species of sunfishes, *Lepomis*. The individuals of each species had greater growth rates when individuals of the other species were absent (Table 3.1a). This is an indication that food was more available and easier to obtain. When stocked separately, each species preferred vegetation for habitat and consequently fed primarily upon vegetation dwelling organisms (Table 3.1b). When the species were stocked together, habitat and food resource utilization patterns of two of the species changed. While *L. cyanellus* continued to inhabit the vegetation and prey upon vegetation dwelling organisms, *L. gibbosus* preyed more heavily on benthic food organisms and *L. macrochirus* on zooplankton in the open water. Thus, individuals of these species have the capacity to shift habitat and food resource utilization, this favoring the coexistence and persistence of the competing species.

A second and more widely employed means of studying the effects of interspecific competition is by comparing the resource utilization patterns of closely-related species in areas where they occur separately (allopatrically) with the patterns in areas where they occur together (sympatrically). The inference is then made that the observed differences in resource use are a result of interspecific competition. Some ecologists, however, have objected to this approach because some observed differences in resource use could also be accounted for by differences, between areas, in factors such as predator and parasite densities (Connell 1975; Wiens 1977). In a study of the marine snails, *Hydrobia ulvae* and *H. ventrosa*, Fenchel (1975) found that in areas where the species were allopatric, they had approximately the same body size distribution (Fig. 3.9). In areas where they were sympatric, however, *H. ulvae* was larger and *H. ventrosa* was smaller than when they were allopatric. The existence of morphological differences such as this in sympatric species is termed character displacement (Brown and Wilson 1956). Body size differences in these snails were related to differential food resource use, snails of different body sizes feeding upon different sizes of food particles. Thus, species have capacities to persist because their populations have evolved the capacities to exhibit different morphological patterns associated with different patterns of resource utilization.

To this point differences in utilization along a single niche dimension have been considered. Populations usually differ in their use of more than one resource. We have seen, in the case of *Lepomis* (Table 3.1), that differences between species in the use of one resource (habitat) may result in differences in the use of another (food). Some species may have a common range of utilization of one resource and reduce the intensity of competition by differing in their use of another resource. For example, species may occupy the same habitat but do so at different times; or they may feed upon different types of prey in the same habitat. Schoener (1974) discusses several examples of resource partitioning along multiple niche dimensions.

(*d*) *Commensalism and Mutualism.* Along with predation and competition, commensalism and mutualism are two other important processes that organize systems of populations. Commensalism is a relationship between two populations in which one is benefited and the other is assumed to be unaffected. Mutualism is a relationship between two populations in which both are benefited. Perhaps the most well known mutualistic relationship is that between flowering plants and their insect pollinators. As a result of this interaction, insects receive nutrients from plants and at the same time aid in the plant's pollination. Small marine fish and shrimp prey upon ectoparasites on the body, mouth, and gills of larger fish. Thus, while obtaining food from the larger fish these 'cleaner' fish and shrimp reduce the degree of parasitic infestation.

Too often commensalism and mutualism are viewed as rather special processes having only very limited roles in the organization of systems of populations. In terrestrial environments, however, trees provide habitat and

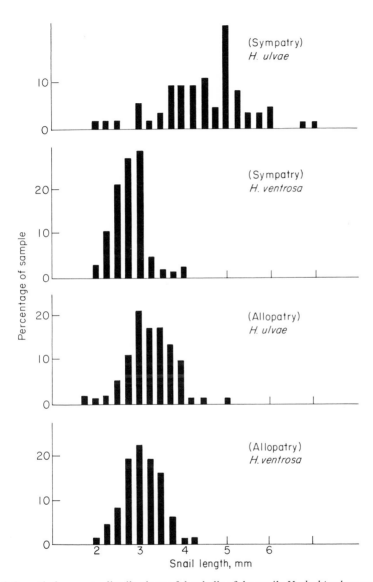

Fig. 3.9. Length-frequency distributions of the shells of the snails *Hydrobia ulvae* and *H. ventrosa* at an area where they occur together (sympatry) and at two areas where they occur separately (allopatry). After Fenchel (1975).

support the food of insect-eating birds. This commensalistic-like relationship is of unquestionable importance in the organization of terrestrial systems. A similar sort of relationship must exist in lakes between aquatic plants and organisms such as fish that forage upon invertebrates that inhabit the vegetation. In the next section, an example of the importance of mutualistic interactions in organizing systems of blue-green algae populations will be presented.

3.2.2 Population distribution and abundance and the structure and organization of systems of populations

We have examined how predation, competition, mutualism, and commensalism are each organizing processes in systems of populations and thus underlie and determine the structure and other performances of these systems. Now we can illustrate, with examples of different kinds of systems in different habitats, how these processes interact and together with physical environmental conditions determine population distribution and abundance and the structure of systems of populations.

Sessile and slow-moving organisms that inhabit the rocky intertidal zone of coastlines have provided unique opportunities for examination of the structure and organization of systems of populations. The distribution of adult and larval barnacles *Balanus balanoides* and *Chthamalus stellatus* and the factors determining their distribution on the coast of England are shown in Fig. 3.10a (Connell 1961b). The lower limit of distribution of *Chthamalus* on the intertidal rocks was determined by interspecific competition for space with *Balanus*. Connell demonstrated this by showing that *Chthamalus* were able to persist in the lower tidal zones when *Balanus* were removed from among the settled *Chthamalus* larvae (Fig. 3.10b). *Balanus* when present were able to undercut or grow over *Chthamalus* and thus severely reduce their survival. The upper limits of distribution of both species appear to be determined by physical environmental factors. *Chthamalus* exists higher on the shore than *Balanus* apparently because it is more resistant to dessication and can thus tolerate longer periods of exposure to air. This adaptation enables *Chthamalus* to persist in an area of high intertidal rocks from which *Balanus* is absent. Below midtidal level, densities of the predatory snail *Thais lapillus* increase and predation becomes an important process affecting the lower limits of distribution of *Balanus*. Thus the distributions of these species are determined by both physical factors (dessication) and biological interactions (competition, predation).

Predation and competition have been shown to be important in determining the structure of systems of populations of intertidal organisms also on the coast of New England. Menge (1976) examined the structure and organization of rocky intertidal systems in areas that were exposed to severe wave action and in relatively more protected areas. In the mid-intertidal zone at exposed areas, barnacles (*Balanus balanoides*) and mussels (*Mytilus edulis*) occupy most of the

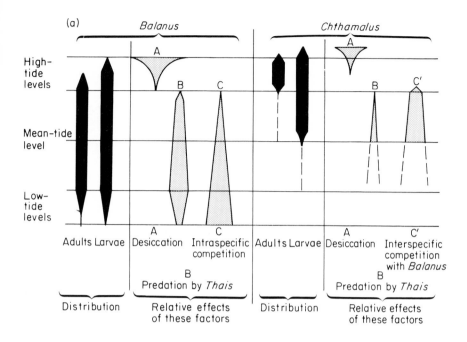

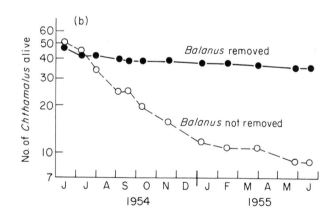

Fig. 3.10. (a) Intertidal distribution of adults and newly settled larvae of the barnacles *Balanus balanoides* and *Chthamalus stellatus* at Millport, Scotland, with a diagrammatic representation of the relative effects of the principal factors determining their distribution. After Connell (1961b). (b) Survival of cohorts of *Chthamalus* at Millport, Scotland when *Chthamalus* grew in contact with *Balanus* and when *Balanus* had been removed. After Connell (1961a).

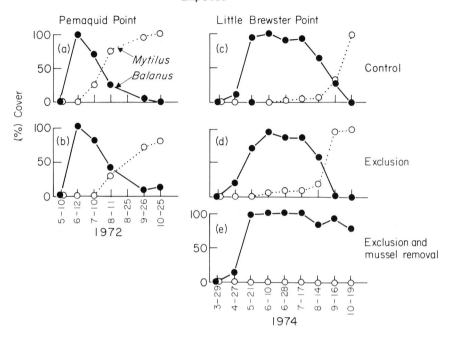

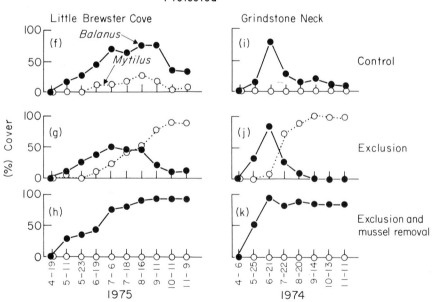

available primary space on the intertidal rocks. Larvae of the barnacle settle first in the spring (Fig. 3.11a, c). But as the mussel larvae settle and begin to grow among the barnacles, the abundance of the barnacles declines. If mussels are removed (Fig. 3.11e), barnacles are able to persist throughout the season. The mussel removal experiments indicate that *Mytilus* outcompetes the *Balanus* for space and, later in the season, may effectively exclude the barnacles from many areas.

Some areas of the mid-intertidal zone were caged to exclude the predator *Thais lapillus* and thus determine the effect of predation on the competitive interaction between barnacles and mussels. In the exposed areas, *Thais* were not abundant and thus the results of predator exclusion experiments were similar to controls (Fig. 3.11b, d). In protected areas where physical conditions are less severe, a somewhat different system structure develops (Fig. 3.11, *bottom*). Here the density of *Thais* is relatively high, and these predators are capable of reducing the densities of both barnacles and mussels to low levels toward the end of the season. When *Thais* are excluded by cages, mussels increase in density and outcompete barnacles for space (Fig. 3.11g, j), much as they did in the exposed areas where *Thais* abundance was low.

Predation by *Thais* reduces the intensity of competition between mussels and barnacles by reducing their densities. Thus, in the presence of *Thais*, *Balanus* is able to persist for longer periods of time (Fig. 3.11f, i). Moreover, by removing barnacles and mussels from the substrate, predation by *Thais* permits the successful colonization and persistence of competitively inferior species such as limpets and the alga *Fucus* (Menge 1976). In the New England rocky intertidal zone, predation can reduce the densities of the dominant competitors and, thus, a greater number of species are capable of coexisting at protected areas than at exposed areas (Fig. 3.12)—that is, the systems in protected areas have greater species diversity.

The intensity of predation has significant effects on the persistence of barnacles and mussels and the structure of this mid-intertidal system of populations. Because the harshness of the physical environment in part determines the abundance and effectiveness of predators such as *Thais*, it has indirect as well as direct effects on the structure and organization of this system of

---

Fig. 3.11. Relative abundance, as measured by percentage of rock substrate covered, of the barnacle *Balanus balanoides* and the mussel *Mytilus edulis* in the mid-intertidal at four areas in New England. Graphs a–e represent areas exposed to severe wave action, and graphs f–k represent areas which were protected from waves. Some areas of the rock substrate were covered with stainless-steel mesh predator exclusion cages to assess the impact of the absence of predators on *Balanus* and *Mytilus* abundance (b, d, g, j). On other areas of substrate, the effects of both predator exclusion and removal of mussels on the abundance of *Balanus* was determined (e, h, k). After Menge (1976) and Menge and Sutherland (1976). Reprinted from *American Naturalist* by permission of The University of Chicago Press. © 1976 by The University of Chicago Press.

Chapter 3

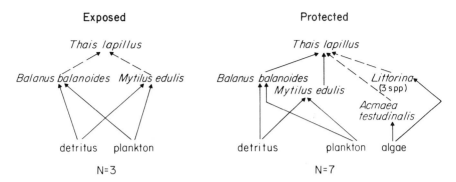

Fig. 3.12. Representation of the structure of rocky intertidal systems at areas in New England that are protected from and exposed to severe wave action. Number of species in each system is indicated below each food web. After Menge and Sutherland (1976). Reprinted from *American Naturalist* by permission of The University of Chicago Press. © 1976 by The University of Chicago Press.

populations. Again it is evident that the structure and other performances of a system of populations are determined by the interactions between all of the populations—the organization of the system—together with the environment of the system. When environmental conditions change—with the environment becoming more or less 'harsh'—the structure of the system changes as a result of direct as well as indirect effects on the populations.

The work of Castenholtz and his colleagues (Castenholtz 1968, 1973; Wickstrom and Castenholtz 1973), summarized by Colwell and Fuentes (1975), is an exceptionally fine example of the interaction of predation, competition, and mutualism or commensalism in determining the distribution of blue-green algae along a temperature gradient in a hot spring channel (Fig. 3.13). One of the species of algae *Synechoccus lividus* is composed of four genetic strains. The potential range of temperatures at which each strain was capable of growing was determined in pure culture. There is some difference in potential growth range between strains, but yet there is a great deal of overlap. The potential growth range of the species extends from 30° to 73°C. In the hot springs channels, the range of temperatures occupied by each of the potentially highly competitive strains is considerably reduced so that at any temperature no more than two strains of *Synechoccus* are abundant.

The potential growth range of *Oscillatoria terebriformis* extends from 30°C to 53°C, but it is killed by higher temperatures. In the hot spring channels where *Oscillatoria* and *Synechoccus* are sympatric competitors, development of an *Oscillatoria* mat prevents growth of *Synechoccus* as a result of shading and thus effectively excludes *Synechoccus* from temperatures below 53°C. Ostracods of the genus *Potamocrypsis* are predators on *Oscillatoria* and are capable of surviving in temperatures up to 48°C. Grazing by ostracods resulted in

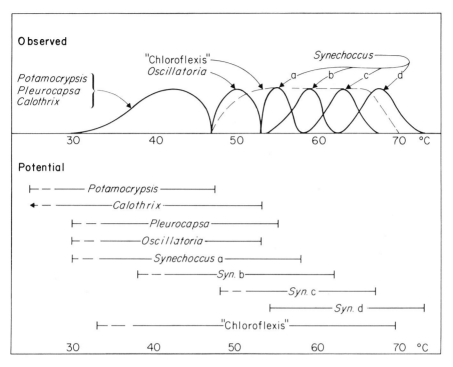

Fig. 3.13. Observed and potential distributions of species along a temperature gradient in an outflow channel of a hot springs in Oregon. The 'observed' frequency distributions are intended to show approximate temperature limits only, and do not represent actual measurements of density. The 'potential' distributions show the approximate thermal limits for growth in laboratory culture. After Colwell and Fuentes (1975). Reproduced, with permission, from the *Annual Review of Ecology and Systematics*, volume 6, © 1975 by Annual Reviews, Inc.

restriction of the utilizable temperature range of *Oscillatoria* to 48°–54°C. Thus the lower limit of *Oscillatoria* distribution was determined by predation and the upper limit by physiological tolerance to heat.

Blue-green algae of the genus *Pleurocapsa* inhabit substrate that is kept free of *Oscillatoria* by grazing ostracods. *Pleurocapsa* cells are thick-walled and form crusts and nodules that protect it from ostracod grazing. A mutualistic relationship is thought to exist between *Pleurocapsa* and algae of the genus *Calothrix*. Filaments of *Calothrix* are embedded in the *Pleurocapsa* mass and thus receive protection from grazing. In turn, there is evidence that *Calothrix* fixes free nitrogen which is required by *Pleurocapsa*. Besides the interaction between *Pleurocapsa* and *Calothrix* there is also a mutualistic relationship between a mat of microorganisms called 'Chloroflexis' and *Synechoccus* and *Oscillatoria*. Evidently these algae require Chloroflexis as a substrate and provide organic matter that is necessary for the persistence of the microorganisms.

The processes of predation, competition, commensalism, and mutualism organize systems of populations. The way a system is organized—the interrelationships between its parts—determines its capacity to manifest different structures, where we take system structure to be the species composition and abundances and distributions of the species populations in time and space. The organization of the system of populations, together with factors in the environment of the system such as fishing effort and physical factors, determine the particular system structure that exists at each particular time. The populations in systems are coadapted so that the capacity of each population is concordant with the capacities of all other populations with which it interacts, either directly or indirectly. Distribution, abundance, resource utilization, and other population performances are determined not simply by predation, competition, mutualism, or commensalism but by all of these processes acting together. Thus, we come to better understand population performances within the context of systems of populations.

3.3   THE STRUCTURE AND ORGANIZATION OF BIOLOGICAL COMMUNITIES

3.3.1   On so conceiving of communities as to make them understandable
There has developed increased appreciation of biological communities, which together with their physical environments form ecosystems. With this appreciation and concern has come the expectation that resource management agencies give more regard to these natural systems.

Over the past 50 years, community conceptions, insofar as they have been expressed at all, have unfortunately been polarized between the *metaphorical-organism view* of Clements (1916) and the *individualistic-population view* of Gleason (1917, 1926) and others. These views were taken to be mutually exclusive hypotheses somehow amenable to empirical testing. The difficulty is that each view makes apparent different characteristics of a community, neither view is falsifiable, but then neither alone is adequate. We need somehow to unify the strengths of these views in an extended and more adequate conception of communities.

Clements (1916) saw the biological community as a continuum of successional states, as a developing system, as a thing extending in time as well as space. But his metaphor of a community being an organism developing and reproducing with fidelity was an overstatement that troubled many ecologists. Gleason (1917, 1926) and his followers believed that communities are more nearly aggregations or, at most, very loose systems of populations. These species populations, in this view, are not highly coadapted in an evolutionary sense and thus can and do distribute themselves quite independently in the colonization of communities. In the individualistic conception, community kinds are not clearly distinguishable and do not exist as natural entities. This individualistic conception is inadequate, but it emphasized another important aspect of communities: there is a high element of probability as to the sequence of species appearing in community colonization and development.

We take a biological community to be an organismic system—that is, a system conforming to the generalizations about capacities, environments, and performances first considered in Chapter 2. A biological community develops through time as a recognizable kind of thing, as Clements suggested. It is individualistic in the sense that it will not be exactly like any other community, because its habitat will in some degree be different and because there is a large element of probability in when and what species will arrive to colonize. But the species populations that become established form systems of closely interacting populations, similar to those considered in earlier sections. These systems of populations are coupled one to another finally to form a community. The community has incorporated these lower level systems and thus has come to have its own level specific capacities and performances. It is to such community organization, capacities, and performances that the remainder of this chapter will be devoted.

3.3.2 The apparent structure of communities

The organization of a system was defined as the ways its subsystems are interrelated in determining its capacities and its performances such as structure, development, and persistence. The structure of a biological community not only is a performance of a community at any time but also is an important aspect of its organization. Community structure is more apparent—is more observable— than is organization. We will take the *structure* of a biological community to be the kinds of organisms present in a community, their relative abundances, their distributions in time and space, and the superficial appearance that these give to the community.

The structure and organization of any community is very much determined by the medium and substrate forming its habitat. The structure or form of aquatic communities appears very different from that of terrestrial communities. A lake community (Fig. 3.14b) does not have the physical structure of trees (Fig. 3.14a) bearing 'photosynthetic surface' and providing habitats for other species. And yet the phytoplankton of a lake—its 'photosynthetic surface'—is supported by the physical structure of the water, which also physically supports the organisms preying on the phytoplankton. The emergent vegetation along the shores of the lake and the shrubs at the edge of the forest each provide further structural similarities between a lake and forest community. In near shore and intertidal communities, the foundation of marine community structure in different respects resembles that of both forests and lakes. Kelp, the giant red alga, extends from its seabed holdfasts to its flotation devices and fronds at the surface (Fig. 3.14c), this providing structure for the community in ways not unlike the trees of the forest. The marine community, like the lake, has its plankton subsystem. Rocks provide the basic structure for the rocky intertidal sub-community. Rocky riffles and carved-out pools provide for the community structure of a small stream (Fig. 3.14d). Upon these surfaces are supported the organisms of the stream community.

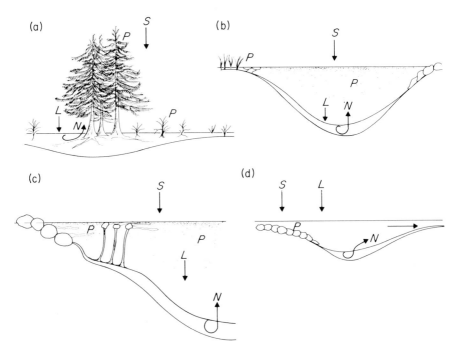

Fig. 3.14. Idealized or simplified representations of analogous features of (a) forest, (b) lake, (c) marine, and (d) stream communities, where *S* represents solar radiation, *P* represents areas of photosynthesis, *L* represents litter or organic debris generation and path, and *N* represents nutrient generation.

Apparent structure provides the basis of community organization, the habitat aspect or 'template' of organization. Another facet of community structure is *species composition*. Representation of the structure of systems so extensive, diffuse, and heterogeneous as biological communities seems to demand simplification. Various forms of graphs of *species-abundance relations* are simplified representations of the species composition aspect of community structure. One such form of graph is obtained when some measure of species relative abundance or dominance is plotted against species rank positioned along the x-axis from most to least abundant species, as shown in Fig. 3.15. When a

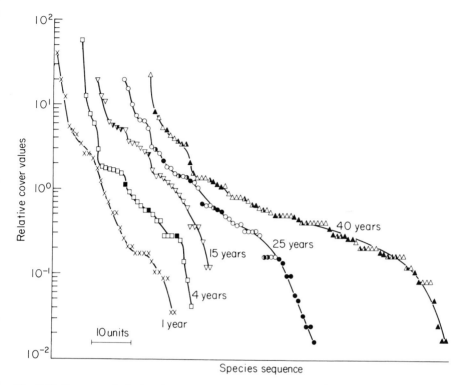

Fig. 3.15. Patterns of relative abundance of plant species in old fields in southern Illinois after 1, 4, 15, 25, and 40 years of abandonment. Relative cover values are the percentages that particular species contribute to total area covered by all species. Values are plotted against the ranks of the species, ordered from most to least abundant. The symbols are open for herbs, half-open for shrubs, and closed for trees. After Bazzaz (1975).

large and rather heterogeneous collection of species is represented in this way, a descending curve results. In the plant community shown in Fig. 3.15, after 40 years of succession, there are a few species that are relatively abundant, a few that are relatively rare, and most that are of intermediate abundance. In the early years of succession, the species-abundance relation appears more linear. The communities of streams that have become polluted tend to be dominated by relatively few tolerant species and thus have species-abundance relations more like those shown for the early successional stages of the plant community in Fig. 3.15. Before pollution these streams tend to have more sigmoidal species abundance relations (Patrick 1972, 1975). Some ecologists have claimed the biological basis of the more linear relations to be resource partitioning; nothing more than chance may determine the form of the more sigmoid curves (May 1975; Pielou 1975). We should not expect to find biological explanation in the form of such graphs; we have little reason to suppose them to be more than convenient summaries of data.

Species diversity indices have been of interest to ecologists, sometimes as an index of the status (the structure) of communities in polluted waters (Warren 1971). Diversity can be thought of as having two components: a *species richness* component and a *species evenness* component. The number of species in a community is essentially what is meant by species richness, and thus it is the simplest expression of this aspect of diversity. And a community will seem to be more diverse if the numbers of individuals in the community are distributed more evenly among all its species. As succession progresses from 1 to 40 years, the plant community shown in Fig. 3.15 becomes more diverse and would have higher diversity indices, because the number of species increases and the distribution of individuals among the species becomes more even. Many different sorts of diversity indices have been proposed. Although any one of them may be convenient and adequate for some purposes, none is a very good summary of the underlying species abundance relation (May 1975).

Community diversity has continued to excite ecological interest of two sorts: the physical and biological reasons for high or low diversity; and whether or not more diverse communities are more stable than less diverse communities. Several reasons have been proposed for differences in community diversity (Pianka, 1966). It has been supposed that communities that have had long periods of time for colonization and evolution will have the highest species diversity, especially if there has been much *climatic stability*. Old-world tropical communities are usually given as an example. *Habitat diversity* is sometimes supposed to be another reason for species diversity. And systems that have the greatest capacity to support plant production, the greatest *productivity*, could support more species. In addition, the kinds and degrees of *competition* and *predation* would be involved in determining species diversity as discussed in Section 3.2.2. All of these 'hypotheses' are importantly involved in determining the species diversity of communities. Each is but a partial representation of interactive processes in which all are importantly involved and from which none can be meaningfully separated.

3.3.3   Habitat, life history, and trophic organization of communities

Understanding of the structure, development, and persistence of biological communities must be based on comprehension of their organization. The energy and material relations that interconnect the parts of a community form its *trophic organization*. This trophic or nutritional aspect has occupied much of the interest of ecologists for many years. But the trophic relations of communities are immensely complex, and this has made it difficult to achieve much understanding of their trophic organization. *Habitat organization, life history organization*, and other equally important aspects of communities tended to be neglected as studies

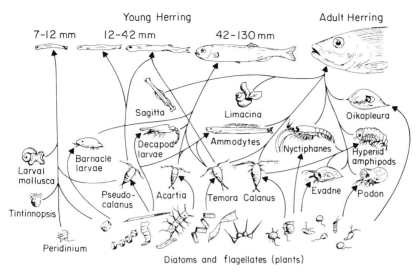

Fig. 3.16. Simplified representation of the food web of the herring. Changes in food relations occur as the herring develops. The herring feeds upon at least two trophic levels. Data from Hardy (1959), after Warren (1971).

of trophic relations became popular. Let us begin with some of the ideas, observations, and difficulties of trophic ecology.

In Fig. 3.16 a greatly simplified though still complex representation of the food web of herring is shown. The food habits of the herring change with life history development, as do those of most animals. Not shown are the seasonal changes in species relationships within the food web. Dependence of the herring on many species of food organisms favors the persistence of predator and prey populations and even the fishery on herring, because environmental changes and consequent changes in the abundance of some prey species need not endanger the food resources of the herring. Darnell (1961) studied the food relations of the fish and larger molluscs in an estuarine community and concluded that most of the species change their food habits as they develop, most are nutritionally opportunistic, most depend for food on many species of different trophic kinds, and organic detritus is the most common food item. One cannot help but to conclude that trophic relations are so universally complex that it may be difficult to see them in some essentially simple, ordered, and unified way, as required for understanding.

Lindeman (1942) proposed that communities could be understood by organizing their species populations into *trophic levels* and emphasizing *energy and material transfer* among these levels. In this view, there is a plant trophic level on which all plants, regardless of species, are placed. The next trophic level includes those animal species that consume plants, all the herbivores. Then there is a level for carnivores that eat herbivores and one for those that consume only carnivores. Finally, there is a group of organisms that utilize and decompose organic detritus. Thus a complex community could, in theory, be reduced to perhaps only five trophic levels—five parts or objects. The relationships between trophic levels are represented as rates of energy and material transfer and utilization, as shown in Fig. 3.17.

One problem with such a view of trophic level organization is that often, because of their complex feeding habits, species cannot be assigned to a single trophic level. The complexity of the feeding relations of the herring and the organisms in Darnell's estuarine community illustrate the difficulties of assigning organisms to particular trophic levels.

What are some of the observational requirements of the trophic dynamic view of biological communities and their physical-chemical environments? For at least the common species of any community, the energy and material relations—the food web—of each ought to be qualitatively known. This, of course, is a tremendous task, so much so that few have even attempted it. Now if the food organisms of the common species in a community were known, we would need to be able to estimate for each species the quantity of each prey species it consumes, say in calories of energy per day per unit area. And of the total energy consumed by a particular species, the amount that is assimilated would need to be estimated, and, of this, the amount allocated to production or growth of that species and respiration and maintenance would need to be determined. For the animals, this must be done species by species, for many species, if we are indeed to be able to reliably estimate energy and material transfer and utilization among the trophic levels of a community. In short, the kinds of information required for trophic dynamic analysis are extremely difficult to obtain. For some kinds of animal populations, reliable production estimates can be obtained with considerable effort. But estimates of energy and material intake and respiration for populations in nature are probably no better than order of magnitude ones. All of this has repeatedly been emphasized (Ivlev 1945; Darnell 1961; Warren 1971; Rigler 1975).

It has long been recognized that energy and material transfer rates are of interest only if a system is in a steady state. But it has not been generally recognized that even then they would not be of much interest, because there are an indefinite number of sets of steady states, each determined by a particular set of environmental conditions such as plant resource input and level of fishing effort, as previous examples illustrate (Figs. 3.3 and 3.17).

Biological communities have many structures and incorporate many life history patterns and many species populations and their interrelations. How can all of this be ordered to achieve the simplicity and unification that must exist if we are to understand communities? One way is to conceive of a community as a hierarchy successively incorporating interesting subsystems. At the top of this

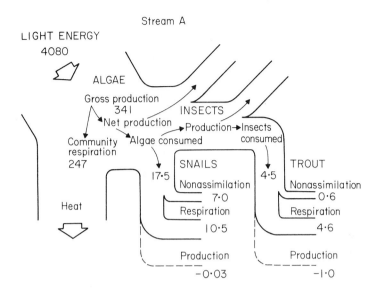

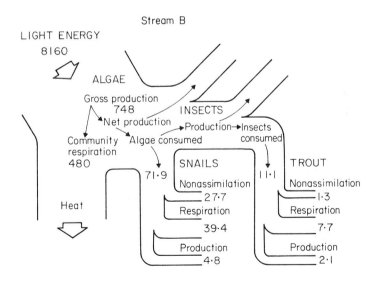

Fig. 3.17. Energy transfer in laboratory stream communities exposed to different light levels. Energy values are given in kilo-calories per square meter for an 80 day period. Differences in light energy result in differences in energy transfer rates between trophic levels. After Warren (1971).

hierarchy we want only a few major subsystems, so as to be able, at this level, to think about a community. Each of these major subsystems can then be decomposed or reduced to its subsystems. The important thing is to choose or conceive of objects or subsystems on each level so as to make then conceptually rich and interesting.

A *habitat organization* view of biological communities can provide us with major community subsystems having successive levels of habitat structure and life history as well as trophic organization.

Thus in a lake the littoral habitat, the pelagic habitat, and the benthic habitat support the three major subsystems of the lake community (Fig. 3.18). Within the

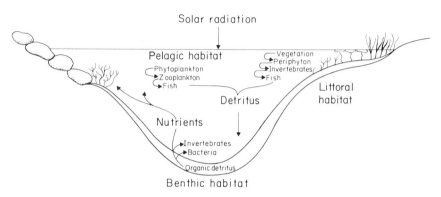

Fig. 3.18. An idealization of the relationships between the subhabitats and subcommunities involved in the organization of a lake community.

littoral subsystem there is much habitat substructure, provided by emergent and submerged vascular plants, by rocks, by sand and silt and, for parasites, on and within some of the animals. Life history and trophic relations are quite close and concordant in the littoral subsystem. Within the pelagic and benthic subsystems there is perhaps less physical habitat differentiation, but the life history and trophic relations of these subsystems are no less closely coupled. Between these community subsystems there are physical and biological relations—'girder systems' (Elton 1966)—that make them together a community. The habitat organization of marine communities and stream communities (Fig. 3.14) could be similarly analyzed, but perhaps the point is made: habitat organization provides community subsystems as objects of understanding.

Such major community subsystems are maintained by the *life history organization* of communities. Most simply, this means that each of these major habitats can be populated only by species having life history patterns concordant with them. Within these habitats, there must be concordance of the life history patterns of the different species: among the phytoplankton, zooplankton, and fish, in the pelagic subsystem; or among the vascular plants and the organisms dependent on them in the littoral subsystem.

The *trophic organization* of communities must, along with habitat and life history organization, be fundamental to our understanding of community organization. Trophic organization is not simply a matter of trophic levels and rates of energy and material transfer. We must consider trophic structure—the kinds of organisms present in the community, their life history patterns, and their distributions and abundances. But trophic organization entails not only the kinds of organisms present in the community but also the relationships between populations of these organisms such as predation, competition, commensalism and mutualism, as discussed in Section 3.2. These relationships provide the explanation of food web structure. The basis of a lake economy is determined by the production of phytoplankton in the pelagic subcommunity, the production of algae and macrophytes in the littoral subcommunity, and the deposits of organic detritus in the benthic subcommunity. Each subcommunity has trophic specialists and trophic generalists. Trophic specialists among the zooplankters determine the structure or species-abundance relation in the phytoplankton (Wetzel 1975) much as fish determine the structure of the zooplankton (Brooks and Dodson 1965).

Finally, community organization determines the capacity of a community for performances such as structure, development, and persistence. A lake community as a whole *incorporates* its subsystems and their level-specific environments (Fig. 2.3). By incorporating the pelagic subcommunity, the lake community as a whole incorporates or makes part of itself the capacity of the pelagic subcommunity. It also incorporates the capacities of the littoral and benthic subcommunities. Thus, in a very important sense, the lake community as a whole has the combined capacities and all emergent possibilities of its pelagic, littoral, and benthic subcommunities together. The performances of the lake community as a whole—its structure, development, and persistence—result from the interactive performances of its subcommunities. And these subcommunity performances are in large part determined by the community as a whole, because the community incorporates the environments of its subcommunities. Thus, the nutrients needed by the pelagic subcommunity are largely generated from organic deposits by the benthic subcommunity. And these deposits derive from organic production in the pelagic and littoral subcommunities. Incorporation and concordance of capacities and environments occur through all the levels of the community hierarchy: subcommunities, systems of populations, populations, and life history patterns, each with its level specific capacities and environments, all organized to form communities as we understand them.

3.4 THE ORIGIN, DEVELOPMENT, AND PERSISTENCE OF COMMUNITIES

3.4.1 On development of a community of a kind

It is not enough to know the structure and organization of communities, how they operate, or even what their functions in still more incorporating systems may be. We must also have some understanding of how communities come into being. There are, perhaps, two aspects of the origin and development of communities. The first aspect is known as community succession or *community development*. This is the development, at some location and over some period of time, of a particular community of a recognizable kind or class of communities. The other aspect of the origin of communities pertains to how such a kind or class of communities came into being in the first place. To the extent that there are recognizable kinds of communities, there must have been some sort of *evolutionary origin* of communities.

Clements' *metaphorical-organism* view has given us a conception of communities as definite things developing through time as well as extending in space. The *individualistic-population* view emphasized other aspects of communities: their colonists come from other communities, perhaps of different kinds, and with a considerable element of chance; and communities are not easily classifiable into kinds. Both views should be taken into account in explanation and understanding of the origin and development of communities. Community development is a performance made possible by suitable community habitat and the 'pool' of species from which the colonists come. Such a *species pool* is the potential capacity making possible the development of a community.

Probably the most familiar example of development of aquatic communities is that occurring as a lake becomes shallower with deposition of organic materials and smaller with encroachment of littoral vegetation, eventually to become a marsh, then a meadow, and perhaps finally a forest. This is commonly known as a *hydrarch succession*. A newly formed lake habitat is usually first colonized by small organisms such as algae, rotifers, cladocera, and insects from nearby bodies of water. As time goes on, conditions change in the lake and colonization by additional species becomes possible. Some of the earlier inhabitants may no longer find conditions suitable. Physical conditions and the composition of the biological community thus change through some sequence of states to some much more persistent state, a *climax*.

The nature and development of stream communities have been difficult for ecologists to conceptualize. Margalef (1960) emphasized the downstream export of information and materials keeping upstream communities pioneer-like and leading to more 'mature' downstream ones, a sort of succession through space. Small stream communities together with their stream habitats are embedded in their watershed environments and develop with them through time. In forested watersheds, the removal of trees dramatically changes the nature of the stream communities. Then, as forest succession progresses, the stream communities develop through recognizable stages until the forest climax is again reached. Stream community development ceases only when the environment of the stream system, the forested watershed, reaches its climax or most persistent state. During the development of a particular stream community, the colonizing organisms come from other stream communities in different stages of their own development, that is, from the species pool of the watershed and its biogeoclimatic zone. Such stream community development is largely externally determined by forest succession. In our example of lake development, the sequence was more internally determined by the production of organisms and the accumulation of organic materials within the lake.

In a review of empirical work relevant to development of intertidal communities, Connell (1972) concluded that the species sequence during colonization may be quite predictable. This results from superior colonizing abilities of the earlier appearing species, not from any necessary organismic preparation of the habitat for later arrivals. This view reflects the individualistic-population aspect of community development. Out of any given species pool, the sequence of colonization of a given sort of habitat will depend on the characteristics of the different species. And although different roles within a community must be filled, the species filling them may be substitutable if there are similar species available in the species pool.

But even with such possible substitution of species in communities, there are distinguishable or recognizable kinds of communities—there are classes of communities into which particular communities can be placed (Warren 1979). Such a recognizable kind of community will develop on a site providing a suitable habitat. On an adjacent site a different kind of community may develop, because there habitat conditions are different. Over a large area, there will be a pattern formed by such communities, some different because their primary habitats are of different kinds, some the same because their habitats are of the same kind. A *biogeoclimatic zone* is a system of such communities and community habitats. These communities share a common species pool, all the species inhabitating the area occupied by the communities. The *potential capacity* for different kinds of communities to develop on different kinds of habitats within the biogeoclimatic zone resides in this species pool. For particular habitat and environmental conditions then, a particular kind of community will develop within a given biogeoclimatic zone.

3.4.2 On historic origin and change in community kinds

Neither of the major community conceptions has led to an adequate account of origin and change in community kinds (Whittaker and Woodwell 1972). The metaphorical-organism view emphasized a fidelity of replication and development that individual communities do not have. Those holding the individualistic population view of communities argued that since communities do not reproduce with fidelity, they cannot be 'natural kinds' of things in the sense that individuals and populations of the same species are natural. If communities are not such kinds of things, their evolution cannot—indeed need not—be explained. But any science including community ecology must have classes of things, for otherwise it cannot generalize. Community ecology must seek to explain and understand kinds or classes of communities. An important part of this understanding would be how such kinds of communities could originate and could change: how communities could evolve. Communities could not evolve in the same way as species populations, because communities are things of a different sort.

To understand the origin and change of community kinds, we must keep in mind not simply particular communities of a kind but also the entire system of communities within a biogeoclimatic zone. A particular community of a kind develops on suitable habitat as a result of colonization by species coming from other communities in the zone, that is, from the species pool. For the kinds of habitats available in the zone, a limited number of recognizable kinds of communities can develop. The development of any such community is an expression or performance of the potential capacity of the species pool of the biogeoclimatic zone.

If any of these kinds of communities were to change considerably but remain recognizably of that kind, or if an entirely new kind of community were to appear in the zone, we would be justified in saying that community evolution had occurred. But how could this take place? There would need to be a considerable change in the potential capacity or species pool of the biogeoclimatic zone, for it to express itself in changed communities or new community kinds. *Community evolution*, most fundamentally, would be a change in the potential capacity—the species pool and all its possible interactive performances—of a biogeoclimatic zone. This could come about in many ways, including new and very different species colonizing a zone, coevolution of species, and vast geoclimatic changes with their consequent effects on entire biotas.

3.4.3   On matters pertaining to community persistence

Scientifically, practically, and aesthetically, man is interested in the persistence of particular individual communities, in the persistence of particular community kinds, and in the persistence of zonal and regional community complexes or systems of communities. Some shift of scientific and management focus to the organizational levels of communities and systems of communities makes apparent scientific and management problems and possibilities that are not visualizable in the traditional population view. Among these is the concern for the understanding and maintenance of species pools and the diversity of habitats in which this potential finds its expression in communities of different kinds. Our society has greatly altered many stream, lake, wetland, and estuarine habitats. And the potential capacities of species pools have been drastically altered by introduction of exotic species, overexploitation of species of interest, destruction of populations supposedly not of immediate concern, and the intentional application of pesticides whose effects are little known. In the narrowest view, not even particular populations of interest—particular resource uses—have always benefited. In a broader view, without understanding, we hardly know what the losses may be. The persistence of communities depends upon our understanding and maintenance of their potential capacities—their species pools—and their kinds of habitats.

3.5 CREATIVITY IN ECOLOGICAL OBSERVATION AND THOUGHT

Ecological science is a human endeavor, its knowledge a human product. It is an endeavor we should seek to understand, if we are to do it well and with enjoyment. Most scientists and many philosophers, in their concern that scientific knowledge be objective, have supposed science to be very different from other human endeavors. It is almost as if they were trying to take the nature of man out of science, as though this were possible. We think this to be most unfortunate, because it diminishes scientific knowledge as well as the vitality and enjoyment of scientific practice.

Man seeks understanding and is creative in all his endeavors. To separate the essential nature of science from art, music, literature, and the practical affairs of man is to lose a unified way of comprehending and practicing science. Whatever else scientific knowledge may be, it is a creative product of the human mind, the human imagination in its search for understanding. This creative product must conform broadly with human experience, or it would not be explanation and understanding.

At its best, ecology is the creative endeavor to understand ecological systems, and it should be practiced and appreciated as such. Out of theory, explanation, and observation, ecology creates objects of understanding that accord with our ideals of understanding and criteria of explanation. Creativity in the search for understanding must be emphasized, not only that ecological science would be better but also that it would be unified with other human creativity and understanding.

3.6   REFERENCES

Bazzaz F.A. (1975) Plant species diversity in old-field successional ecosystems in southern Illinois. *Ecology,* **56**, 485–488.

Booty W.M. (1976) *A Theory of Resource Utilization.* M.S. Thesis, Oregon State University, Corvallis.

Brocksen R.W., G.E.Davis, and C.E.Warren (1968) Competition, food consumption, and production of sculpins and trout in laboratory stream communities. *J. Wildl. Manage.* **32**, 51–75.

Brocksen R.W., G.E.Davis, and C.E.Warren (1970) Analysis of trophic processes on the basis of density-dependent functions. *In* J.H.Steele (ed.), *Marine Food Chains,* pp.468–498. Berkeley: University of California Press.

Brooks J.L. and S.J.Dodson (1965) Predation, body size, and composition of plankton. *Science,* **150**, 28–35.

Brown W.L. and E.O.Wilson (1956) Character displacement. *Syst. Zool.* **5**, 49–64.

Castenholz R.W. (1968) The behavior of *Oscillatoria terebriformis* in hot springs. *J. Phycol.* **4**, 132–139.

Castenholz R.W. (1973) The ecology of blue-green algae in hot springs. *In* N.G.Carr and B.H.Whitten (eds.), *The Biology of Blue-Green Algae,* pp.379–414. Oxford: Blackwell.

Colwell R.K. and E.R.Fuentes (1975) Experimental studies of the niche. *Annu. Rev. Ecol. & Syst.* **6**, 281–310.

Connell J.H. (1961a) Effects of competition, predation by *Thais lapillus*, and other factors on natural populations of the barnacle *Balanus balanoides. Ecol. Monogr.* **31**, 61–104.

Connell J.H. (1961b) The influence of interspecific competition and other factors on the distribution of the barnacle *Chthamalus stellatus. Ecology,* **42**, 710–723.

Connell J.H. (1972) Community interactions on marine rocky intertidal shores. *Annu. Rev. Ecol. & Syst.* **3**, 169–192.

Connell J.H. (1975) Some mechanisms producing structure in natural communities: a model and evidence from field experiments. *In* M.L.Cody and J.M.Diamond (eds.), *Ecology and Evolution of Communities,* pp.460–490. Cambridge: Belknap.

Clements F.E. (1916) *Plant Succession.* Washington: Carnegie Institute of Washington Publication 242.

Darnell R.M. (1961) Trophic spectrum of an estuarine community, based on studies of Lake Pontchartrain, Louisiana. *Ecology,* **42**, 553–568.

Elton C.S. (1966) *The Pattern of Animal Communities.* London: Methuen.

Enright J.T. (1976) Climate and population regulation: the biogeographer's dilemma. *Oecologia,* **24**, 295–310.

Fenchel T. (1975) Character displacement and coexistence in mud snails (Hydrobiidae). *Oecologia,* **20**, 19–32.

Gause G.F. (1934) *The Struggle for Existence.* Baltimore: Williams and Wilkins. Reprinted, 1964 New York: Hafner.

Gleason H.A. (1917) The structure and development of the plant association. *Bull. Torrey bot. Club,* **44**, 463–481.

Gleason H.A. (1926) The individualistic concept of the plant association. *Bull. Torrey bot. Club,* **53**, 331–368.

Hardy A. (1959) *The Open Sea: Its Natural History*. Part 2, Fish and Fisheries. Boston: Houghton Mifflin.

Howard L.O. and W.F.Fiske (1911) The importation into the United States of the gipsy-moth and the brown-tailed moth. *Bull. Bur. Ent. U.S. Dep. Agric.* **91**. 344pp.

Ivlev V.S. (1945) The biological productivity of waters. (Translated from Russian by W.E.Ricker. *J. Fish. Res. Board Can.* **23**, 177–1759.)

Lindeman R.L. (1942) The trophic-dynamic aspect of ecology. *Ecology*, **23**, 399–418.

Margalef R. (1960) Ideas for a synthetic approach to the ecology of running waters. *Int Revue ges. Hydrobiol.* **45**, 133–153.

May R.M. (1975) Patterns of species abundance and diversity. *In* M.L.Cody and J.M.Diamond (eds.), *Ecology and Evolution of Communities*, pp.81–120. Cambridge: Belknap.

Menge B.A. (1976) Organization of the New England rocky intertidal community: role of predation, competition, and environmental heterogeneity. *Ecol. Monogr.* **46**, 355–393.

Menge B.A. and J.P.Sutherland (1976) Species diversity gradients: synthesis of the roles of predation, competition, and temporal heterogeneity. *Am. Nat.* **110**, 351–369.

Patrick R. (1972) Use of algae, especially diatoms, in the assessment of water quality. *Am. Soc. Testing Mater. spec. tech. Publ.* **528**, 76–95.

Patrick R. (1975) Structure of stream communities. *In* M.L.Cody and J.M.Diamond (eds.), *Ecology and Evolution of Communities*, pp.445–459. Cambridge: Belknap.

Pianka E.R. (1966) Latitudinal gradients in species diversity: a review of concepts. *Am. Nat.* **100**, 33–46.

Pianka E.R. (1975) Competition and niche theory. *In* R.M.May (ed.), *Theoretical Ecology: Principles and Applications*, pp. 114–141. Philadelphia: Saunders.

Pielou E.C. (1975) *Ecological Diversity*. New York: Wiley.

Randall J.E. (1967) Food habits of reef fishes of the West Indies (Proc. Int. Conf. on Trop. Ocean, 1965), *In Stud. in Trop. Ocean Miami*, **5**, 665–847.

Rigler F.H. (1975) The concept of energy flow and nutrient flow between trophic levels. *In* W.H.vanDobben and L.H.Lowe-McConnel (eds.), *Unifying Concepts in Ecology*, pp.15–26. The Hague: W. Junk.

Roughgarden J. (1974) Species packing and the competitive function with illustrations from coral reef fish. *Theor. Populat. Biol.*, **5**, 163–186.

Schoener T.W. (1974) Resource partitioning in ecological communities. *Science*, **185**, 27–39.

Silliman R.P. (1968) Interaction of food level and exploitation in experimental fish populations. *Fishery Bull. Fish Wildl. Serv. U.S.* **66**, 425–439.

Smith H.S. (1935) The role of biotic factors in the determination of population densities. *J. Econ. Ent.* **28**, 873–898.

Warren C.E. (1971) *Biology and Water Pollution Control*. Philadelphia: Saunders.

Warren C.E. (1979) Toward classification and rationale for watershed management and stream protection. Ecological Research Series, USEPA, EPA-600/3-79-059.

Warren C.E. and W.J.Liss (1977) Design and evaluation of laboratory ecological system studies. Ecological Research Series, USEPA, EPA-600/3-77-022.

Werner E.E. and D.J.Hall (1976) Niche shifts in sunfishes: Experimental evidence and significance. *Science*, **191**, 404–406.

Wetzel R.G. (1975) *Limnology*. Philadelphia: Saunders.

Whittaker R.H. and G.Woodwell (1972) Evolution of natural communities. *In* J.Wiens (ed.), *Ecosystem Structure and Function*, pp.137–195. Corvallis: Oregon State University Press.

Wickstrom E.E. and R.W.Castenholz (1973) Thermophilic ostracod: highest temperature tolerant aquatic metazoan. *Science*, **181**, 1063–1064.

Chapter 4
Biology of Fishes

DORA R. MAY PASSINO

4.1 DIVERSITY OF AQUATIC ANIMAL AND PLANT RESOURCES IN FISHERIES

Nearly every nonpoisonous aquatic animal and plant has served as food to peoples somewhere on earth. In addition, many nonfood items are taken, such as shells, ivory, bone, plant fibers, sponges, pearls, lubricants, toxins, and medicinal chemicals. Major groups of economically important organisms are listed in Table 4.1. To adequately discuss the classification, anatomy, physiology, behavior, and life histories of all these groups would obviously fill a thick textbook, much less a chapter. Hence, this chapter describes the biology of only those animals of major economic importance, i.e., shellfish (mollusks, crustaceans), fish, and marine mammals.

4.2 SHELLFISH – MOLLUSCA

4.2.1 Characteristics and systematic résumé
The phylum Mollusca includes valuable shellfish such as clams, oysters, scallops, and abalone, in addition to edible forms less familiar to the western palate, viz. octopus, squid, and land snails. The phylum Mollusca is second only to the Arthropoda in abundance, having over 80,000 living species and 35,000 fossil species. Although the diverse body forms and variety of habitats may give members of this phylum seemingly little similarities, yet certain features set them apart from other phyla. The mantle, which is a fold of body wall that secretes the shell, and the radula, a flexible tongue-like or file-like structure in the pharynx, are unique to this phylum. Other characteristics of the mollusks are a coelom, bilateral symmetry, which may be secondarily asymmetrical; generally a lack of metamerism or segments; a calcareous shell of one or more pieces; a ventral muscular organ of locomotion, the foot; a distinct head; gills; heart; open circulatory system; and metanephridia for excretion (Hyman 1967).

A systematic résumé of mollusks including only those classes important in fisheries is shown in Table 4.2.

Table 4.1. Classification of Commercially Important Aquatic Plants and Animals
(Rounsefell 1975, Bold and Wynne 1978, Barnes 1974, Villee *et al.* 1968)

Phylum or division	Subgroup	Chief commercial items
Chlorophycophyta		Food supplement for cattle, poultry
Phaeophycophyta		Kelp for food, agar
Chrysophycophyta	Bacillariophyceae	Diatomaceous earth
Rhodophycophyta	Bangiaceae	*Porphyra* ('Nori') for food
Porifera	Demospongiae	Sponges
Coelenterata	Anthozoa	Precious corals
Annelida	Polychaeta	Bait worms, Palolo worms
Mollusca	*	Food, shells, pearls
Arthropoda	Merostomata	Horseshoe crab
	Crustacea*	Food
Echinodermata	Stelleroidea	Starfishes
	Echinoidea	Sea urchins
	Holothuroidea	Sea cucumbers
Chordata	Urochordata	Sea squirts
	Vertebrata	
	Cephalaspidomorphi*	Lampreys
	Chondrichthyes*	Sharks, rays
	Osteichthyes*	Most fish
	Amphibia	Frogs
	Reptilia	
	Chelonia	Turtle, tortoises
	Squamata	Sea snake leather
	Crocodilia	Crocodiles, alligators
	Aves	Guano
	Mammalia	
	Carnivora*	Seals, walrus, sea otter
	Sirenia	Manatees, dugong
	Cetacea*	Whales, dolphins

*See this chapter.

4.2.2   Biology of selected mollusks

Most members of the class Gastropoda have a shell and a visceral mass, which is
coiled in a right-handed spiral and is carried dorsally. Most crawl on a flattened
ventral foot. In a bilaterally symmetrical gastropod, the head is anterior and the
pallial cavity containing the gills, nephridia and anus is posterior. The majority
of gastropods have apparently undergone torsion so that the pallial cavity is also
located anteriorly, dorsal to the head (see discussions in Morton 1967 and Stasek
1972). The resulting sanitation problem is overcome in a gastropod like the

Table 4.2. Classification of selected examples of phylum Mollusca (Barnes 1974)

Taxonomic level	Genus Species	Common name
Class Gastropoda		
Subclass Prosobranchia		
Order Archaeogastropoda	*Haliotis rufescens*	Red abalone
Order Mesogastropoda	*Strombus gigas*	Conch
Subclass Pulmonata		
Order Stylommatophora	*Helix aspera*	Escargot
Class Bivalvia		
Subclass Lamellibranchia		
Order Anisomyaria	*Mytilus edulis*	Bay mussel
	Crassostrea gigas	Japanese oyster
	Placopectin magellanicus	Deep sea scallop
Order Heterodonta	*Venerupis japonica*	Manilla clam
Order Adapedonta	*Siliqua patula*	Razor clam
Class Cephalopoda		
Order Decapoda	*Sepia esculenta*	Cuttlefish
	Loligo opalescens	Squid
Order Octopoda	*Octopus hongkongensis*	Octopus

abalone *Haliotis* by the presence of perforations in the shell and ventilating water currents that flow up and out of the holes in the shell. Wastes from the anus and nephridia are thus carried out. In the land snails of the order Pulmonata the mantle cavity has been converted into a lung. The mantle cavity no longer contains gills but is highly vascularized. An opening in the edge of the mantle cavity (the pneumostome) is used in ventilation (Barnes 1974).

Gastropods include herbivores, carnivores, scavengers, deposit feeders, and suspension feeders. The feeding organ or radula is a ribbon-shaped membrane beset on one side with transverse rows of chitinous teeth. Muscles move the radula in a licking motion for grating, rasping, brushing, and cutting food and conveying it into the pharynx. Extracellular digestion occurs in the stomach with the aid of enzymes produced by salivary glands, esophageal pouches, the digestive diverticula, or a combination of these structures. Most pulmonates such as *Helix* are herbivores. The conch *Strombus*, which is a deposit feeder, has a large mobile proboscis that sweeps across the bottom like a vacuum cleaner.

Most prosobranchs, such as the abalone, are dioecious (i.e., sexes are separate). Gametes are usually discharged into a short gonoduct that opens into the right nephridium. Gametes are shed from the nephridiopore and fertilization is external (Meglitsch 1972). Eggs develop into free-swimming trochophores, which are transformed into veliger larvae (Fig. 4.1). The European edible snail *Helix aspera* is hermaphroditic with an ovotestis, which produces both eggs and

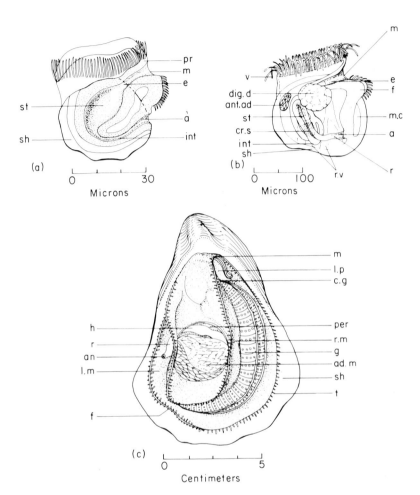

Fig. 4.1. (a) Trochophore larvae of *Ostrea edulis*. (b) Early free-swimming veliger of *O. edulis*. a.—anus; ant.ad.—anterior adductor muscle; cr.s.—crystalline style sac; dig.d—digestive diverticula; e.—esophagus; f.—rudiment of foot; int.—intestine; m.—mouth; m.c.—mantle cavity; pr.—prototroch; r.—rectum; r.v.—velar retractor muscles; sh.—shell; st.—stomach; v.—velum. (c) Organs of *Crassostrea virginica* seen after the removal of right valve. ad.m.—adductor muscle; an.—anus; c.g.—cerebral ganglion; f.—fusion of two mantle lobes and gills; g.—gills; h.—heart; l.m.—left mantle; l.p.—labial palps; m.—mouth; per.—pericardium; r.—rectum; r.m.—right mantle; sh.—shell; t.—tentacles. The right mantle contracted and curled up after the removal of the right valve, exposing the gills. Portion of the mantle over the heart region and the pericardial wall were removed. Drawn from live specimen (Galtsoff 1964).

sperm. Snails have both a vagina and a penis and cross-fertilization is the rule (Hickman 1967). After copulation, each individual deposits a small number of large, separate, yolky eggs in shallow burrows in the ground. Development is direct, and the young emerge as small snails.

The class Bivalvia or Pelecypoda includes clams, mussels, cockles, oysters, and scallops. Bivalves are laterally compressed and have a calcareous shell with two valves, hinged dorsally. The muscular foot is wedge or hatchet shaped and the head is greatly reduced. The spacious mantle cavity encloses the foot, visceral mass, and gills. One gill hangs from the mantle-cavity roof on each side of the body. Each gill consists of two lamellae and serves for filter feeding as well as respiration. Essentially in a clam, water is circulated in the ventral or incurrent siphon, through the pores or ostia in the gill lamellae to a suprabranchial chamber or cloaca, and then out the dorsal or excurrent siphon. As water passes through the gills, microscopic food particles (bacteria and phytoplankton) are filtered out by cilia. Mucus also plays a role in trapping the particles, which are then moved to the labial palps, which aid in selecting particles that enter the mouth. Larger particles rejected by the labial palps become part of the pseudofeces. The problem of possible clogging of the gills with mud is obvious, and differences in the ability of oysters to handle sediment determines culture methods for *Ostrea* and *Crassostrea*. The anatomy of *Crassostrea virginica* is shown in Fig. 4.1. In many intertidal bivalves such as *C. virginica* and the cockle *Cardium edule*, feeding occurs only at high tide and digestive processes exhibit a tidal rhythm (Barnes 1974, Fretter and Graham 1976). Gas exchange takes place both within the gills and the inner mantle surface.

The edge of the mantle is the principal location of bivalve sense organs. The mantle edge often bears tentacles, which are both tactile and chemosensory, and simple pigment-spot ocelli, which detect changes in light intensity. In the swimming scallop, *Pecten*, the ocelli are well developed with cornea, lens, and retina. Nearly all bivalves have a pair of balancing organs, the statocysts, in the foot.

Most bivalves are dioecious, and the gonoducts are simple, since fertilization is usually external. Some species of cockles, oysters, and scallops are hermaphroditic. During its life time, the European oyster, *Ostrea edulis*, shifts from being a male to a female and then back to being a male again. Eggs of *Ostrea edulis* are fertilized within the mantle cavity by sperm entering the incurrent siphon. The fertilized eggs are brooded within the gills of *O. edulis* and then released as veliger larvae (Fig. 4.1). In most marine bivalves, eggs hatch as trochophore larvae, which later develop into veliger larvae. As the young settle to the bottom, they select specific substrates, often the shells of the adults.

The members of the class Cephalopoda, such as squid and cuttlefish, are generally adapted for swimming although the octopods lead less active lives as bottom-dwellers. The shells of squids and cuttlefish are reduced and internal;

however, cuttlefish shells provide buoyancy by the presence of fluid and gas in the spaces between thin septa walls. Cuttlefish shells are used as a dietary supplement for caged birds. The contractions of the mantle walls of squids and cuttlefish force water out the funnel and propel the animal like a jet. The octopods have no shell. Although they can swim by water jets, they usually crawl on the rocky bottom where they live. The highly developed swimming ability and numerous tentacles of cephalopods enable them to capture prey, which are then bitten and torn by powerful beak-like jaws. Chunks of food are pulled into the buccal cavity by the radula. Food of cephalopods includes fish, pelagic shrimp, benthic shrimp, crabs, and snails. Associated with the cephalopod's mobility and carnivorous habit is a highly developed nervous system with a complex brain and eyes that are strikingly similar in structure to vertebrate eyes. Cephalopods are usually dioecious. After courtship display, the male cephalopod seizes the female head-on and uses his specially modified tentacle to deposit a spermatophore, or packet of sperm, in the female's mantle cavity. Fertilized eggs are usually deposited by the female in a gelatinous mass, and embryonic development is direct.

4.3   SHELLFISH-CRUSTACEA

4.3.1   Characteristics and systematic résumé

The Crustacea include some of the world's most valuable fisheries resources, on a price per pound basis. The dockside value of penaeid shrimp is greater than any other U.S. fishery. The class Crustacea are members of the Arthropoda, as are horseshoe crabs, spiders, ticks, mites, insects, millipedes and centipedes.

The 26,000 species of the class Crustacea are primarily aquatic, have gills for respiration, and are distinguished from insects by possessing two pair of antennae (Fig. 4.2). The class Crustacea contains seven subclasses of smaller types of crustacea (brine shrimp, ostracods, copepods, barnacles) that are important in food chains and were formerly grouped as Entomostraca (Barnes 1974). The eighth subclass, the Malacostraca, includes the order Decapoda (8500 species), which will be the subject of the remainder of our discussion on crustaceans. The body plan of decapods is typified by the crayfish (Fig. 4.2), which has 19 segments: 5 head segments; 8 thoracic segments with the first 3 pair of appendages (maxillipeds) modified for feeding and the last 5 pair (pereiopods) for walking (hence the name decapod); and 6 abdominal appendages, 5 bearing pleopods. The first pair of pereiopods is frequently heavier and modified as a claw or cheliped. The head and thoracic segments are fused dorsally and covered with a well-developed carapace, which encloses the gills.

A systematic résumé of decapods with examples of species important in fisheries is shown in Table 4.3.

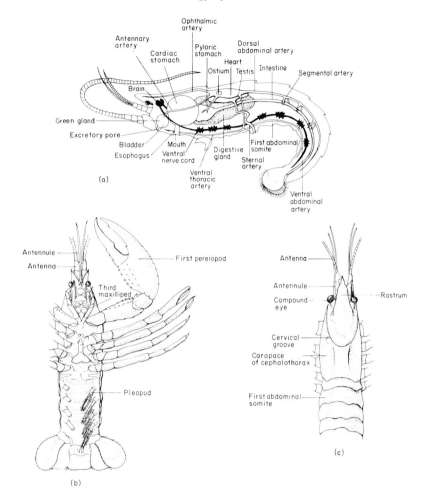

Fig. 4.2. Anatomy of a crayfish. Crayfish are typical of the macruran reptants, with a large abdomen extending back from the thorax. Lobsters and spiny lobsters have essentially the same structure. (a) Scheme of internal organization; (b) Ventral view; (c) Dorsal view (Meglitsch 1972).

4.3.2   Biology of decapods

Commercially important decapods are found in marine, estuarine, and fresh-water habitats from arctic to tropical waters. Marine species are primarily in littoral or shallow-water areas of the continental shelf. Although shrimp are good swimmers, they are generally found on or near the bottom, especially in the day time. *Penaeus*, the most important commercial shrimp throughout the world, usually lives on sand bottoms in 5 to 30 meters of water. The Gulf of Mexico white shrimp (*P. setiferus*) and brown shrimp (*P. aztecus*) prefer muddy

Chapter 4

Table 4.3. Classification of selected examples of order Decapoda (Barnes 1974).

Taxonomic level	Genus Species	Common name
Suborder Natantia		
Section Penaeidea	*Penaeus monodon*	Sugpo prawn
Section Caridea	*Macrobrachium rosenbergi*	Giant freshwater prawn
	Pandalus borealis	Rough pink shrimp
Suborder Reptantia		
Section Macrura		
Superfamily Scyllaridea	*Panulirus japonicus*	Spiny lobster
Superfamily Nephropidea	*Homarus americanus*	American lobster
	Astacas astacas	Crayfish
	Procambarus clarkii	Red crayfish
Section Anomura		
Superfamily Paguridea	*Paralithodes camtschatica*	King crab
Section Brachyura		
Superfamily Brachyrhyncha	*Cancer magister*	Dungeness crab
	Callinectes sapidus	Blue crab
Superfamily Oxyrhyncha	*Chionoecetes bairdi*	Tanner or snow crab

bottom, while the pink shrimp (*P. duorarum*) lives predominantly on sand, shell-sand or coral-mud bottom (Williams 1965). Availability of suitable vegetative cover may be important as in *Macrobrachium rosenbergi*, which establishes territories and defends them by cannabilizing intruders. The blue crab (*Callinectes sapidus*), which ranges from Cape Cod, Massachusetts, to northern South America, inhabits shallow, brackish water estuaries as juveniles. Dungeness crabs (*Cancer magister*) prefer sandy bottom apparently due to their habit of lying almost entirely buried in the sand with only the stalked eyes and antennae visible (Rounsefell 1975). Lobsters inhabit crevices in rocky or coralline bottoms.

The Natantia are designed for pelagic existence as may be seen by examining shrimps and prawns. Their laterally compressed bodies have a full set of well-developed, fringed swimming pleopods and their slender thoracic segments are not well suited for walking. Rapid ventral flexion of the abdomen is used for quick backward darts.

The Reptantia or 'crawlers' have heavier thoracic legs and dorsoventrally flattened bodies. The first pair of legs are usually chelipeds, sometimes large and meaty as in the lobster (*Homarus americanus*). The Macrura, represented by lobsters and crayfish, have a large extended abdomen with a full complement of appendages. Macrurans can swim rapidly backward by flexing the abdomen ventrally (Barnes 1974). Brachyurans or true crabs have a greatly reduced abdomen which fits tightly beneath the cephalothorax. The center of gravity is shifted forward beneath the legs. Crabs can walk forward slowly, but commonly walk sideways, especially when running rapidly. The blue crab, a powerful and agile swimmer, has its last pair of legs shaped as broad flattened paddles. These limbs move in figure eight patterns, essentially like propellers, enabling the animal to swim sideways, backwards, and sometimes forward.

Most decapods are either scavengers or omnivores. 'Prey or other food is caught or picked up by the chelipeds and then passed to the third maxillipeds, which push it between the other mouth parts. While a portion is bitten by the mandible, the remainder is torn away by the maxillae and maxillipeds. The severed piece is then pushed into the pharynx and another bite is taken' (Barnes 1974). The crustacean digestive tract (Fig. 4.2), like that of other arthropods, consists of a foregut, lined with cuticle and shed at each molting, a midgut lined with endoderm, and a hindgut also lined with cuticle.

Although some crustacea utilize the body surface for respiration, decapods respire with gills housed in the branchial chamber. A ventilating current is produced by the beating of a paddle-like projection of the second maxilla. Water is pulled forward, and the exhalent current flows out anteriorly in front of the head. The path of the water varies in different groups depending on the tightness of the fit of the carapace. The blood of decapods contains hemocyanin for oxygen transport.

The excretory glands of decapods are the antennal (or green) glands. The end of the gland lies in front of and to both sides of the esophagus (Fig. 4.2) and leads into a bladder by excretory tubules. From the bladder a short duct leads to an opening in a papilla on the base of the second antenna. Most nitrogenous waste is in the form of ammonia, which diffuses from the body surface, especially across the gills. The antennal glands appear to function in controlling internal fluid pressure and salt content.

The nervous system of decapods consists of a dorsal brain, circumesophageal connectives, and a double ventral nerve cord with segmental ganglia. Of the different sensory receptors, an important and common type is one connected with hairs, bristles, or setae (Barnes 1974). The bristle is designed so that when it is moved, the receptor end of the shaft or base is stimulated. Most decapods, such as shrimps and crabs, have compound eyes composed of many cylindrical units called ommatidia. These units have a lens at the outer end and pigmented retinular cells at the basal end. The many ommatidia of one compound eye can form crude images and are especially useful for detecting moving objects (predators or prey) around them. A pair of statocysts is located in the basal segment of the first antennae of nearly all decapods.

The male decapods have paired but connected testes lying in the thorax and the anterior portion of the abdomen (Fig. 4.2). The terminal end of the sperm duct is a muscular ejaculatory duct that opens on the eighth thoracic segment. In males the first two pair of pleopods are modified to aid in sperm transfer and a single penis or pair of penes are usually present. The paired female ovaries empty into simple oviducts that end in gonopores on the third thoracic segment. Many decapods exhibit precopulatory courtship behavior. Male *Cancer* and *Callinectes* crabs carry the adult female about until she molts, after which copulation takes place. In most decapods the fertilized eggs are attached and brooded on the pleopods of the female and hatch as zoea larvae (Fig. 4.3).

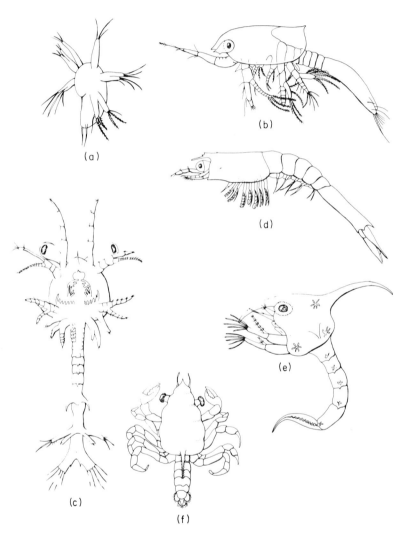

Fig. 4.3. Representative crustacean larvae. (a) A penaeid nauplius. (b) A penaeid protozoea. (c) A penaeid zoea. (d) A mysis larva. (e) A crab zoea. (f) A crab megalops (a, b, and d, after Gurney; c, after Claus; e and f, after Poole—see Meglitsch 1972).

Penaeid shrimp, however, shed their eggs directly into the water and the eggs hatch as nauplius larvae (Fig. 4.3).

4.4 FISH

4.4.1 Cephalaspidomorphi

The class Cephalaspidomorphi, of superclass Agnatha (Cyclostomes) and phylum Chordata, consists of the lampreys and hagfish and has little positive economic value (Norman and Greenwood 1975). Rounsefell (1975) notes that lampreys were eaten by aborigines of the British Columbia area, the colonists on the Atlantic seaboard of the United States, and by certain English kings. Generally this group is noted for destructive predatory attacks by lampreys, *Petromyzon marinus*, on economically important fish, such as the lake trout, *Salvelinus namaycush*, in the Laurentian Great Lakes (Fig. 4.4) and attacks by hagfish, *Myxine glutinosa*, on fish immobilized in fishermen's nets.

The Cephalaspidomorphi are characterized by lack of jaws (i.e., are

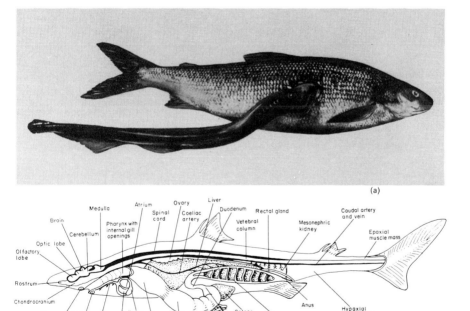

Fig. 4.4. (a) Sea lamprey, *Petromyzon marinus*, attacking a whitefish (Photo courtesy of the U.S. Fish and Wildlife Service). (b) Internal anatomical features of a spiny dogfish shark, *Squalus acanthus* (Lagler *et al.* 1977).

agnathous), and paired pectoral and pelvic fins. The main axial skeleton is cartilaginous or fibrous and the notochord is unconstricted; they have a single median nostril; instead of true gill arches, they have the gills in pouches with a branchial basket external to the gills; they have a neurocranium, i.e., the branchial basket is united firmly to the brain case; two semicircular canals in the ear are on each side of the head of lampreys but only one on each side in hagfishes (Lagler *et al.* 1977).

The body of hagfish is soft-skinned and nearly cylindrical with a low caudal fin around the tail. The eyes are vestigial and covered with skin. The hagfish, *Myxine*, has 6 pairs of gill sacs but only one opening. Hagfish are able to produce copious amounts of slime. The common or North Atlantic hagfish, *Myxine glutinosa*, lives on soft bottoms. Boring into the ocean floor is accomplished by powerful undulations of the posterior portion of the body while in a vertical position (Grzimek 1973). Hagfish feed on benthic invertebrates such as annelids, echiurid worms, snails, mussels, shrimp, and hermit crabs. In addition they scavenge dead fish and other carrion. Hagfish are hermaphroditic but can produce only one kind of gamete at a time (Hickman 1961).

Lampreys have smooth eel-shaped bodies with well-developed dorsal and caudal fins. The seven paired gill sacs of *Petromyzon* have seven external openings. In contrast with hagfish, adult lampreys have well-developed eyes. Parasitic lampreys, such as *Petromyzon marinus*, have sucker-like mouths and muscular protrusible tongues provided with horny, epidermal teeth (Hickman 1961). The lamprey attaches to the host fish by its sucker mouth and rasps off small bits of flesh with its teeth. The tongue creates suction by acting like a plunger. The lamprey then injects an anticoagulant and sucks blood from the host fish. When gorged, the lamprey releases its hold. Such attacks are often fatal to the host, although many lake trout are found with healed lamprey scars. When the lamprey is attached, water passes into as well as out of the gill slits.

Lampreys are dioecious and spawn in the spring on shallow gravel beds in streams. Males clear away pebbles with their buccal funnels and form a nest or pit. When a female anchors herself to a pebble over one of these pits, a male seizes her, winds his tail around her, and discharges sperm over the eggs as they are extruded from her body into the nest. The adults die soon after spawning. The eggs hatch in two weeks into ammocoete larvae, which burrow in sediment and filter feed on organic ooze for 3 to 7 years. The ammocoete has a fleshy oral hood at the anterior end and skin covering its eyes. Control of lampreys in the Great Lakes is accomplished by adding to streams a chemical that is selectively toxic to the ammocoetes. Metamorphosis, which occurs in the fall, involves replacement of the oral hood with the sucker mouth containing teeth, development of larger uncovered eyes, shifting of the nostril to the top of the head, and development of a rounder but shorter body (Hickman 1961). Parasitic lampreys then migrate to the lake or sea to prey on fish.

4.4.2 Chondrichthyes

The class Chondrichthyes of the superclass Gnathostomata (jaw-bearing fishes) consists of sharks, rays, and chimaeras. The oldest fossils of this class are found in Lower Devonian rocks (Norman and Greenwood 1975). Since members of this class have cartilagenous skeletons, only fragmentary remains such as spines, teeth, and dermal denticles have been found and the origins of the class are obscure. Although no other group of fish has such an evil reputation or causes such fear as the sharks, actually they serve several useful purposes. Sharks yield excellent leather, polishing agents, marketable flesh for human consumption, and liver oil, which is used as a source of vitamin A or for other commercial uses (Gilbert 1963). The meat of the pectoral fins or 'wings' of flapper skates (*Raja batis*) and thornback rays (*Raja clavata*) is considered a delicacy in France and Germany (Grzimek 1973). Liver of the ratfish (*Chimaera monstrosa*) yields an oil that is highly prized as a lubricant for precision equipment (Herald 1972). The classification of Chondrichthyes is shown in Table 4.4.

Characteristics of the class Chondrichthyes are as follows: (1) jaws (gnathos-

Table 4.4. Classification of selected examples of the class Chondrichthyes (Lagler *et al.* 1977)

Taxonomic level	Genus Species	Common name
Subclass Elasmobranchii (Selachii)		
Order Heterodontiformes	*Heterodontus japonicus*	Japanese horn shark
Order Hexanchiformes	*Hexanchus griseus*	Sixgill shark
Order Squaliformes		
Family Lamnidae	*Carcharodon carcharias*	Great white shark
	Isurus oxyrinchus	Mako shark
Family Rhincodontidae	*Rhincodon typus*	Whale shark
Family Cetorhinidae	*Cetorhinus maximus*	Basking shark
Family Carcharhinidae	*Galeorhinus zyopterus*	Soupfin shark
	Prionace glauca	Great blue shark
Family Sphyrnidae	*Sphyrna zygaena*	Hammerhead shark
Family Squalidae	*Squalus acanthias*	Spiny dogfish
Order Rajiformes (Batoidei)		
Family Rhinobatidae	*Rhinobatos rhinobatos*	Guitar fish
Family Rajidae	*Raja batis*	Flapper skate
Family Dasyatidae	*Dasyatis pastinaca*	Stingray
Family Mobulidae	*Manta birostris*	Manta ray
Order Torpediniformes	*Torpedo torpedo*	Electric ray
Subclass Holocephali		
Order Chimaeriformes	*Hydrolagus colliei*	Ratfish

tomous), (2) notochord constricted by vertebrae, which are cartilaginous, (3) paired pectoral and pelvic fins, (4) paired nostrils, (5) three semicircular canals in each ear, (6) gill arches cartilaginous and internal to the gills, and (7) gill arches not firmly united to the brain case, but joined to it by connective tissue (Lagler *et al.* 1977). In addition the subclass Elasmobranchii has five to seven pairs of gills and gill clefts, a spiracle opening on the dorsolateral surface of the head behind each eye, placoid scales usually present, a cloaca, and pelvic intromittent organs in males. In the order Squaliformes (sharks), the gill clefts are positioned on each side of the head, in front of the pectoral fins, whereas in the order Rajiformes (skates and rays), the five gill clefts are on the ventral side of the pectorals, which are greatly enlarged and expanded. The subclass Holocephali has four pairs of gills with a single gill cleft covered by a dermal opercle, naked skin in adults, separate anal and urogenital openings instead of a cloaca, and both a clasping organ on the forehead of some (*Chimaera*) and pelvic intromittent organs in males.

A distinguishing feature of the Chondrichthyes is the placoid scale (Fig. 4.5), which is similar in structure to a vertebrate tooth. The placoid scale consists of a basal plate and a projecting spine composed of dentine and containing a pulp cavity. In most elasmobranchs the placoid scales or denticles are small and give a rough texture to the skin, but in skates, large denticles are present (Hyman 1942).

Sharks normally swim with undulating motions of their body, the caudal fin providing the main propulsion (Grzimek 1973). The pectoral fins elevate the anterior of the body and the heterocercal tail generates lift for the posterior of the body thus counteracting the negative buoyancy of the body, which lacks a gas bladder. Nevertheless, because of their negative buoyancy, sharks must swim continuously to avoid sinking. The flattened sides of skates and rays have become fused with the large pectoral fins and generate vertical undulations from front to rear. The sides of eagle rays and especially manta rays are elongated like wings and have a beating motion like birds' wings (Grzimek 1973).

The teeth of predatory sharks are enlarged formidable cutting weapons with one or more sharp points and often serrated edges (Hyman 1942). The teeth are arranged in rows, of which only one or a few rows are standing upright and in use, whereas the others lie flat along the inner surfaces of the jaws. Sharks do not keep the same teeth throughout a lifetime but rather have a gradual replacement of older, functional teeth by the younger teeth lying behind them. Predatory sharks eat porpoises, water birds, turtles, other sharks, fishes of all kinds, sea lions and even man. Other sharks, such as basking sharks (Cetorhinidae) and whale sharks (Rhincodontidae), strain small fish and plankton from large volumes of water by gill rakers. Bottom feeding sharks such as the nurse sharks (*Ginglymostoma*) and rays such as guitar fish (Rhinobatidae) and sting rays (Dasyatidae) have small, blunt teeth for crushing and grinding. Their diets include mollusks, crustaceans, sea urchins and smaller fishes (Norman and

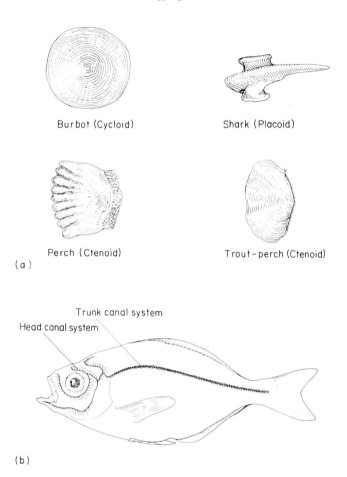

Burbot (Cycloid)

Shark (Placoid)

Perch (Ctenoid)

Trout-perch (Ctenoid)

(a)

Trunk canal system

Head canal system

(b)

Fig. 4.5. (a) Types of fish scales (Lagler *et al.* 1977). (b) Lateral line system of the walleye surfperch, *Hyperprosopon argenteum* (Cahn 1967).

Greenwood 1975). The jaws of chimaeras are armed with large flat plates studded with hardened points or 'tritors'. Their varied diet consists of seaweeds, fishes, worms, echinoderms, mollusks, and crustaceans. The large intestine of Chondrichthyes contains a spiral valve, which is a large twisted fold of absorptive tissue (Fig. 4.4).

During the respiratory cycle of sharks, water is taken in the mouth and expelled over the gills and out the gill clefts. When skates (*Raja*) are on the bottom, there is danger of introducing sediment into the gills. Skates thus inhale water through the spiracles, which are on the dorsal surface of the head, and expel water out the gill clefts (Norman and Greenwood 1975).

Marine sharks, skates, and rays have an unusual means of adjusting to the high salt content and osmotic pressure of sea water. Their blood is slightly hypertonic to sea water and about 50 percent of the osmotically active blood solutes are nitrogenous compounds, primarily urea (Lagler *et al.* 1977). Little or no urea is lost through shark gills, and their tissues have a high content of urea, viz., as much as 2 to 2.5 percent.

Chondrichthyes have well-developed nervous systems and sense organs. The olfactory organs, consisting of nostrils lined with sensory tissue, are of primary importance to predatory sharks in locating their food (Gilbert 1963). Sharks follow odor gradients by swimming in circles or in figure-eight patterns (Lagler *et al.* 1977). The corresponding olfactory lobes of the brain of sharks are relatively large compared with other vertebrates (Fig. 4.4). Taste buds in the mouth and pharynx of sharks enable them to discriminate between food and nonfood items, although this sense is not always used when sharks are in a 'feeding frenzy.' Although movement, which is detected by the lateral line system, and olfaction play primary roles in attracting sharks to food from a distance, in closer quarters the rush at living prey is mainly optically oriented (Gilbert 1963). Prominent extrinsic muscles rotate the eyeball in such a manner that a shark maintains a constant visual field while it is turning, twisting, or moving forward.

Internal fertilization is the rule for Chondrichthyes. The modified pelvic fins or claspers of males each have an internal cartilaginous skeleton and a lengthwise groove leading from a glandular sac at the base (Norman and Greenwood 1975). During copulation the claspers are thrust into the cloacal aperature of the female, and seminal fluid is introduced into the oviducts. In some sharks and rays, development takes place in the oviduct and the young are born in an advanced state of development. In the sand sharks, the larger embryos cannibalize the smaller embryos (Herald 1961). In egg laying species, the eggs are encased in a tough, chitinous shell, in which the embryo develops. The rectangular, smooth capsule of sharks such as the spotted dogfish (*Scyliorhinus*) has long tendrils at the four corners, which coil around rocks or other objects and anchor the egg capsule (Norman and Greenwood 1975). The upper and lower halves of the capsule are united at first by a membrane, but at the time of hatching they separate at one end, leaving an opening through which the young fish can make its escape.

4.4.3   Osteichthyes

(*a*) *Origins and classification.* The class Osteichthyes or bony fish has a rich fossil history reaching back to the Lower Devonian with many extinct species (Lagler *et al.* 1977). The subclass Crossopterygii with one living species, *Latimeria chalumnae*, was thought to have been extinct since the Cretaceous until a specimen was caught off South Africa in December 1938 and identified by Professor J.L.B.Smith (Smith 1956). The subclass Dipnoi or lungfish has six living species, including *Protopterus aethiopicus*, which is commercially important in Lake Victoria, the world's second largest freshwater lake.

The third subclass, Actinopterygii (ray-fins), with over 20,000 species, is characterized by maxilla and premaxilla in the upper jaw, muscles not extending into fin base, palatoquadrate not fused to the cranium, no movable joint between anterior and posterior parts of the skull, and the absence of internal nares and cloaca. The Actinopterygii includes the majority of economically important fish for commercial fishermen, anglers, and aquarium hobbyists. Table 4.5 gives the classification of class Osteichthyes.

Table 4.5. Classification of selected examples of the class Osteichthyes (modified from Lagler *et al.* 1977)

Taxonomic level	Genus Species	Common name
Subclass Crossopterygii		
Order Coelacanthiformes		
Family Latimeriidae	*Latimeria chalumnae*	Lobefin
Subclass Dipnoi		
Order Dipteriformes		
Family Lepidosirenidae	*Protopterus aethiopicus*	African lungfish
Subclass Actinopterygii		
Order Clupeiformes		
Family Clupeidae	*Clupea harengus*	Herring
	Alosa pseudoharengus	Alewife
	Brevoortia patronus	Gulf menhaden
	Dorosoma cepedianum	Gizzard shad
	Sardina pilchardus	Sardine
Family Engraulidae	*Engraulis mordax*	Northern anchovy
	Cetengraulis mysticetus	Anchoveta
Order Salmoniformes		
Family Salmonidae	*Salvelinus namaycush*	Lake trout
	Salmo gairdneri	Rainbow trout
	Oncorhynchus nerka	Sockeye salmon
		Kokanee
	Salmo salar	Atlantic salmon
		Landlocked salmon
Family Esocidae	*Esox lucius*	Northern pike
	Esox masquinongy	Muskellunge
Order Gonorynchiformes		
Family Chanidae	*Chanos chanos*	Milkfish
Order Cypriniformes		
Family Characidae	*Cheirodon axelrodi*	Red neon tetra
Family Cyprinidae	*Cyprinus carpio*	Carp
	Carassius auratus	Goldfish
	Pimephales promelas	Fathead minnow
	Notropis hudsonius	Spottail shiner

Table 4.5. Continued.

Taxonomic level	Genus Species	Common name
Family Ictaluridae	*Ictalurus punctatus*	Channel catfish
	Pylodictis olivaris	Flathead catfish
	Ictalurus furcatus	Blue catfish
Order Cyprinodontiformes		
Family Poeciliidae	*Poecilia reticulata*	Guppy
Order Perciformes		
Family Percichthyidae	*Morone saxatilis*	Striped bass
	Morone chrysops	White bass
Family Centrarchidae	*Micropterus dolomieui*	Smallmouth bass
	Lepomis macrochirus	Bluegill
	Lepomis cyanellus	Green sunfish
Family Percidae	*Perca flavescens*	Yellow perch
	Stizostedion vitreum	Walleye
Family Pomatomidae	*Pomatomus saltatrix*	Bluefish
Family Sparidae	*Stenotomus chrysops*	Scup
Family Sciaenidae	*Cynoscion nebulosus*	Spotted seatrout
Family Cichlidae	*Tilapia mossambica*	Java tilapia
	Tilapia nilotica	Nile tilapia
Family Scombridae	*Thunnus thynnus*	Bluefin tuna
	Euthynnus pelamis	Skipjack tuna
	Scomberomorus maculatus	Spanish mackeral
Order Pleuronectiformes		
Family Pleuronectidae	*Pleuronectes platessa*	Plaice
	Hippoglossus stenolepis	Pacific halibut
Order Gadiformes		
Family Gadidae	*Gadus morhua*	Atlantic cod
Family Merlucciidae	*Merluccius productus*	Pacific hake

(*b*) *External anatomy.* The skin of fish is composed of an outer layer, the epidermis, which is constantly worn away, and an inner layer, the dermis, which is composed primarily of connective tissue with some nerves, muscles, blood vessels and glands such as the mucus glands. The latter bring about the slimy feeling of fish. Although some fish, such as catfish (Ictaluridae), are 'naked,' most have scales (Lagler *et al.* 1977). The Actinopterygii in Table 4.5 have two scale types, namely, cycloid and ctenoid (Fig. 4.5). The thin bony-ridge scales of cycloid type, which are smooth and approximately circular, are characteristic of soft-rayed fish such as Clupeidae and Salmoniformes. The bony-ridged ctenoid scale type, which are toothed or comb-like on the posterior margin, are characteristic of spiny-rayed fish such as Perciformes.

The appendages of Osteichthyes are median fins (dorsal fin, the tail or caudal fin, and the anal fin) and paired fins (pectoral and pelvic fins). In most soft-rayed fish (Clupeidae, Salmonidae, Cyprinidae) the pectorals are in a thoracic position and the pelvics in an abdominal position. However, in the spiny-rayed fish (Percidae), the pelvic fins are located far forward on the belly just below the pectoral fins.

The principal gross openings of bony fishes include a single gill opening on each side of the head covered by an operculum. In most fishes the openings of the urogenital ducts are at the surface, behind the anus (Lagler *et al.* 1977). In certain herrings, the gas bladder opens through a pore near the anus. Sense organ openings include the nares and other small sensory pores, e.g., the pores of the lateral line system.

(c) *Form and locomotion.* Fish are ideally torpedo-shaped (fusiform) and most often ovoid in cross section as exemplified by fast-swimming tunas (Scombridae). However, many exceptions to the fusiform shape include the flattened flounder (Pleuronectidae), depressed catfish (Ictaluridae), and attenuated eels (Anguillidae) (Lagler *et al.* 1977). Besides bone and cartilage, the skeleton also includes notochord, connective tissue, non-bony scale and tooth elements (enamel and dentine), supporting cells of the nervous system, and fin rays. The three main regions of the skeleton are the skull, vertebral column, and fin skeletons (Fig. 4.6). The skull is made up of two parts—the neurocranium, enclosing the brain and sense organs; and the branchiocranium, i.e. the upper and lower jaws, hyal bone (the jaw-supporting hyoid arch and the bones of the gill-covering opercular series), and the arches supporting the gills. Many bony fishes such as herrings, salmons, suckers, and carps have small splint bones of assorted shapes in the myosepta, and these bones can be most annoying when fish are eaten by people (Jollie 1962).

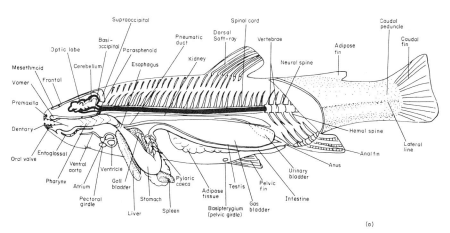

Fig. 4.6. Internal anatomical features of a soft-rayed bony fish, the brook trout, *Salvelinus fontinalis* (Lagler *et al.* 1977).

The muscles of fishes, as other vertebrates, are of three types, i.e., skeletal or striated muscles, smooth or involuntary muscles of the gut, and cardiac or heart muscle. The musculature of the head is primarily for movement of the jaws and gill arches. The trunk, median fin, and paried fin muscles are primarily for locomotion. The membranous skeleton of connective tissue divides the lateral trunk muscles into bundles called myomeres, which are readily seen on the side of a skinned fish (Lagler *et al.* 1977). When the myomeres on one side of a fish contract and shorten, the head and tail bend toward that side and the myomeres on the opposite side stretch. Alternating serial contractions of the myomeres on the two sides of the body flex the fish or successive parts of the fish into a wave form. In the most common type of locomotion, carangiform or jack-like, the fish drives itself forward by side-to-side sweeps of the tail region. The vertebral column whips or undulates from side to side, in contrast with whales where it moves up and down. The paired and median fins are used as stabilizing and maneuvering organs. The median fins serve as keels. The principal functions of the pectorals and pelvics are for climbing, diving, banking, turning, and stopping. In addition to swimming, some fish can burrow (flounders), crawl (morays-Muraenidae), jump (mudskipper-Gobiidae), leap (tarpon-Megalopidae), and fly (flying fishes-Exocoetidae).

(*d*) *Feeding, digestion, and growth.* The digestive tract consists of the following features: mouth (with toothed jaws), oral cavity with teeth on the roof and tongue with teeth on the floor, pharynx with pharyngeal tooth pads on the throat sides of the gill arches and with gill rakers on the internal branchial openings, esophagus or gullet, stomach, pylorus (pyloric valve followed by openings into pyloric caeca), small intestine with openings of ducts bringing in bile and pancreatic secretions, large intestine, and anus (Fig. 4.6) (Lagler *et al.* 1977). The feeding preferences of economically important fish may be grouped into the following types: predators, grazers, strainers, and suckers. Predatory fishes feed on macroscopic animals either by actively hunting their prey or lying in wait for their prey. Predators include bluefish (Pomatomidae), salmon and trout (Salmonidae), pikes (Esocidae), and groupers (Serranidae). These fish have well-developed grasping and holding teeth, and, in common with other carnivores, they have strong acid secretions in the stomach and a relatively short intestine. Grazing fishes may feed on bottom organisms or plankton, taking food in bites. Bluegills (Centrarchidae) graze on the bottom of a lake. Parrotfish (Scaridae) graze or browse on coral reefs. Cichlids of Africa with chisel-like incisors cut weeds or scrape algae from rocks. Strainers or filter feeders strain plankton from the water based generally on size, not species of food organism. Herring and menhaden (Clupeidae) are plankton feeders. Both suckers (Catostomidae) and sturgeons (Acipenseridae) feed by sucking a combination of food and bottom sediments into their mouths. They have toothless jaws, protrusible sucking mouths, and large fleshy lips (Herald 1972).

The growth rate of fishes is determined by genetic factors inherent in each species, the amount of food, temperature (higher temperatures up to the optimum temperature result in faster growth), hormonal cycles, and density-dependent factors such as crowding (crowded fish have depressed growth). Age and growth of fish may be measured by examining annuli or year marks in scales, fin rays, otoliths or opercular bones of temperate-zone fish.

(*e*) *Circulation and respiration.* As in other vertebrates, the circulatory system of fish carries oxygen and assimilated food to the body tissues and carries waste products of nitrogen and carbohydrate catabolism from the tissues to the gills and kidneys. The heart is situated immediately behind the gills in a cavity called the pericardium (Fig. 4.6). In mammals blood is circulated from the right ventricle to the lungs, where blood is oxygenated, and then back to the heart, where the contractions of the left ventricle send the blood forcefully on its way. However in fish, which have single chambered auricles and ventricles (with the exception of lungfish), blood is circulated from the ventricle to the gills for oxygenation and thence throughout the body. Consequently, fish have lower blood pressure than mammals (Curtis 1949). Fish blood has red corpuscles with hemoglobin for binding and transporting oxygen.

The respiration cycle begins when the gill covers are closed and water is drawn into the mouth, which is expanded, creating negative pressure. The mouth is then closed and negative pressure is created in the space between the gills and the operculum, drawing water into the gill cavity. When the opercula open, water is expelled outward from the gill cavity. Many fish possess a gas bladder which is variously used in respiration (gars-Lepisosteidae), sound reception (minnows-Cyprinidae), sound production (drums-Sciaenidae), and as a hydrostatic organ (many species).

(*f*) *Osmoregulation.* Osmoregulation is the maintenance of proper internal salt and water concentrations. Fish face three different osmotic situations (Lagler *et al.* 1977). Freshwater fish are hyper-osmotic to the environment and tend to lose salts and to gain excess water. Marine fish are usually hypo-osmotic to the environment and tend to be dehydrated and to gain excess salts. Estuarine fish may be iso-osmotic to the environment. Fish meet the challenge of osmotic differences by adjusting the volume of urine excreted and the salt concentration of the urine, by eliminating salts through the gills, by eliminating salts in the feces, and by other mechanisms such as drinking water. Fish such as salmon and eels, which migrate between fresh and sea water, must make major osmotic adjustments. Knowledge of osmotic requirements of fish is necessary for successful aquaculture.

(*g*) *Nervous system, endocrine system, sense organs, and behavior.* Successful living of fish in their environment depends in part on their ability to perceive their environment accurately and their ability to react appropriately (Gosline 1971). The fish's nervous system consists of a brain and spinal cord, cranial and spinal nerves, and an autonomic system of ganglia and nerve fibers (Fig. 4.6). Stimuli are relayed by nerves to the spinal cord and brain which in turn relay impulses to muscles and glands. The fish's brain is relatively smaller than mammals.

The endocrine glands, together with the nervous system, play an integrating and regulatory role. Although much less is known about endocrine function of fish than mammals, the major actions of these ductless glands are known. The pituitary gland, located at the base of the brain, is the master gland. The pituitary and sex organs control the reproductive cycle. The pituitary and thyroid glands control growth and metabolic rate. The pituitary, the adrenal cortical tissue and possibly the thyroid control osmoregulation. The suprarenals control blood pressure and pulse rate. Metabolism is affected by hormones from the pancreatic islets, adrenal cortical tissue, corpuscles of Stannius, and ultimobranchials (Lagler *et al.* 1977).

Fish detect most odors through the olfactory epithelium, which lines the nasal sac and is connected to the olfactory nerve. The detection of odors is important in finding food and mates. Odors also play a role in the location of home streams and spawning sites of salmon. Fish eyes have no lids, since their eyes are continually bathed in water. Fish also have no lachrymal glands and hence cannot cry. Instead of accommodating the eyes to vision at varying distances by altering the convexity of the lens, fish change the position of the lens with regard to the retina (Norman and Greenwood 1975). The inner ear serves both for hearing and for balance. The inner ear consists of a membranous sac and three semicircular canals enclosed in a chamber in the skull. Another sensory system in fish is the lateral line system, which has connections in the brain with afferent nerve pathways from the auditory system of the inner ear (Lagler *et al.* 1977). This system is typified by lateral-line canals and their pores in skin or scales on the head and body of bony fish (Fig. 4.5). This system functions in the detection and location of disturbances in the water and aids in finding food, in avoiding enemies, in social behavior, in echo location, and as an accessory organ for orientation in flowing water (Jollie 1962).

How intelligent are fish? Can they think? Do they feel pain? Do they sleep? These are some of the questions dealt with by behavioral scientists. Much research remains to be done on fish behavior. Fish can associate stimulus and response, and their behavior can be conditioned by both positive reinforcement (food) or negative reinforcement (electrical shock) (Adler 1975). Under natural conditions, fish are capable of learning and remembering quite complex relationships. For example, the tide-pool dwelling goby *Bathygobius soporator* (Gobiidae) jumps from one tide pool to another when the tide is out, although it cannot possibly see the location of the target pool at the time of jumping. These fish survey the location of the pools during high tide and remember this information for use at low tide. Tests have shown that their memory persists for at least a month.

Many fish display territoriality at some time in their lives, especially during courtship and breeding. The need for individual space refers to the space that a fish will attempt to keep clear around itself and is exemplified by a school of characins with each individual holding his own position in the group. Dominance hierarchies develop within any group of fish. These relationships determine who is boss and who follows next in line, and so on until a definite order has been established. Adding a fish to an established tank carries the risk that the resident population will collectively attack the new arrival (Adler 1975).

(*h*) *Reproduction, Development, and Life History.* Although most Osteichthyes are dioecious, examples of hermaphroditic (*Serranellus subligarius*-Serranidae) and parthenogenetic (*Poecilia formosa*-Poeciliidae) fish exist (Lagler *et al.* 1977, Norman and Greenwood 1975). Fecundity, i.e., the number of eggs produced, is related to the parental care given the eggs and young. Female cod (*Gadus*), which merely emit their eggs in the vicinity of a male, produce as many as 9 million eggs per female per season. Sticklebacks (*Gasterosteus*) produce 30 to 100 eggs and provide parental care for the eggs and fry. Guppies (*Poecilia*) bear alive a brood of less than two dozen. Fish eggs vary with regard to their buoyancy and surface characteristics. The pelagic eggs of cod have neutral buoyancy. Most stream fishes have negatively buoyant eggs that sink into the stream bed rather than being swept downstream. In addition, trout and salmon eggs are temporarily adhesive during water hardening after they are extruded. Adhesive eggs pick up little particles of sand that help them to sink to the bottom and anchor there during development.

In addition to producing gametes, ovaries and testes secrete hormones that influence reproductive cycles, courtship and breeding behavior, and secondary sexual characteristics. Pearl organs or nuptial tubercles appear on males of smelt (*Osmerus*), minnows (Cyprinidae), and suckers (Catostomidae). Enlarged fins or bright coloration are characteristic of some male fishes.

The characteristics and timing of developmental stages of embryos of hatchery fish are well known. Meroblastic cleavage is the prevalent form of egg division in Osteichthyes (Lagler *et al.* 1977). Larval fish often lack the distinguishing characteristics of their parents, so that identification keys for larval fish have been published (Mansueti and Hardy 1967) (Fig. 4.7).

4.5   MAMMALS

Marine mammals have long been of interest to mankind, not only for meat, oil, fur, and other commercial products, but also as objects of curiosity because of the similarity of their behavior to humans. Marine mammals are classified into three orders, i.e., Carnivora, Cetacea, and Sirenia. The first two orders, which

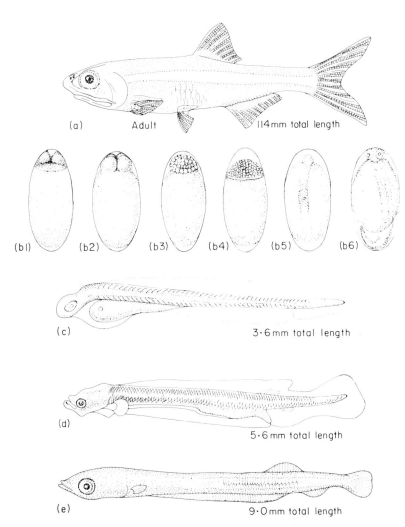

Fig. 4.7. The striped anchovy, *Anchoa hepsetus*, showing developmental stages. (a) Adult, 114 mm TL. (b) Development of egg. Major axis ca. 1.5 mm; minor axis ca. 0.8 mm. (b1) Two-cell stage. (b2) Four-cell stage. (b3) Early morula. (b4) Late morula. (b5) Early embryo. Eight somites visible; Kupffer's vesicle present. (b6) Embryo just before hatching. (c) Yolk-sac larva, 3.6 mm TL, newly hatched. (d) Larva, 5.6 mm TL; few chromatophores along anterior dorsal half of intestine not illustrated. (e) Larva, 9.0 mm TL. (a, Hildebrand; b–e, Hildebrand & Cable—see Mansueti and Hardy 1967).

include seals, whales and dolphins, will be discussed below, but the Sirenia will not be described because of the low economic value of manatee and dugong.

4.5.1   Pinnipedia

Seals, sea lions, and walruses are classified by most taxonomists in the order Carnivora as shown in Table 4.6, although some taxonomists elevate Pinnipedia to ordinal status.

The teeth of pinnipeds are nearly homodont (teeth of uniform shape) and are adapted for grasping and tearing, not chewing. Adult walruses lack lower

Table 4.6. Classification of selected examples of class Mammalia (Scheffer and Rice 1963)

Taxonomic level	Genus Species	Common name
Order Carnivora		
Suborder Pinnipedia		
Family Otariidae		Eared seals
Subfamily Otariinae	*Zalophus californianus*	California sea lion
Subfamily Arctocephalinae	*Callorhinus ursinus*	Northern fur seal
Family Odobenidae	*Odobenus rosmarus*	Walrus
Family Phocidae		Earless seals
Subfamily Phocinae	*Phoca vitulina*	Harbor seal
	Pagophilus groenlandicus	Harp seal
Subfamily Monachinae	*Monachus schauinslandi*	Hawaiian monk seal
Subfamily Cystophorinae	*Cystophora cristata*	Hooded seal
Order Cetacea		
Suborder Odontoceti		Toothed whales
Family Platanistidae	*Inia geoffrensis*	Amazon dolphin
Family Ziphiidae	*Hyperoodon ampullatus*	Bottlenose whale
Family Physeteridae	*Physeter catodon*	Sperm whale
Family Delphinidae	*Delphinapterus leucus*	Beluga or white whale
	Orcinus orca	Killer whale
	Delphinus delphis	Common dolphin
	Phocoena phocoena	Harbor porpoise
Suborder Mysticeti		Baleen whales
Family Eschrichtiidae	*Eschrichtius glaucus*	Gray whale
Family Balaenidae	*Balaena mysticetus*	Greenland right whale
Family Balaenopteridae	*Balaenoptera musculus*	Blue whale
	Balaenoptera physalus	Fin whale
	Balaenoptera borealis	Sei whale
	Balaenoptera acutorostrata	Minke whale
	Megaptera novaeangliae	Humpback whale

incisors but the upper canines of both sexes are modified to form tusks. Although the walruses are bottom feeders, taking largely sessile organisms such as mollusks, they occasionally eat small whales, narwhals, and seals. The harp seal feeds primarily on krill. Some seals such as the leopard seal (Monachinae) feed on penguins, other birds that land on the water, and other seals.

The pinnipeds are actually amphibious since they obtain food from the sea and use the land for mating, bearing young, and resting (Stains 1976). Pinnipeds are well-suited for an aquatic existence. The body is streamlined with ears reduced or absent and mammary teats and external reproductive organs constructed so that they can be withdrawn beneath the skin. The limbs are enclosed within the body skin to or beyond the elbows and knees. The ends of the limbs are flattened to form paddles. The pelvic girdle is more nearly parallel to the vertebral column than in terrestrial carnivores enabling the hind limbs to be held in a trailing position and to be more efficient in propulsion through the water. In addition to having a tough, thick skin, they have a subcutaneous layer of fat or blubber, which insulates them from the cold, increases buoyancy, and provides them with a reserve source of energy during lactation or times of fasting.

Although human divers can only dive to about 100 meters and hold their breath for $2\frac{1}{2}$ minutes, the Weddell seal (*Leptonychotes weddelli*) dives as deep as 330 meters with dives lasting up to 43 minutes (Stains 1976). Seals adjust to diving conditions by reducing their metabolism, by their large volume of blood compared with humans, and by shunting the blood to the heart and brain from the extremities. The muscles function anaerobically during diving. Deep dives are possible since many structures of their bodies are built to yield to pressure rather than to resist pressure. Some of these features are smaller lungs, elimination of most air from the lungs before the dive, a thorax that permits the lungs to collapse, incomplete rings in the trachea permitting complete collapse, and driving air out of the middle ear by partly filling the ear cavity with blood.

Seal's vision is good under water, in air, and at night. Hearing and smell are probably poorly developed since ears and nostrils are closed under water. Their well-developed facial whiskers and system of nerves in the facial area aids in locating prey and avoiding underwater collisions.

Pinnipeds are gregarious during the breeding season with more than a million individuals congregating on an island. Each male walrus or otarid seal sets up a territory and gathers as many females as it can. It defends the harem by loud bellows, roars, and honks, and rushes at any approaching rival male. The male goes without food for as long as 2 months while defending the harem. After a gestation period of 8 to 12 months, the young are born and nursed on land or ice. Female pinnipeds are sexually mature at about 3 to 4 years old. Males of the polygamous species do not mate until several years after reaching maturity when they can successfully defend a territory against other males. Phocid seals, however, are usually monogamous in their breeding activities (Stains 1976).

4.5.2   Cetacea

Whales of the North Pacific provided Eskimos with food, fuel, light, cordage from sinews, and tools from the skeleton. The modern whaling industry makes similarly complete use of whales, obtaining the following products: meat for animal and human consumption, bone meal for fertilizer, and oil for soap, cosmetics, glycerin, varnish, linoleum, and watch oil (Purrington 1976). The blue whale is 30 meters long on the average and reaches a maximum size of 38 meters and 130,000 kilograms, being the largest animal that has ever lived on earth. Classification of the two major groups of whales is shown in Table 4.6.

The two suborders of Cetacea are distinguished on the basis of their teeth. The Odontoceti are mostly homodont and most lack milk teeth. Adult baleen whales lack teeth and feed by straining plankton from the water by baleen plates made of keratin (Norris 1976). The teeth of odontocetes are used for seizing prey, which is not chewed before being swallowed. Carnivorous cetaceans usually have a large crop before their main stomach where food is likely pulverized (Slijper 1976).

Cetacean body shape is fusiform or spindle-shaped with the tail ending in a horizontal blade (the flukes). A dorsal fin is usually present. Cetaceans have flipper-shaped forelimbs, but lack hindlimbs, having the pelvic bones represented by small bones remote from the vertebrae. No ear pinnae are present and the mammary teats are buried in longitudinal folds. Whales and dolphins swim by vertical undulations of the tail. The swimming animal describes a roughly sinusoidal path as it swims, coming to the surface for a breath at the top of each curve and descending below the surface in between. Breathing occurs through the paired or single blowhole on top of the head. Similar to the pinnipeds, cetaceans can also dive to great depths. A sperm whale was found entangled in a telephone or telegraph cable on the ocean floor at 988 meters depth (Slijper 1976).

Smell is reduced in cetaceans. The lens of the cetacean eye is modified for underwater vision. Hearing is a major sense. In addition to producing a wide variety of sounds for communication, odontocetes produce brief clicks of high-intensity, high frequency sounds used for echolocation.

Most odontocetes spend their lives in highly organized schools whereas baleen whales only congregate for breeding and sometimes feeding. Schools of toothed whales show a variety of swimming patterns. Groups of mothers and young are usually found in the center of the school. The mother-young relationship may persist for several years, with the young returning to its mother in times of distress even when full grown. Playful behavior occurs mainly in the odontocetes. Whales also show epimeletic behavior, i.e., assisting another individual that is in trouble.

Cetaceans are seasonal breeders. Gestation periods are usually 11 to 12 months, although the sperm whale has a gestation period of 16 months. Birth of whale calves takes place under water and growth is rapid. Whale milk is very high in fat content and almost lacking in milk sugar. Young baleen whales nurse for 7 to 10 months. Most baleen whales undergo long seasonal migrations between the feeding and calving grounds (Norris 1976).

4.6   REFERENCES

Adler H.E. (1975) *Fish Behavior: Why Fishes Do What They Do*. Hong Kong: T.H.F. Publications.

Anderson H.T., (ed.) (1969) *The Biology of Marine Mammals*. New York: Academic.

Barnes R.D. (1974) *Invertebrate Zoology*, 3rd ed. Philadelphia: Saunders.

Bold H.C. and M.J.Wynne (1978) *Introduction to the Algae: Structure and Function*. Englewood Cliffs, New Jersey: Prentice-Hall.

Cahn P.H., (ed.) (1967) *Lateral Line Detectors*. Bloomington: Indiana Univ. Press.

Curtis B. (1949) *The Life Story of the Fish: His Morals and Manners*. New York: Dover.

Fretter V. and A.Graham (1976) *A Functional Anatomy of Invertebrates*. New York: Academic.

Galtsoff P.S. (1964) The American oyster, *Crassostrea virginica* Gmelin. *U.S. Fish Wildl. Serv. Fish. Bull.* No. **64**, 1–480.

Gilbert P.W., (ed.) (1963) *Sharks and Survival*. Boston: Heath.

Gosline W.A. (1971) *Functional Morphology and Classification of Teleostean Fishes*. Honolulu: Univ. Press of Hawaii.

Grzimek H.C.B., (ed.) (1973) *Grzimek's Animal Life Encyclopedia*, Vol. 4 *Fishes I*. New York: Van Nostrand Reinhold.

Grzimek H.C.B., (ed.) (1974) *Grzimek's Animal Life Encyclopedia*, Vol. 5 *Fishes II and Amphibians*. New York: Van Nostrand Reinhold.

Herald E.S. (1961) *Living Fishes of the World*. New York: Doubleday.

Hickman C.P. (1961) *Integrated Principles of Zoology*, 2nd ed. St Louis: Mosby.

Hickman C.P. (1967) *Biology of the Invertebrates*. St Louis: Mosby.

Hyman L.H. (1942) *Comparative Vertebrate Anatomy*, 2nd ed. Chicago: Univ. of Chicago Press.

Hyman L.H. (1967) *The Invertebrates*, Vol. 6 *Mollusca I*. New York: McGraw-Hill.

Jollie M. (1962) *Chordate Morphology*. New York: Reinhold.

Lagler K.F., J.E.Bardach, R.R.Miller, and D.R.M.Passino (1977) *Ichthyology*, 2nd ed. New York: Wiley.

Mansueti A.J. and J.D.Hardy Jr. (1967) *Development of Fishes of the Chesapeake Bay Region: An Atlas of Egg, Larval, and Juvenile Stages*. Baltimore: Natural Resources Institute, Univ. of Maryland.

Meglitsch P.A. (1972) *Invertebrate Zoology*, 2nd ed. London: Oxford Univ. Press.

Morton J.E. (1967) *Molluscs*, 4th (rev.) ed. London: Hutchinson Univ. Library.

Norman J.R. and P.H.Greenwood (1975) *A History of Fishes*, 3rd ed. London: Ernest Benn.

Norris K.S. (1976) Whale. *In Encyclopedia Britannica*, 15th ed., vol. 19, pp.805–809. Chicago.

Purrington P.F. (1976) Whaling. *In Encyclopedia Britannica*, 15th ed., vol. 19, pp.811–813. Chicago.

Rounsefell G.A. (1975) *Ecology, Utilization, and Management of Marine Fisheries*. St Louis: Mosby.

Scheffer V.B. and D.W.Rice (1963) A list of the marine mammals of the world. *U.S. Fish Wildl. Serv. Spec. Sci. Rep. Fish.* **431**, 1–12.

Schmitt W.L. (1965) *Crustaceans*. Ann Arbor: Univ. of Michigan Press.

Slijper E.V. (1976) *Whales and Dolphins*. Ann Arbor: Univ. of Michigan Press.

Smith J.L.B. (1956) *Old Four-Legs*. London: Longmans.

Stains H.J. (1976) Carnivora. *In Encyclopedia Britannica*, 15th ed., vol. 3, pp.927–944. Chicago.

Stasek C.R. (1972) The molluscan framework. *In* M.Florkin and B.J.Scheer (eds.), *Chemical Zoology*, vol. 3, pp.1–44. New York: Academic.

Villee C.A., W.F.Walker Jr., and F.E.Smith (1968) *General Zoology*, 3rd ed. Philadelphia: Saunders.

Williams A.B. (1965) *Marine Decapod Crustaceans of The Carolinas. U.S. Fish Wildl. Serv. Fish. Bull. (U.S.)*, **65**, 1–298.

Chapter 5
Dynamics of Fished Stocks

A. V. TYLER AND V. F. GALLUCCI

5.1 INTRODUCTION

5.1.1 The Stock Concept

The term 'stock' is used in a special way in fisheries management. There are distinctions between the fisheries concept of a stock and the biological concept of population that should be made clear here. A population is the breeding unit of a species. Over a species' geographic range there tend to be clusters of individuals with relatively rapid gene flow among all members of the group. 'Relatively rapid' means that genetic exchange within the group is faster than genetic exchange between groups. Members of a population would tend to have a common spawning ground, and larval fish of the population would develop in the same geographic area. Because of the relative lack of mixing between populations, and because populations are breeding units, gene frequencies within one population tend to be different from other populations. Biologists can use the frequency of occurrence of blood proteins (types of haemoglobin and transferin) among fishes to detect gene frequency shifts from area to area.

We use the term 'stock' in an operational sense, following Gulland (1969). Essentially a stock is defined so that fisheries yield models will work when they are applied. Any statistical sample of individuals from a stock must have similar production characteristics to any other sample of the stock. The basic production characteristics incorporated into yield models are number of births in a given year (cohort strength), growth rate, natural mortality rate, and fishing mortality rate. If geographic clusters of a species differ in these characteristics, more than one stock would be set up for management purposes. In this way a basic fish sampling unit is established that is characterized by homogeneity of natural production parameters. A stock may be a portion of a population, or include more than one population.

How do stocks and populations differ in terms of the basic production characteristics? Cohort strength is influenced by both density-dependent responses to the fish stock's own biomass density, and by density-independent weather and hydrographic factors. Density-dependent responses tend to operate differently and independently on separate populations. Therefore if there were only density-dependent responses, two populations would likely have to be categorized as two fishery stocks. However, weather factors are very wide-ranging and could be influencing the two populations equally. If variations in cohort strength were influenced more by wide-ranging weather factors than by density-dependent factors, then the two populations could be treated as one stock, all other things being equal.

It is sometimes true that different groups of juvenile fish of a single population grow to maturity in different areas, and do not mix until they spawn for the first time several years later. They could easily be exposed to different temperatures and feeding conditions and so form two or more units in terms of growth rate. During this period the two groups of juveniles could be subjected to different rates of predation, fishing, disease, and parasitic exposure, and so the mortality rates of the two groups would be distinct. The two parts of the population would have to be treated as two stocks. On the other hand it could happen that two populations with similar, weather influenced, cohort-strength trends, would mix on a feeding ground through much of their life history. They would have such similar growth rates and natural mortality rates that they would be indistinguishable in this respect. They could be classed as one stock even though migration pathways, physical characteristics, and separate spawning grounds showed they were two populations. It is interesting that males and females of the same population often have different mortality and growth rates and so must be managed separately.

If two year-by-year abundance trends were made for a single stock by sampling two areas occupied by the stock, the trends would be parallel in time. Such trends can be helpful in stock identification. Relative abundance trends over years can sometimes be measured as catch per unit effort (CPUE) data generated by either a fishery or research sampling. This is because average catch rates are usually proportional to the density of the fish in an area. The more fish there are, the more a unit of fishing gear catches per hour, at least when these CPUE data are averaged over the fishing season or the year. There are of course sampling problems that may be insurmountable, and so in any particular case CPUE may not be a valid measure. But if it is, then long-term, annual trends in CPUE within one area of a stock should be correlated with CPUE trends in any other area occupied by the stock. Stock biomass trends are completely determined by the combined rates of recruitment, mortality, and growth. Homogeneity of these rates within a stock must lead to correlated CPUE trends if sampling can be properly carried out.

5.1.2   Types of Rates

All stock processes, e.g., natality and mortality, have rates associated with them. An introduction to stock-process rates is necessary since the field of stock dynamics is the study of rates and how they change. We will discuss three classes of rates in this chapter: absolute rates, relative rates, and instantaneous rates. Absolute rates express the change in a dependent variable, Y, against the change in the independent variable, X. Usually X is time, and the absolute rate, r, is expressed as:

$$r_1 = \frac{Y_2 - Y_1}{t_2 - t_1}$$

where Y_2 and Y_1 are two values of variable Y and t_2, t_1 are the corresponding times at which Y had those values. Thus,

$$r_1 = \frac{5 \text{ kg} - 3 \text{ kg}}{\text{day 3} - \text{day 1}} = \frac{1 \text{ kg}}{1 \text{ day}}$$

r_1 is the absolute rate of 1 kg per day.

Relative rates are absolute rates expressed in terms of a standardizing value. These rates may also be termed proportionate rates, such that

$$r_2 = \frac{(Y_2 - Y_1)/(t_2 - t_1)}{Z}$$

where Z is the value of any selected variable. Examples of a standardizing value for a change in the dependent variable Y would be the starting value of Y, or else the mean value of Y, for the interval, $t_2 - t_1$, during which the measurements of Y were taken. An example of this kind of rate is growth (change in body size) as a proportion of mean body weight ($\overline{Wt}$),

$$\frac{Wt_2 - Wt_1/t_2 - t_1}{\overline{Wt}}$$

Wt_2 is weight at time, t_2. Wt_1 is weight at some earlier time, t_1. Changes of Y per unit time can be expressed in terms of an independent variable. Examples of this kind of relative rate are food conversion efficiency (change in body weight per mean ration per unit time) and individual or per capita birth rate (number of young added per individual spawner per unit time).

Instantaneous rates are rates of change of curvilinear functions. The term 'instantaneous' is an unfortunate one that tends to confuse. One should not try to imagine a rate that is applied at an instant. The term is just jargon that refers to an algebraic exponent of a relationship. The curvilinear functions most commonly used in fisheries modeling are exponential functions. The instantaneous rate, k, is a constant and an exponent of the relationship:

$$Y_2 = Y_1 e^{-k(t_2 - t_1)}$$

where the dependent variable Y_2 is a function the starting Y (Y_1) and the time interval ($t_2 - t_1$), and where e is the base of the natural logarithm. Often

t is taken as one time unit, say one year. Then where k is determined on a year basis,

$$Y_2 = Y_1 e^{-k}$$

Note that as long as k does not change, e^{-k} does not change. Y_2 is a constant proportion of Y_1, and the relationship between Y_2 and Y_1 may be expressed as a straight line with intercept of zero. The term, e^{-k}, is seen to be a proportionate, or relative rate.

5.2   SURPLUS PRODUCTION MODEL

5.2.1   Dynamics of density dependence

The general question that fisheries managers ask is, how much fishing will the stock stand and also maintain high, long-term yields? Long-term persistence means that the birth rate equals the death rate on balance during course of the year, or a period of a few years. Under completely natural conditions a stock either maintains this equality, or completely disappears. If on balance the birth rate already equals the death rate, how can we expect to catch large quantities of fish and still hope for stock persistence?

There are two kinds of stock responses that allow us to maintain fisheries. Some years may be better than others for the stock due to natural environmental fluctuation, i.e., the birth rate may be higher than the death rate in those years. A fishery can then operate by cropping off this production that otherwise would have increased, perhaps temporarily, the size of the stock. Obviously, if the fishery is dependent on changes in the weather for its allowable catch, the concept of a sustainable yield is only a pleasant fiction. A constant removal of fish biomass based on what fishermen should remove in 'good' years would deplete the stock.

The second type of stock response allows some approximation of constant catch removal. When fish stocks are 'thinned out' their rate of productivity increases. In fact, within bounds, the harder a stock is fished, the harder it 'fights back.' This type of response is termed compensatory density-dependence. As density is reduced by whatever cause, the productivity rate of an individual fish increases (see Fig. 3.1). It grows faster in body, it increases its reproductive output, and often its probability of death decreases. These responses occur for a variety of reasons, depending on the species of fish. The general cause is that the resources an individual draws upon are less taxed when the stock has fewer fish in it. There tends to be more food per individual, more space per individual, and less interference among individuals in spawning and feeding activities. For those species of fish exhibiting parental care, there is less interference in finding and preparing sites for egg deposition, defended territories tend to be larger, only the best areas are used, and less energy needs to be expended in territory defense. As a result growth rate increases, survival rate increases, disease and parasite interference decreases, size at first spawning decreases, life expectancy and the number of times that an individual spawns increases, the number of yolk-filled eggs carried by a female increases, the quality of the yolk-filled eggs increases, and survival rate of young increases.

These changes in productivity must be expressed in quantitative relationships if yield estimates are to be made for the stock. Let us suppose that the proportion of the stock that dies each year increases as stock size (or stock density) increases. This relationship can be expressed as a curvilinear relationship between number dead per year and number of fish in the stock (Fig. 5.1a). The

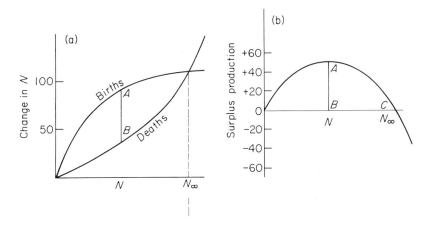

Fig. 5.1. (a) Numbers of fish added due to births, and missing due to deaths, as a function of the stock size in numbers (N). N_∞ is the carrying capacity. For each N, the difference between the births and deaths curves gives the stock increments termed 'surplus production.' Height AB is the maximum increment in N. (b) Surplus production as a function of N. Height AB is the same as AB in panel a. Notice that negative surplus production is possible to the right of N_∞.

number of births also increases with increase in stock size, rapidly at first, and then slowly. At the stock size where the two lines cross (N_∞) births equal deaths. At stock sizes smaller than this point there are more births than deaths, and the stock tends to increase. At stock sizes above N_∞ the stock decreases. N_∞ is called the equilibrium stock size or the carrying capacity. The difference between the birth and mortality curves to the left of the equilibrium stock size is the source of increase from the stock. This quantity, which gradually increases with increase in stock size, and then decreases, is termed 'surplus production.' It is surplus production that is available for harvest without cutting into the size of the generating stock. The surplus production in Fig. 5.1a is re-plotted in Fig. 5.1b.

Fluctuations of a stock may be due to density-dependent responses. These responses can occur under rather constant physical environmental conditions. That is, fluctuating physical factors are not always the sole cause of observed stock fluctuations. This principle may be understood in terms of the difference between the births and deaths curves. The maximum distance between the births and deaths curves is a measure of the stock's ability to return to the equilibrium stock size when a fishery is applied. The measure of that quality, called the resilience of the stock, is expressed by the ratio of distance AB to BC (Fig. 5.1b). Suppose the stock replaces itself each generation, the second generation completely replacing the first, say, each year. Also, suppose a large number of animals were removed by human exploitation during one year only. The stock will respond to this removal in different ways depending on the ratio of $AB:BC$. If the ratio is one or less the stock will approach N_∞ along an increasing curve over time (Fig. 5.2a). This type of population increase is known as logistic population growth. The successive increments in stock size are drawn as vertical lines in Fig. 5.2b, with the height of the line at A being equal to the stock size increase between A and B. If the ratio is increased so that its value lies between 1 and 2 (Fig. 5.2c), the time course will show a series of oscillations (Fig. 5.2d). If the ratio is increased still further so that the value of the ratio is greater than 2 (Fig. 5.2e), the oscillations will either diverge wildly, then converge, or diverge to the point where the stock becomes extinct (Fig. 5.2f).

5.2.2   Surplus production and fisheries yield

Up to a point there are as many sustainable yields as there are levels of fishing effort. To illustrate this we have arranged the birth and mortality curves so that the resulting surplus production shows symmetry with increasing stock size (Fig. 5.3). We have drawn the mortality line so that a constant percentage of the stock dies each year due to natural causes no matter what the stock size.

For this example, we express level of fishing effort in units of boat-hours (10 similar trawlers fishing for 10 hours would produce 100 boat-hours). We will allow that a set level of fishing effort means removal of the same percentage of the stock as stock size increases. That is the number of fish taken would be

---

Fig. 5.2. (a) Change in stock numbers over time corresponding to surplus production additions in panel b, where height at $N = A$ is stock increment 0 to A over time in panel a. (c) Surplus production plot showing resilience greater than in panel b. The time course resulting is in panel d. (e) Further increase in resilience brings divergent stock fluctuation in panel f, resulting in stock extinction.

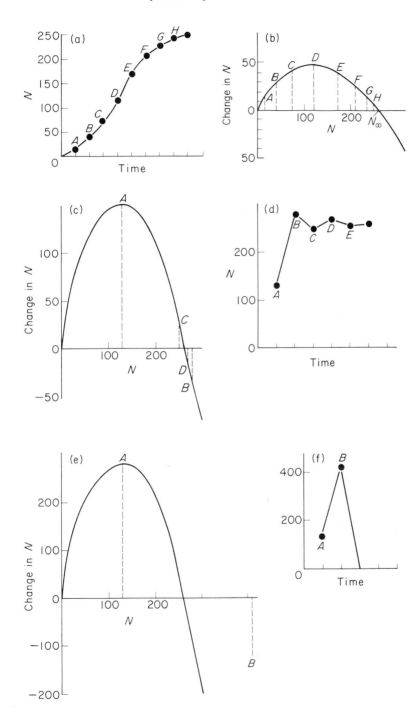

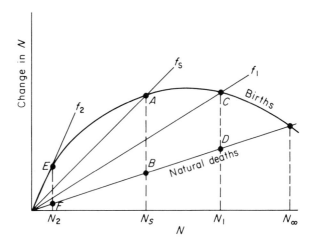

Fig. 5.3. Annual births and deaths curves (change in N) as a function of stock size N. Line f_1 represents increase in deaths due to a small, constant percentage fishing mortality. CD is the sustainable yield. Lines f_s and f_2 are further increases in fishing mortality. f_s takes the maximum sustainable yield, AB; f_2 takes reduced sustainable yield, EF. N_1, N_s, N_2 are equilibrium stock sizes for the various fishing rates.

in direct proportion to the stock size. This constant proportional catch can be added to the natural mortality, and total number of deaths per year can be expressed as a function of stock size (line f_1 in Fig. 5.3). The equilibrium yield in numbers of fish is line segment CD. As fishing effort increases, equilibrium yields also increase until effort level f_s is reached. Line segment AB is the maximum surplus production and maximum sustainable yield (MSY). Further increases in fishing effort will still produce sustainable yields, but they will be less than MSY.

  If the stock is fished down to a very low level, what will be the time course of the recovery if fishing is completely stopped? Suppose fishing effort level f_2 is completely stopped and the stock will reach its new equilibrium level in 10 years. At first the surplus production additions will be small, corresponding to EF (Fig. 5.3). By year 5, surplus production will be equal to AB, i.e., stock increase will have maximized. Over the next 5 years surplus production will decrease progressively as stock size increases. The resulting time course is a sigmoid curve, ending at N_∞ (Fig. 5.2a).

5.2.3   Yield calculations from the surplus production model

The algebra of yield calculations from the surplus production model is not difficult. Most fisheries are conducted on a weight basis rather than by numbers. The graphical presentation above would not be at all different had we developed it in biomass terms, though surplus production then would subsume somatic growth as well as reproduction.

From the foregoing graphical presentation it should be clear that the rate of change of the stock biomass (ΔB) depends on the size of the stock itself and also on the distance between the biomass and the maximum equilibrium stock size, B_∞. This can be expressed algebraically as:

$$\Delta B = kB_E \frac{(B_\infty - B_E)}{B_\infty} \tag{5.1}$$

where the distance between B_E and B_∞ is written as a proportion of B_∞. The constant of proportionality, k, is termed the intrinsic rate of population growth. Elsewhere k has been written as r and B_∞ as K, but we follow the notation of Ricker (1975) here. Quantity ΔB is the surplus production that may be taken in a unit time as equilibrium yield, which we will designate Y_E. If the surplus production curve is symmetrical, then the stock size that produces maximum sustainable yield, B_s, is one half B_∞,

$$B_s = \frac{B_\infty}{2} \tag{5.2}$$

Expanding eq. 5.1,

$$Y_E = kB_E - \left(\frac{k}{B_\infty}\right)B_E^2 \tag{5.3}$$

To find the maximum sustainable yield in terms of B, we substitute B_s from eq. 5.2 into eq. 5.3 in place of B_E:

$$Y_s = k\frac{B_\infty}{2} - \left(\frac{k}{B_\infty}\right)\left(\frac{B_\infty}{2}\right)\left(\frac{B_\infty}{2}\right)$$

$$Y_s = k\frac{B_\infty}{4}$$

In words, the annual maximum sustainable yield represents removal of one quarter of the maximum stock size each year, multiplied by the rate of increase.

It must be emphasized that the surplus production model allows calculation of yield under equilibrium fishing conditions only. The level of fishing effort must be relatively constant over a minimum time interval equal to the time it takes the stock to completely replace itself. The yield from the final year of that interval is Y_E. If the effort then jumps to another level and again remains constant for the stock replacement time interval, a second Y_E and associated B_E can be estimated at the end of that interval. Eventually a plot of Y_E against B_E can be made. Since there is a one to one correspondence between equilibrium stock size and steady effort (Fig. 5.3), effort can be substituted for N (Fig. 5.4).

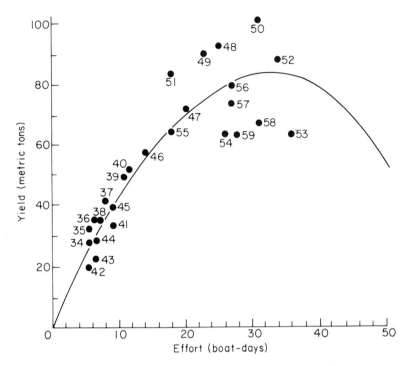

Fig. 5.4. Equilibrium catch and effort (fitted line), and annual catch and effort data points for Pacific yellowfin tuna. Drawn from Pella and Tomlinson (1969).

If the equilibrium yield has not been reached yet (because the time interval for complete stock replacement has not come to completion), both yield and stock size will change each year. M.B. Schaefer invented a method of dealing with this situation. We will designate this non-equilibrium yield as Y_t. We will call the biomass at the beginning of the year B_t, and the biomass at the beginning of the following year B_{t+1}. Then Y_E may be calculated as:

$$Y_E = Y_t + (B_{t+1} - B_t)$$

When effort is applied to an unfished stock, or when effort is increased from previous levels, the term $(B_{t+1} - B_t)$ will be negative because biomass will be decreasing during the stock replacement time interval. Other methods for estimating values for the application of surplus production models may be found in Ricker (1975). Also, see Chapter 7 for comments and other formulation of the Schaefer model.

5.2.4 Assumptions of the surplus production model
Mathematical models are consciously simplified representations of the real

world (see Chapter 7). Simplifications are developed as contingencies. If a given circumstance is true, then the model can be applied. Or if the condition is roughly true, then the estimate made from the model is an approximation of reality. Because models developed to make yield estimates incorporate simplifications of reality, a particular model cannot be universally applied. The user of a model should always point out the likely biases or discrepancies in the resulting estimate.

One of the contingencies of the surplus production model is that there have been slow and regular changes in levels of fishing effort. The model deals only with equilibrium yield, and so large and rapid changes in fishing effort levels will invalidate application of the model, even with the Schaefer method. The pulse fishing techniques of the large, Soviet, distant water trawlers must be dealt with using non-equilibrium models. Rapid growth of domestic fishing fleets, as is currently occurring in the U.S. West Coast shrimp fleet, does not allow application of the surplus production model. On the other hand, the historical slow growth of the demersal fish industry in the U.S. Northwest Coast has allowed estimates from surplus production models (Hayman *et al.* 1980).

The model also assumes that natural production rates change slowly, and that environmental conditions do not incorporate long-term, unidirectional changes in physical factors that influence rates of productivity. One of the interesting areas of research in fisheries modeling is evaluation of simple models, such as the surplus production model, by using complex computer simulations that take account of physical factor changes. It is becoming possible to discover how 'robust' the simpler models really are in the face of the variable real world. Surplus production models have built into them a host of density-dependent rate changes, yet their data requirements are easily attainable by fisheries agencies. If a surplus production model has been developed over a period of 30 to 40 years, there may be a degree of uncertainty associated with estimates made from it. These estimates would be more acceptable for stock management if some estimate of the degree of uncertainty could be made.

The model is useful because of its simple data requirements. Exact aging of fishes is not necessary, nor are estimates of growth rate, mortality rate, or reproduction rate. One need deal only with net biomass increments and associated yields. The model has been useful in calculating yields on tuna stocks where accurate aging of individuals is not possible (Schaefer 1957). It has also been applied to Pacific halibut (Ricker 1975).

5.3   YIELD PER RECRUIT MODEL

5.3.1   The negative exponential depletion

Suppose a stock is depleted over time by a constant percentage, or constant

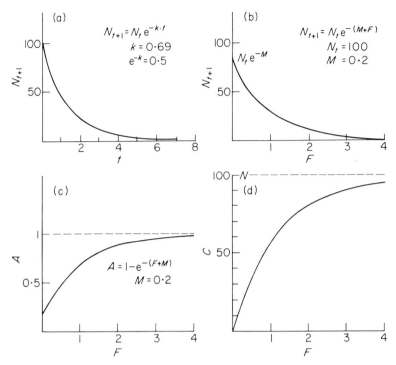

Fig. 5.5. (a) Exponential decrease of stock numbers over time. (b) Response of stock numbers to instantaneous fishing mortality (F) as expressed in the exponential depletion function. (c) Change in total annual proportionate mortality rate (A) as a function of F. Note that the Y-intercept is greater than zero. (d) Plot of Baranov catch function with catch in numbers (C) from a stock of size N individuals, with instantaneous fishing mortality (F) as independent variable.

proportionate natural mortality rate, and there are no new individuals added through reproduction. The time course of depletion would be the progressively less steep curve in Fig. 5.5a. Algebraically, we could express this relationship as:

$$N_{t+1} = N_t S \tag{5.4}$$

where S is the proportionate rate of survival, sometimes called the individual, or per capita, survival rate, since it is multiplied by the number of individuals at time t (N_t) to calculate the number remaining one time unit later, N_{t+1}.

S can be expressed in terms of e, the base of natural logarithms, and the instantaneous rate of mortality, Z,

$$S = e^{-Z}$$

Z is the sum of instantaneous natural mortality, M, and instantaneous fishing mortality, F. Substitution into eq. 5.4 gives:

$$N_{t+1} = N_t e^{-(F+M)} \qquad (5.5)$$

One can solve eq. 5.5 for F to find how the stock size changes as F increases. When F goes to zero, $e^{-(F+M)}$ goes to e^{-M}, and so N_{t+1} goes to $N_t e^{-M}$. When F goes to infinity, $e^{-(F+M)}$ goes to zero, and N_{t+1} goes to zero. The curve of eq. 5.5 is plotted for a range of values of F in Fig. 5.5b.

By tautology the proportionate mortality rate, A, and the proportionate survival rate sum to one,

$$A + S = 1$$

By substitution of eq. 5.5 in the above equation,

$$A = 1 - e^{-F+M} \qquad (5.6)$$

If A is calculated as a function of F where F ranges from zero to infinity, an upside down, mirror image of the curve in Fig. 5.5b results (Fig. 5.5c), with A ranging from $1 - e^{-M}$ to one. As F is increased the increase in the proportionate mortality rate toward one becomes less rapid for a given stock size.

5.3.2   The Baranov catch function

The catch, C, in numbers of fish is directly proportional to the number of fish in the stock, N. As stated above it is also directly proportional to F. But it is inversely proportional to M. Natural mortality is in a sense competing with fishermen for fish. If we multiply the proportionate total mortality rate, A (eq. 5.6) by N we would calculate the number of deaths from both fishing and natural sources. If we would like to calculate the catch alone, we would be interested in only some portion of A. That proportion is the ratio of $\dfrac{F}{F+M}$. So the catch C from stock N is:

$$C = N \frac{F}{F+M} A$$

$$C = N \frac{F}{F+M} (1 - e^{-(F+M)}) \qquad (5.7)$$

This function is referred to as the Baranov catch function after its originator.

Since A, when plotted as a function of F, is a curve increasing to an asymptote equal to one, then C must also follow this shape as a function of F (Fig. 5.5d). Its asymptote, however, is N, since as F goes to infinity, both

$\dfrac{F}{F+M}$ and $1 - e^{-(F+M)}$ go to one.

5.3.3 Production from growth

In many fisheries the basic management computations are in terms of weight or biomass. Factors such as the speed at which weight is added, the average maximum weight under various environmental conditions, and body length at different weights are important. In addition to fulfilling a management need, basic ecological questions may be addressed by growth studies, such as the role of density dependence or of abiotic environmental factors, or the study of extremely variable patterns of growth exhibited by a single species in apparently similar environments. For example, Fig. 5.6 depicts haddock growth curves in

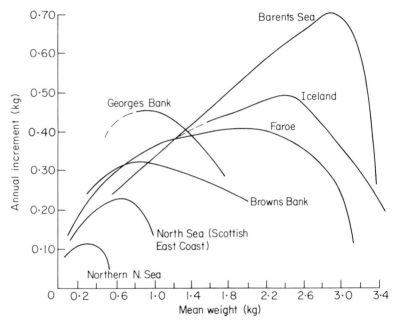

Fig. 5.6. The relationship between annual growth increments (kg) and mean weights (kg) for north Atlantic haddock. Adapted from Jones and Johnson (1977), p.39.

terms of the weight added as a function of the weight of the animals, each curve from a different habitat in the north Atlantic. The extreme variability indicates the importance of identifying stocks and of managing them as separate units.

A pattern or method of organism growth may be represented most directly by plotting the length or weight of a fish at different times (t) during its life, i.e., plotting $l(t)$ for length at t and $w(t)$ for weight at t. However, it is rarely practical to wait years to see this pattern. The alternative is to imagine an instantaneous photograph of the sizes or weights of the whole stock with animals of all ages, from which an average length or weight for each year class is found. Exploratory fishing will provide animals that may be aged by independent methods such as reading growth rings on scales, otoliths, shells or teeth and then length-at-age or weight-at-age can be plotted. The resulting smooth curves, however, can be deceptive (1) because of variability in growth between similar habitats as seen in Fig. 5.6, (2) because even within one environment there is variability among individuals, and (3) because the environmental conditions experienced during the early years of a ten-year-old may not be the same as those experienced by a three-year-old. These and other sources of variability all contribute to the standard error that should be imagined around each datum.

The simple addition of biomass that we see as growth is the result of many physiological processes. Naturally, the general nutritional level in the environment is also important but density-dependent and resultant behavioral characteristics within the population are frequently of central importance, as well. In these complex interactions may be the explanations for such diverse observations as why laboratory and field experiments often yield different results; why the numerical abundance of halibut appears to be decreasing, the weight at age is increasing, but so is the time to first reproduction; or why the size of Pribilof Island subadult male seals on a crowded unharvested island are smaller than on a nearby island where these males are sparser because they are harvested. Such observations raise difficult questions that require further study.

(a) The model of von Bertalanffy

Forewarned now about the seductive power of smooth curves, we proceed to consider the most useful of smooth curve growth models. The rate of change of length (dl/dt) is a function of the actual length at time t, $l(t)$ and how far that length at t is from some hypothesized maximum length denoted by L_∞. One of the simplest functions for the rate of change of length is a differential equation which says that the rate of change of length will get smaller and eventually be zero as the length gets close to its maximum possible, *viz.*,

$$\frac{dl}{dt} = K(L_\infty - l(t)) \tag{5.8}$$

This is the well-known Bertalanffy differential equation whose solution is

$$l(t) = L_\infty (1 - e^{-K(t - t_0)}) \tag{5.9}$$

L_∞ is the asymptotic length ($L_\infty = \lim_{t \to \infty} l(t)$), K a growth constant (not a rate) and t_0 the time at which $l(t) = 0$. Parameters L_∞ and K and t_0 are all estimated from the growth data. Alternatively, eq. 5.8 can be solved in the form

$$l(t) = L_\infty - (L_\infty - l_0)e^{-Kt} \tag{5.10}$$

In this case, l_0 is the length when time equals zero (t_0). The paradox that length is not zero at time zero occurs because these equations do not apply to the growth of very young fish and do not pass through the origin (Fig. 5.7a). Gallucci and Quinn (1979) review the history behind this function and present a view of how the function is used and misused and how its parameters are estimated.

A sketch of the von Bertalanffy length model (Fig. 5.7a) may first appear similar to that of the logistic (Fig. 5.2a) but they are quite different. The von Bertalanffy model for length lacks an inflection point and has a different type of slope at low values near zero. However, if eq. 5.8–5.10 were converted from length units into weight units the new von Bertalanffy curve would have an inflection point at about $\frac{1}{3}$ of the asymptotically maximum weight size, W_∞ (Fig. 5.7b).

The standard way to convert from length to weight is to use the allometric equation

$$w = al^b \tag{5.11}$$

where $w = w(t)$ and $l = l(t)$ are weight and length at the same t-value. Parameters a and b are usually estimated by regressing weight on length. Linear regression can be used if both the weight and length data are transformed logarithmically,

$$\log w = \log a + b \log l$$

Nonlinear regression is also possible and is likely to be more accurate (Gallucci and Quinn 1979; Glass 1967; Zar 1968). The value of b is usually near 3 and is often set equal to 3, and only a is estimated.

If the allometric equation with $b = 3$ is substituted into the von Bertalanffy length equation the result is the von Bertalanffy model in biomass

$$w(t) = W_\infty (1 - e^{-K(t - t_0)})^3 \tag{5.12}$$

where W_∞ ($\propto L_\infty^3$) is the asymptotic weight and everything else is as defined for eq. 5.9. This new function $w(t)$ is the one used in the biomass yield model of Beverton–Holt to be described later. Figure 5.7b shows the shape of the growth curve defined by this equation.

(b) Other models

Despite the widespread use of the von Bertalanffy model, several other models are used in fisheries. One model used for halibut management is the generalized von Bertalanffy model (Southward and Chapman 1965) which includes precisely the same parameters as the standard version but has an additional parameter

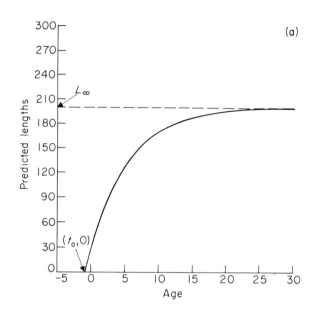

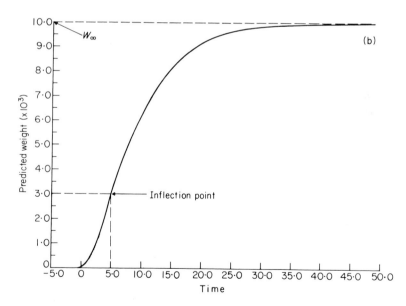

Fig. 5.7. (a) The von Bertalanffy equation in terms of length, with $K = 0.17$ and $t_0 = -1.0$. (b) The von Bertalanffy equation in terms of weight, with $K = 0.17$ and $t_0 = -1.000$. The inflection point is at 2960.

'*m*,' which adds an inflection point to the model. This function, written from eq. 5.10,

$$l(t)^{1-m} = L_\infty^{1-m} - (L_\infty^{1-m} - l_0^{1-m})e^{-K(1-m)t} \tag{5.13}$$

It is generally true in mathematics that use of more parameters permits better agreement between the data and the model. Thus, the use of the additional fitting parameter '*m*' naturally results in better agreement with the data.

Another variation of the standard von Bertalanffy model results from using the allometric equation, eq. 5.11, without pre-setting $b = 3$, but evaluating b from the regression of weight on length data. The resulting $w(t)$ function does not lend itself as readily to yield computations because the resulting yield equation involves an incomplete beta function (Gulland 1969).

A characteristic of most growth functions is a rapid biomass increase for low values of biomass (of an individual or a population) and thus, the popularity of the exponential growth model for populations. An analogous situation exists for organism growth; the model is known as the Ricker growth model (Ricker 1975) in weight

$$w(t) = w_0 e^{Gt} \tag{5.14}$$

where w_0 is an initial weight and G is the instantaneous growth rate. The model is quite useful if growth is over a short interval where it is accurately 'exponential-like.' The major advantages of this model are the ease of estimation of parameters G and w_0 (by fitting a regression line to a semi-log transformation of the data) and that a simple yield expression can use this formulation directly. The model is less frequently used in these days of computers because it has a limited range of biological usefulness and computational ease of the type provided is usually unnecessary.

The model of Richards (1959) is the most general of growth models because all of the models discussed above can be shown to be special cases. This model is used in the growth literature, but not frequently seen in management work.

5.3.4   Beverton–Holt yield per recruit model

Yield per recruit models are among the most commonly used estimation procedures for stock assessment. Unlike the surplus production model, the yield per recruit model treats somatic growth, mortality, and natality as separate processes. Though more explicit estimates can be made from yield per recruit models, there are more kinds of data used than in the surplus production model. Consequently it is usually more difficult to get data for this model.

We will first approach the concepts involved from the standpoint of a single cohort of fish—those individuals produced by the stock in a single spawning season. A cohort may also be termed a brood, or a year class. During the course of its lifespan a cohort is subjected to natural mortality that decreases the numbers of individuals in the cohort, and so decreases the biomass of the cohort (Fig. 5.5a). The individuals grow in body size, and that tends to increase the biomass of cohort (Fig. 5.7b). The interplay of the two processes acting simultaneously over the life span of the cohort produces an increase in cohort biomass during youth while growth dominates, and a decreasing biomass at older ages while mortality dominates. So, the change in biomass during the lifespan can be represented by a dome shaped curve (Fig. 5.8a). Clearly, maximum yield from the cohort would result if all individuals were caught at the intermediate age where biomass is at a maximum.

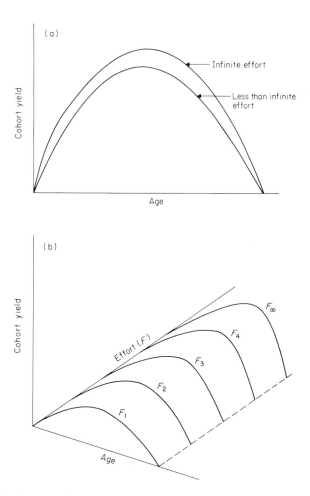

Fig. 5.8. (a) Yield or catch from a cohort as a function of age at first capture. (b) Yield from a cohort as a function of age at first capture and four levels of fishing effort. Intensity of effort increases from level 1 through level 4.

To catch all fish at a given age would require an enormous amount of fishing effort. At practical levels of fishing effort, some individuals of the cohort would escape the fishery and survive to more advanced ages. Suppose we apply a series of effort levels to the biomass curve in Fig. 5.8a. If we applied infinite effort, then the total biomass curve represents the various yields that could be attained if we could choose the age to apply the effort to. If we applied a reduced level of effort that missed some of the fish, then the yield would be less at any single application of the effort. The new dome shaped curve would be contained within the infinite effort curve. We could then develop yield curves for a series of effort levels. We could now plot effort as a third axis to accommodate the family of curves in Fig. 5.8b. The result of this would be a hill-like surface. If we contoured the plot in Fig. 5.8b in the same way that topographical maps are contoured, the plot would take a new appearance (Fig. 5.9). Whereas on a map, the lines of equal land height are drawn, in Fig. 5.9 lines of equal yield are drawn. These lines are termed yield isopleths. They are used to find the combinations of fishing effort and age at which the cohort becomes vulnerable to fishing (age at first capture) that gives equal yield

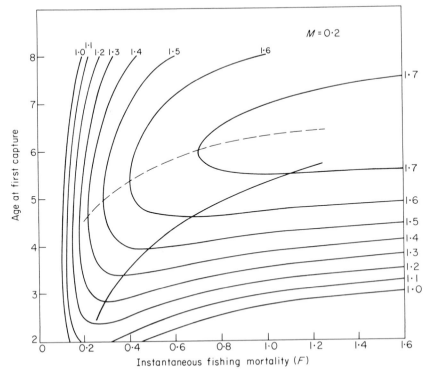

Fig. 5.9. Yield isopleth plot in kg for pollock of the southwest Scotian Shelf Georges Bank. Redrawn from Clark, Burns, Halliday (1977) *ICNAF Selected Papers No. 2*; 15–32.

levels. Age at first capture is a management option achieved by regulating mesh size of nets, hook size, or retention size.

To estimate the total yield from a cohort it is necessary to account for the fish that are not caught during the first and succeeding years of fishing. The lower the effort level, the more fish will get by the fleet. As a consequence, more will be lost due to natural mortality. The number of survivors after a year of fishing, for fish that have just entered the fishery, is calculated as product of the survival rate (S) and the number of fish starting the year (N_t).

$$N_{t+1} = N_t S$$

The number of fish caught from N_t is calculated by the Baranov equation (eq. 5.7). Weight of the fish is calculated by the von Bertalanffy size at age function (eq. 5.12). The catch in numbers is multiplied by the weight at age to arrive at catch weight at the age of first capture. These calculations are repeated for the second year, the third year, to the nth year after the cohort enters the fishery. The nth year is the year that the last individual fish disappears from the stock. This calculation procedure may be approached as an iterative flow that is easily adaptable to computer programming. The flow from one functional relationship to the next in a loop is outlined in Fig. 5.10. The iterative solution

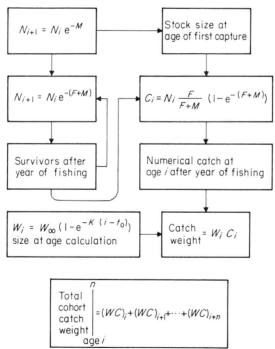

Fig. 5.10. Iterative flow of calculations for Beverton and Holt yield per recruit calculations.

follows the actual time course of the cohort, and so is easier to grasp than the original formulation by Beverton and Holt (1957), reviewed by Royce (1972).

Transfer of the calculations from a cohort to a stock requires only the addition of a contingency. If the number of young entering the fishery each year (recruitment to the fishery) is a constant, then a graph of the number of fish in a single cohort plotted year by year is the same as a graph of the number of fish in each cohort in the stock in any one year. Over a period of years, if the number of new fish entering the fishery varies about a mean within bounds, and if deviations above the mean roughly equal deviations below the mean without a unidirectional time trend, then the averaged cohort trends over time are the same as the average cohort structure of the stock in any year. If one can estimate the average recruitment level, and it is often possible to do so (Gulland 1977), then the model gives yield estimates at different ages of first capture and different effort levels (Fig. 5.9).

The beauty of the three dimensional, mathematical surface, however, is that the shape is independent of the number of recruits. The rates alone in the functions in Fig. 5.10 determine the shape. That means if just the natural mortality rate, present fishing mortality rate, and growth rate can be estimated, a stock assessment can still be made. Yield per 1000 recruits, or more usually, yield per recruit is calculated as the vertical axis. If instantaneous natural mortality is roughly estimated within a range of M's, M_1 and M_2, then a family of yield curves is usually generated with three or four M values (Fig. 5.11). Plotting yield against a range of F values shows whether an empirically estimated fleet F is near the maximum yield per recruit. Each of the yield curves in Fig. 5.11 can be interpreted as a cut through an isopleth diagram at a constant age of first recruitment. These particular curves were calculated with data from a complex of eastern Canadian cod stocks. There are three curves at that age because M was varied from 0.10, to 0.20, to 0.30. The arrow indicates the mean estimated level of fishing between 1960 and 1969, inclusive. At worst the stock is slightly over fished if M is less than 0.20.

5.4   STOCK AND RECRUITMENT

The number of juvenile animals added to a population through natural reproduction is obviously central to the future of a population. Such aspects as the number of animals added relative to (1) the number lost through natural mortality and harvest mortality, (2) the number of reproductive animals that produced these additions and (3) the number of animals present when the juveniles joined the population are especially important. The animals added to the population are called recruits and the process of their addition, recruitment.

It is common in management to distinguish among three stages of recruit-

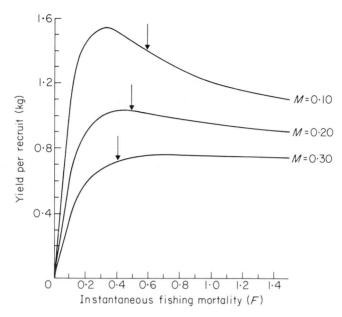

Fig. 5.11. A yield per recruit curves in kg from the eastern Scotian Shelf cod stock complex. Redrawn from Halliday, R.G. (1972), *ICNAF Res. Bull.* **9**, 118–122.

ment: (1) recruited to the harvestable stock, (2) recruited to the harvest, and (3) recruited to the mature (fully reproductive) stock. For example, younger fish may mingle with older fish and be exposed to the harvest operation but not be retained by the gear, or they may be illegal to keep and are returned. They are nevertheless part of the harvestable stock. At a later age, these young fish reach a certain size and are 'fully' recruited (equally catchable or 'retainable' as the older fish) to the harvest. The age of recruitment to the reproductive stock is obviously important also because mature animals produce the greater amount of eggs or offspring and thus have the potential for making the largest contribution to the next generation or cohort. Thus, the crux of the management problem is balancing the social and economic requirements of the harvest of reproductive animals with the biological requirements necessary to keep the stock at a constant or increasing size.

5.4.1   Principles and Models

Theoretical considerations suggest a negative feedback or stabilizing mechanism is active over population sizes the stock may commonly experience. This stabilizing factor would tend to depress the net rate of recruitment as the population reaches the environmental carrying capacity and to increase the net rate of recruitment when the population size is smaller, thus stabilizing the size around the carrying capacity. It is the net reproductive rate (the difference between the total rate of recruitment and the rate of natural mortality on juveniles) which is most important because it is theoretically possible for the total rate of recruitment to be positively correlated with population size if the rate of natural mortality on juveniles is also appropriately positively correlated.

A natural concern is whether there exist conditions that may block the feedback mechanism. In some cases it may be that if the stock size falls below some critical level, the rate of recruitment will not be able to increase (e.g., because a lowered density may make fertilization improbable), but will decrease and eventually result in extinction for the stock. Alternatively, the growth rate might be very fast at very low densities and have no relationship to stock size (density-independence). Then, upon reaching some higher density, the feedback mechanism begins to operate.

Fish larvae move in patches with the currents seeking patches of food. Thus, the patch of larvae is part of a competitive feeding guild in which it is both predator and prey. Most of the attention in stock-recruitment studies focuses on early life and juvenile stages which are more likely to be affected by density-dependent factors such as competition, predation and cannibalism. Density-dependent factors would not greatly affect adults.

Density-independent changes in natural mortality usually imply abiotic factors such as temperature, chemical nutrients in the water, salinity, etc. Abiotic factors are frequently more subtle in their effects than it may first appear. For example, low temperature may act directly and freeze a new set in an intertidal clam fishery or it may act indirectly and mix the upper water layers containing food for larvae, over-dispersing the food and making it inaccessible to larvae.

One way to study the relationship between the spawning stock and the new generation is to use a mathematical formulation. A graph of the number of recruits (R) against the number of spawners (P) is called a *recruitment curve* where recruits are defined in any of the several possible ways noted initially. The two common and famous mathematical models of recruitment used in fisheries, the Ricker model (1954) and the Beverton–Holt model (1957) (Fig. 5.12 and 5.13), have properties consistent with three important theoretical considerations *viz.*,

1. There is an upper limit to recruitment regardless of how large the population of spawners.

2. Eventually, the rate of recruitment R/P (the slope of curve) will decrease continuously or effectively cease to increase with an increase in P.

3. Recruitment R exceeds stock size P over some P-values to compensate for mortality; otherwise the stock would have to decrease in size.

Neither of these models, however, allows for a critical stock size, N_c, below which reproductive processes may be density-independent and result in inadequate recruitment to perpetuate the stock. Nor do these models allow for the possibility of recruitment from the immigration of larvae or juveniles via water currents. That is, a zero *R*-value for positive *P*-values and positive *R*-values for a zero *P*-value are pooled into curves that go through the origin.

The model proposed by Ricker can be written as

$$R = aPe^{-\beta P} \tag{5.15}$$

R is the number of recruits. *P* is the parental stock size measured in numbers, biomass, egg production, etc. a is the dimensionless parameter sometimes associated with the magnitude of the effect of density-independent factors and is sometimes considered to be characteristic of the stock. β is a parameter with dimensions $1/P$ usually associated with the effect of density-dependent factors.

The characteristics of the curve in Fig. 5.12 conform to those described above. The *P*-value where recruitment is maximized $P = P_m$ is found by setting the rate change of *R* to zero and solving for *P*. That is,

$$dR/dP = a\, e^{-\beta P_m}(1 - \beta P_m) = 0$$

$$P_m = 1/\beta$$

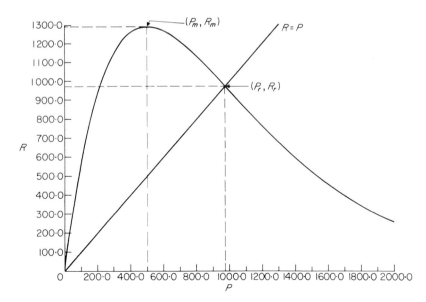

Fig. 5.12. A dome-shaped recruitment curve from the Ricker model with $a = 7.000$, $\beta = 0.002$, $R_m = 1288$, $P_m = 500$, $P_r = R_r = 973$.

The corresponding maximum recruitment $R = R_m$ is found by substituting P_m into eq. 5.15,

$$R_m = a\left(\frac{1}{\beta}\right)e^{-1} = a/\beta e$$

Suppose P and R are measured in the same units, e.g. numbers (or weight) of spawners and numbers (or weight) of four-year-old halibut, respectively. It is then possible to determine the values of P and R, where recruitment just replaces (equals) spawning stock, *viz*, the replacement values $P = P_r$ and $R = R_r$. Algebraically, it is clear that this found by setting $P = R$ in eq. 5.15 to obtain

$$1/a = e^{-\beta P_r}$$

and

$$P_r = (\ln a)/\beta$$

Geometrically, the line $R = P$ in Fig. 5.12 intersects the recruitment curve for eq. 5.15 at the point $(P, R) = (P_r, R_r)$.

The replacement value represents an obviously important concept and also provides a method to express eq. (5.15) in terms of just one parameter, a. This is accomplished by letting $a = \ln a = \beta P_r$ and substituting into eq. 5.15 to obtain

$$R = Pe^{a(1 - P/P_r)} \tag{5.16}$$

where $a = P_r/P_m$ is easily shown. If $a > 1$ the replacement level occurs at sizes in excess of P_m, the stock level maximizing recruitment, and when $a < 1$ the replacement level occurs prior to P_m. Indeed, the higher the value of a, the more steeply R rises in eq. 5.16. The model also predicts that a displacement from the equilibrium will return to the stock-recruitment curve by a damped oscillation if the slope (dR/dP) near the point (P_r, R_r) is greater than -1. If $dR/dP < -1$ the oscillation will not be damped and will lead to biological extinction via increasing fluctuations.

The alternative, Beverton–Holt model is

$$R = \frac{1}{a + \beta/P} \tag{5.17}$$

or, equivalently

$$R = \frac{P}{aP + \beta} \tag{5.18}$$

where R and P are as usual and a and β are new parameters evaluated from data. Again, the replacement value (P_r, R_r) is found by setting $P = R$ to find

$$P_r = \frac{(1 - \beta)}{a} = R_r$$

The Beverton–Holt model is also expressed in terms of a single parameter, $A = (1 - \beta)$ and the replacement value P_r,

$$R = \frac{P}{1 - A(1 - P/P_r)} \tag{5.19}$$

The Beverton–Holt model differs somewhat from the Ricker model in the shape of the curve, as may be seen in Fig. 5.13. In contrast to the Ricker model,

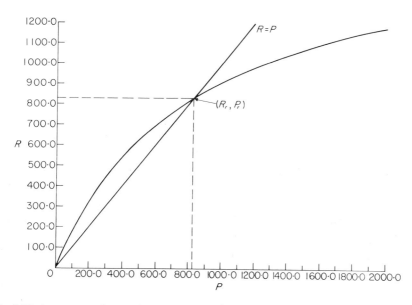

Fig. 5.13. An asymptotic recruitment curve from the Beverton–Holt model with $a = 0.0006$, $\beta = 0.5$, $R_m = 1667$, $P_r = R_r = 833$.

recruitment does not start to decrease after passing through the maximum value, but instead, it remains almost constant. That is, the model asymptotically approaches a maximal recruitment, $R_m = 1/a$, found by taking the limit of R, as $P \to \infty$. A consequence of the dome shape is that the Ricker model predicts recruitment to be near or at zero at a very high spawning stock size, whereas the Beverton–Holt model predicts essentially the same recruitment as it did for a high spawning stock of smaller magnitude. Another difference is that the Beverton–Holt recruitment curve never rises above the replacement value R_r until after intersecting at (P_r, R_r) with the replacement straight line.

Fisheries scientists have generally accepted the observation that gadoid stocks (e.g., cod, haddock, hake) will usually exhibit recruitment curves with well-defined dome shapes. In contrast, some herring stocks have such shallow dome-shaped curves that the Beverton–Holt model appears to fit best, whereas other herring stocks are better described by a Ricker, dome-shaped curve. A last, interesting possibility suggested for flatfish (such as plaice and halibut) is that recruitment may be independent of spawning stock size.

In principle, one does not wish to choose a 'model' on the basis of which curve the data fits, but instead, on the basis of the underlying assumptions of each model (Gallucci and Quinn 1979). However, the state of our understanding of the recruitment process is too primitive and the similarity of the assumptions behind the two models too great for any other clear criteria to be applied.

5.4.2   Processes and patterns of recruitment

It is valuable to consider different patterns of recruitment by examining the processes that act upon invertebrate, fish, and marine mammal populations. Many marine invertebrates of interest to fisheries spawn with fertilization occurring in the water column external to the parents. Many terminate larval development as sedentary or quasi-sedentary organisms of the benthos. These include molluscs such as clams, oysters, and scallops and crustaceans such as lobster and crab. In broad terms, the stages of recruitment include larval production by the spawning stock, larval survival during the planktonic period, and survival after metamorphosis and settlement. Survivors are not usually harvested until a suitable age or size is attained, which varies from two years (e.g., for some clams) to six years (e.g., for some lobsters). In both cases the juveniles may be caught in the commercial fishery after one year or so but usually are rejected and returned, and thus are not recruited to the fishery. Despite the apparently excellent circumstances afforded for studying recruitment in sedentary organisms, invertebrate recruitment patterns are not easily generalized, even within a species. Overall, there is very little evidence to support density-dependent arguments of the standing crop having any major impact upon recruitment. Rather, abiotic density-independent factors seem to dominate. Hancock (1973) provides an excellent overview.

A better argument may be made to support a relationship between recruitment to fish stocks and the standing crop, but there continue to be many cases where the data support density-independent factors. We consider two recent, contrasting examples. Lett and Doubleday (1975) studied recruitment and year-class size in northern Gulf of St Lawrence cod. Two biological processes dependent upon the spawning stock were identified as central to recruitment: the growth rate of adult cod (and the corresponding positive correlation with egg production) and the degree of predation of adult cod upon juvenile cod. In contrast, Sissenwine (1975) analyzed recruitment in the yellow-tail flounder fishery from southern New England and concluded from the analysis that the number of juveniles produced is independent of the size of the spawning stock. His regression analysis suggested that flounder recruitment is independent of interactions with other species, e.g., cod, haddock or blackback flounder, but that abiotic factors, in particular temperature, exert the dominant influence upon recruitment.

Although we might be expected to know more about the stock-recruitment relationship for marine mammals, this is not yet the case. Allen (1974) notes that recruitment to whale stocks is influenced by a number of density-controlled components. Indeed, we might expect mammalian, homeothermic, animals to be influenced somewhat less directly by abiotic factors. These density-dependent components include (i) the age females become mature, (ii) pregnancy rate, (iii) the rate of successful births per pregnancy, (iv) the rate of survival from birth to recruitment to the fishery, and (v) the age of recruitment. The International Whaling Commission reports that female Antarctic fin whales mature at 6 to 7 years of age in heavily exploited stocks and at 10 years of age in unexploited stocks. Here then is a mechanism which fits the compensatory patterns: sexual maturity occurs earlier, more births occur and recruitment increases, all as a consequence of exploitation. Clearly, to have omitted consideration of age structure in analyzing the recruitment pattern would have obscured observation and possible mechanisms behind a stock size which is shrinking further and further 'under the dome' of a Ricker recruitment curve, only resulting in increased recruitment. Ultimately, after getting smaller than P_m, recruitment could start to drop again. Very few studies of stock-recruitment consider the age structure of the spawning stock but it is from this approach that new, interesting results will likely come, and the marine mammals are logical animals to begin with.

5.4.3   Concluding comments on recruitment
Birth and mortality patterns for human and animal populations have been a subject of concern for many years. For example, note the following conversation between Samuel Johnson and Boswell in 1769 (Chapman 1970):

> Boswell: *Have not nations been more populous at one period than another?*
> Johnson: *Yes, sir, but that has been owing to the people being less thinned at one period than another, whether by emigration, war, or pestilence, not by their being more or less prolific. Births at all times bear the same proportion to the same number of people.*

That is, birth rate is a constant and environment regulates population size. Malthus (writing around 1800) would have supported this view whereas Darwin (1868), in writing about domesticated animals, suggested that fecundity is a function of the general health of the breeding stock.

The implications of Darwin's suggestion, applied to human populations in our age of generally excellent nutrition, lead to interesting hypotheses. Since the average age of maturity for men and women in developed countries has dropped by two to three years, there probably have also been downward shifts in the age of maximal female fertility and the age when juvenile survival is maximized. In fact, nutritional data are the basis of the concept that an average minimal critical body weight is associated with the age of menarche (Frisch 1978, Trussell 1978). Speculation about the role of the newly 'discovered' female orgasm suggests that it may be a mechanism by which the female alters the probability of fertilization.

These demographic ideas may be applied directly to mammals that exhibit social structure, such as whales, seals (especially during haul-out periods) and porpoise. A great deal more remains to be learned about the behavioral characteristics within complex school or harem-like social orders.

The application of these lines of thought to fishery management in general requires fishery scientists to concentrate on collecting age-dependent data. Although a suitable theory for stock and recruitment in age-structured populations does not yet exist, it is probably central to understanding the recruitment processes for long-lived fish such as halibut and most marine mammals.

The fisheries literature does contain a minor amount of exploration of the role of nutrition in reproduction. For example, Nikolskii (1969, p.42) reports for several fish species that nutrition affects not only fecundity but the quality of the sex products, by which is meant the likelihood of larval survival. Tyler and Dunn (1975) found that winter flounder fecundity is positively correlated with ration size. Lett and Doubleday (1975) actually suggest that growth rate is 'the homeostatic core of the recruitment mechanism.'

Another avenue of research will probably involve multispecies interactions. Lett and Kohler (1976) argue that predation by mackerel and cannibalism by older herring are the population controls on recruitment to the size of harvestable herring. The focus upon multispecies interactions is representative of an 'across the board' return to this philosophy in fishery research and coincides with the widespread availability of computers to do modeling and multivariate analysis.

Additional reading on stock and recruitment questions might include the seminal work of Ricker (1954), Ricker (1975), the symposium proceedings edited by Parrish (1973), and the lectures by Cushing (1975).

5.5   A SELF-GENERATING STOCK MODEL

At this point in the chapter we can develop a mathematical representation of the life history of a stock with cohort structure, with individuals that grow in size, enter a fishery, reproduce, and die due to natural causes. We have talked about models that set up calculations for recruitment relationships, mortality rates, and body growth. We have dealt with cohort structure as a phenomenon independent of recruitment rate and have described how growth and mortality functions may be coupled with a catch function to give yield per recruit estimates. The surplus production model mimicked the dynamics of the stock, but did not explicitly account for the life history aspects just mentioned. We will now develop a synthesis of stock process mathematics that links up all these life history phenomena into a simple simulation of a stock. Suggestions for such a model were first made by Beverton and Holt (1957). We will develop the model with the functional relationships used by these authors, but develop a logic as might be used by a computer modeler.

A. Von Bertalanffy size-at-age function

$$l_i = L_\infty (1 - e^{-K(i - i_0)})$$

$$W_i = W_\infty (1 - e^{-K(i - i_0)})$$

B. Negative exponential survival

$$N_{t+1} = N_t e^{-z \cdot t}$$

C. Baranov catch function

$$C = N_t \frac{F}{F+M} (1 - e^{-(F+M)})$$

D. Beverton-Holt stock-recruit function

$$R = \frac{1}{a + b/B}$$

Fig. 5.14. A summary of functional relationships used in the self-generating stock model. Symbols are as follows: l_i, length of fish at age i; L_∞, ultimate length; w_i, weight at age i; W_∞, ultimate weight; K, growth parameter; i_0 age at which length $= 0$; Z, instantaneous total mortality rate; F, instantaneous fishing mortality rate; M, instantaneous natural mortality rate; t, years; N_t, number of fish at beginning of year; N_{t+1}, number of fish at beginning of following year; C, catch in numbers; R, biomass of recruits to fishery; B, spawning biomass; a, b, statistically fitted parameters.

There are four functional relationships that will be used in the model, all of which have been previously discussed in this chapter: the von Bertalanffy size-at-age-function, the negative exponential survival decrease function, the Beverton and Holt stock-recruit function, and the Baranov catch function (Fig. 5.14).

We will enter the model logic with the assumption that we have an estimate of the number of young in the age 1 cohort, and then introduce this estimate as N_t in the model (Step 1 in Fig. 5.15). We will create a stock that spawns during a short period (say one month) once per year during late winter, and will

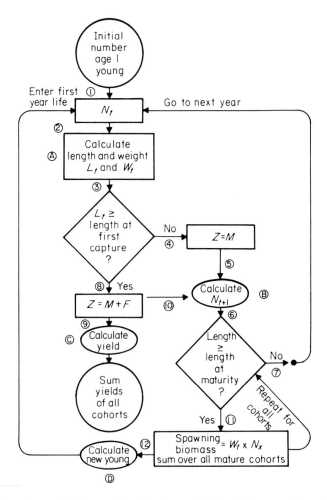

Fig. 5.15. Flow chart of self-generating fisheries model. Capital letters refer to the functional relationships of Fig. 5.14.

allow the fishery to operate during the summer months only. We can calculate the mean length and weight as the average size of the fish in a cohort mid-way through the fishing season by using the von Bertalanffy function (Step 2 and point A in Fig. 5.15). We now ask the question: is this the length at which the fish are first captured by the fishery (Step 3)? We will make the simplification that the length at first capture is the length at which 100 percent of the individuals are available to the fishery. The more realistic situation is that an increasing proportion of the cohort is available as length increases. Since the cohort is only at age 1 and small, we assume that the fishery does not catch them. In that case the total instantaneous mortality rate (Z) equals the natural mortality rate (M) only (Step 4). We move on to calculate the number of survivors at the end of the year using the negative exponential survival function (Step 5, and function B, Fig. 5.14). We check to see whether the fish are large enough to be mature and spawn (Step 6), and since they are only age 1 we pass spawning (let these hypothetical fish be mature at age 6) and progress N_{t+1} to the beginning of the loop in the schematic (Step 7). We can loop the cohort through a second and then a third year, following Steps 2 through 7. During the fourth time through this loop we find that the cohort has reached the length at first recruitment to the fishery, and then we set $Z = M + F$, where F is the instantaneous fishing mortality rate (Step 8). Cohort yield is calculated at Step 9 using the Baranov function. Step 10 moves to calculation of the survivors (N_{t+1}) using negative exponential survival. We check to see whether the fish are large enough to be mature (Step 6), and since they are only age 4 we send them to N_t at the beginning of the loop. Repeating this process until age 6, we find in Step 6 that the fish will spawn and move to Step 11 where the spawning stock biomass is summed over any other cohorts that exist in the model (only one cohort at this point). At Step 12 the Beverton–Holt stock-recruit function is used to calculate the number of age 1 young. Step 13 then sets the new young into the model flow, and also carries the parent cohort through another loop. Eventually there will be a number of cohorts iterating through the model logic. It is usual to use a computer to keep track of these cohorts. The number of cohorts in the stock in any one year depends on the rates of mortality and recruitment.

These next paragraphs are transitional between the simple population models dealt with in this chapter and the systems models discussed in Chapter 7. Since the number of cohorts and the abundance of the cohorts are determined by the rates of production and mortality, extrinsic factors that influence these rates will each partially determine cohort structure and stock size.

There are two modeling approaches to coupling extrinsic factors to self-generating yield models. The simpler is to run the model with a statistical correlation between extrinsic factors and a stock variable such as number of recruits at age 2. The other is to develop time-invariate functional relationships, or rules of change, between those factors and the response rates. These functional relationships can express the behavioral and physiological processes through which the extrinsic factor acts. In this way the modeler can explore alternative biological mechanisms that are the possible causes of the statistical correlation used in the first approach. In both approaches the model may be driven with an empirical record of suspected environmental factors. Since the techniques used in the second type of modeling are described in Chapter 7, we will describe only the first approach here.

The model outlined in Fig. 5.15 uses an estimate of cohort size based on a relationship between cohort size and spawning biomass. To incorporate environmental factors, we could develop a multivariate statistical study to investigate the possible effects of a number of variables on cohort strength. We would need a time series of environmental variables as well as corresponding measures of cohort strength and spawning biomass to develop the statistical study. We would try to avoid the spurious correlations that can result from indiscriminate statistical scanning of a large number of environmental variable measurements selected merely because the data exist. To achieve this discrimination we must have hypotheses regarding how environmental variables might operate on various life stages of the young. Out of these hypotheses we would select the relevant environmental data and carry out a step-wise, multiple correlation to discover which variables, if any, account for significant proportions of the variance in the cohort-strength time-series. The study would be successful if most of the variance could be statistically accounted for. The multiple correlation equation and the environmental factors incorporated can be used to estimate cohort strength. The equation could be used in place of the simple stock-recruitment function in the model of Fig. 5.10. Strictly speaking, the equation applies only to the historical period covered by the time series. But as a model, it applies to the fish stock, and so can be used as an estimator. The contingency of this new model is that the environmental factors that dominated cohort strength in the historic period will also dominate during some finite period in the future. This contingency can be tested. In addition, the model can be used for management decisions on the basis that it is a substantial piece of scientific evidence, giving more relevant estimates than do other alternatives. To be sure, factors not incorporated into the model will influence cohort strength during some periods in the future, and so some estimates will be widely discrepant. Statistical studies must be continued because of the certainty of these environmental changes. Better estimation will then be the result.

A number of multivariate correlation studies have been carried out recently that statistically relate parent stock size, plus hydrographic measurements, to annual cohort strength of a pre-recruit age. For example, in a region on the continental shelf of northwestern United States, 65 percent of the variation in cohort strength of Dover sole (*Microstomus pacificus*), and 84 percent of English sole (*Parophrys vetulus*), was statistically accounted for by three aspects of upwelling (Fig. 5.16). Because of these significant correlations, estimates of cohort strength can be made with upwelling measures, and the influence of upwelling on yield estimated. It is the variation of recruitment due to upwelling that influences the yield estimates. Maximum sustainable yield calculated from the model in Fig. 5.16 on Dover sole was 15 percent less than sustainable yield calculated with a yield per recruit model scaled with the average level of recruitment during the period. In another study, Lett *et al.* (1975) showed that

Fig. 5.16. Correspondence between age 6 abundance of Dover sole estimated by cohort analysis observed and age 6 abundance estimated using a multiple correlation model based on upwelling and offshore divergence of water mass. The sole stock lies near the mouth of the Columbia River on the West Coast of the U.S. Redrawn from Hayman and Tyler (1979).

the temperature at which cod larvae (*Gadus morhua*) developed was related to cohort success. Their model indicated that sea temperature fluctuation could cause the sudden collapse of the cod stock under conditions of heavy fishing that would otherwise be sustainable.

5.6   REFERENCES

Allen K.R. (1974) Recruitment to whale stocks. *In* W.Scheville (ed.), *The Whale Problem*: A Status Report, pp.352–358. Cambridge, Mass.: Harvard University Press.

Beverton R.J.H. and S.J.Holt. (1957) On the dynamics of exploited fish populations. *Fishery Invest., Lond.,* Series 2, **19**.

Chapman D.G. (1961) Population dynamics of the Alaska fur seal herd. *Trans. Twenty-Sixth North American Wildlife and Natural Resources Conference*, pp.356–369.

Chapman, R.W. (ed.) (1970) *Boswell, Life of Johnson*, edition corrected by J.D.Fleeman. New York: Oxford University Press Paperback.

Cushing D.H. (1973) *Recruitment and Parent Stock in Fishes.* Washington Sea Grant Publ. 73–1. Seattle: Univ. of Washington.

Darwin C. (1868) *The Variation of Animals and Plants under Domestication,* 2nd edn. 1894. vol. 2, pp.89–90. New York: Appleton

de Ligny W. (1969) Serological and biochemical studies of fish populations. *Oceanogr. Mar. Biol.* 7, 411–514.

Frisch R.E. (1978) Population, food intake, and fertility. *Science,* **199**, 22–30.

Gallucci V.F., and T.J.Quinn. (1979) Reparameterizing, fitting, and testing a simple growth model. *Trans. Am. Fish. Soc.* **108(1)**, 14–25.

Glass N, (1967) A technique for fitting nonlinear models to biological data. *Ecology,* **48**, 1010–1013.

Gulland J.A. (1969) *Manual of methods for fish stock assessment. Part 1.* FAO FRS/M4.

Gulland J.A. (1977) The analysis of data and development of models. *In* J.A.Gulland (ed.), *Fish Population Dynamics.* p.86. Wiley.

Hancock D.A. (1973) The relationship between stock and recruitment in exploited invertebrates. *Cons. Int, Explor. Mer Rapports et Proces-Verbaus,* **164**, 113–131.

Hayman R.A., A.V.Tyler, and R.Demory (1980) A comparison between cohort analysis and catch per unit effort for Dover sole (*Microstomus pacificus* and English sole (*Parophyrs vetulus*). *Trans. Am. Fish. Soc.,* **109**, 35–53.

Hayman R.A. and A.V.Tyler (1980) Environment and cohort strength of Dover sole (*Microstomus pacificus* and English sole (*Parophrys vetulus*). *Trans. Am. Fish. Soc.,* **109**, 54–70.

Jones R. and C.Johnston (1977) Growth, reproduction and mortality in gadoid fish species. *In* J.H.Steele (ed.), *Fisheries Mathematics.* London: Academic.

Lett P.F. and W.G.Doubleday (1975) The influence of fluctuations in recruitment on fisheries management strategy, with special reference to Southern Gulf of St. Lawrence cod. *Int. Comm. Northwest Atl. Fish. Sel. Pap.* **1.1**, 171–193.

Lett P.F., A.C.Kohler, and D.N.Fitzerald (1975) Role of stock biomass and temperature in recruitment of southern Gulf of St. Lawrence Atlantic cod, *Gadus morhua. J. Fish. Res. Board Can.* **32**, 1613–1627.

Lett P.F. and A.C.Kohler (1976) Recruitment: a problem of multispecies interaction and environmental perturbations, with special reference to Gulf of St Lawrence Atlantic Herring (*Clupea harengus harengus*). *J. Fish. Res. Board Can.* **33**, 1353–1371.

Nikolskii G.V. (1969) *Fish Population Dynamics.* Edinburgh: Oliver & Boyd.

Parrish B.B. (ed.) (1973) Fish stocks and recruitment. *Rapp. P.-V. Reun Cons. Int. Explor. Mer.* **164**.

Pella J.J. and P.K.Tomlinson (1969) A generalized stock production model. *Inter-Am. Trop. Tuna Comm. Bull.* **13(3)**, 421–458.

Richards F. (1959) A flexible growth function for empirical use. *J. Exp. Boit.* **10**, 290–300.

Ricker W.E. (1954) Stock and recruitment. *J. Fish. Res. Board Can.* **11(5)**, 559–623.

Ricker W.E. (1975) *Computation and interpretation of biological statistics of fish populations.* Bull. 191 Canada Fisheries and Marine Service.

Royce W.F. (1972) *Introduction to the Fishery Sciences.* Academic.

Schaefer M.B. (1957) A study of the dynamics of the fishery for yellowfin tuna in the eastern tropical Pacific Ocean. *Inter-Am. Trop. Tuna Comm. Bull.* **2**, 247–268.

Sissenwine M.P. (1975) Variability in recruitment and equilibrium catch of the Southern New England yellowtail flounder fishery. *J. Cons. Int. Explor. Mer.* **36(1)**, 15–26.

Southward G., and C.Chapman (1965) Utilization of Pacific Halibut stocks: study of Bertalanffy's growth equation. *Int. Pac. Halibut Comm. Rep.* **39**.

Trussell J. (1978) Menarche and fatness: re-examination of the critical body composition hypothesis. *Science* **200**, 1506–1509.

Tyler A.V. and R.S.Dunn (1975) Ratio, growth, energy, budget, and measures of condition in relation to meal frequency in winter flounder (*Pseudopleuronectes americanus*). *J. Fish. Res. Board Can.* **33**, 63–75.

Zar J. (1968) Calculation and miscalculation of the allometric equation as a model in biological data. *Bioscience,* **18**, 1118–1120.

Chapter 6
The Human Dimension

MICHAEL K. ORBACH

6.1 INTRODUCTION

The human dimension of fisheries management is an aspect which seems so simple, but which quickly becomes infinitely complex. We often hear the phrase, 'one doesn't manage fish, one manages people'. What does this mean, and how and why must we become acquainted with fisheries in their human dimension?

Any system is a set of interrelated components. In a fishery system, one of the primary, most dynamic components of the system are people and their behaviors. Most fisheries are exploited by groups of individuals who interact closely, perform observable behaviors, and share describable characteristics. The behaviors of these individuals comprise a set of processes which contribute to the extraction or use of fisheries resources. These behavioral processes form the human component of fisheries systems.

There are two primary reasons to manage a fishery. One reason is to make use of a portion of the physical and biotic component: to put food on the table; to produce other products for human use or consumption; to provide a physical environment which satisfies human desires for recreation, leisure, or intellectual or sensual pleasure. The other reason is somewhat more altruistic. As an integral part of the physical ecosystems ourselves, but a part having a moral and aesthetic conscience, we may feel the need to protect the physical and biotic components from the 'exogenous' effects of humans and their manufactured technologies; for example, we should not push species to extinction and we should not produce materials which render natural processes incapable of their normal resiliency.

To accomplish either of these purposes we must have a clear picture of the ways in which people participate in fisheries. We must know the functions of individuals and groups, and the uses and satisfactions which specific individuals derive from fisheries-related activity. The resulting picture is the human dimension of fisheries management.

Any such picture which may emerge, however, has the potential for infinite diversity and complexity. The marine, estuarine, and freshwater environments and the fisheries resources within them are used by people of different national, racial, and ethnic backgrounds; from different social and economic strata; from different cultures with varying tastes, preferences, beliefs, and lifestyles; and from groups with divergent political and economic philosophies. Each of these groups of people derives different benefits from the use of fisheries resources, and each has a different set of preferences concerning the way in which those resources are used.

6.2   THE HUMAN COMPONENT OF FISHERIES SYSTEMS

The human component of a fishery involves more than fishermen themselves, that is, others besides those who harvest the resource from its habitat. These harvesters are only a small part of the total set of people involved in fisheries. For every commercial fisherman, for example, there are three sets of people who are equally a part of the human dimension of his activity: his family and 'community' in the social or political sense; the people in the boatyards, supply stores, and service facilities who are both integral to and dependent upon the harvesting activity; and the distributors, marketers, and consumers who create the demand for his product. These chains of involved individuals often extend thousands of miles, across local, state, and national boundaries (Fig. 6.1).

A Mexican-American shrimp fisherman in the Gulf of Mexico may live with his family in Brownsville, Texas. He may fish, however, off Campeche, Mexico, and sell his catch through a broker in New Orleans, Louisiana, for consumption on dinner tables in California. The skipper of a tuna seiner in San Diego may draw his crew from a Portuguese community in California, yet fish for the tuna off the coast of Central America. He depends on towns in Costa Rica or Panama for supplies, and unloads his catch and has his boat repaired in the Caribbean. A Japanese longline fisherman may be away from home for two years at a time, he and his vessel contributing to the economies of dozens of ports around the world while still being the economic focal point of his family in Japan. A salmon fisherman in the British Isles could well be from Fargo, North Dakota, having disbursed the economic benefits of his quest for a recreational experience over thousands of miles between the two continents.

People in each of these physical locations participate in or are affected—socially, economically, and often politically—by the fishery. Any change in the resource exploitation system, through management or otherwise, will affect the lives of individuals in all of these diverse locations and situations. Each has a family and a community, with unique characteristics, which will be affected by changes in the resource exploitation system. For our purposes, however, let us limit our discussion to those people involved in the harvesting sector of the system; to the fishermen themselves. Who and what is a fisherman?

A public official and ex-fisherman from Alaska is fond of saying of fishermen

Recreational system

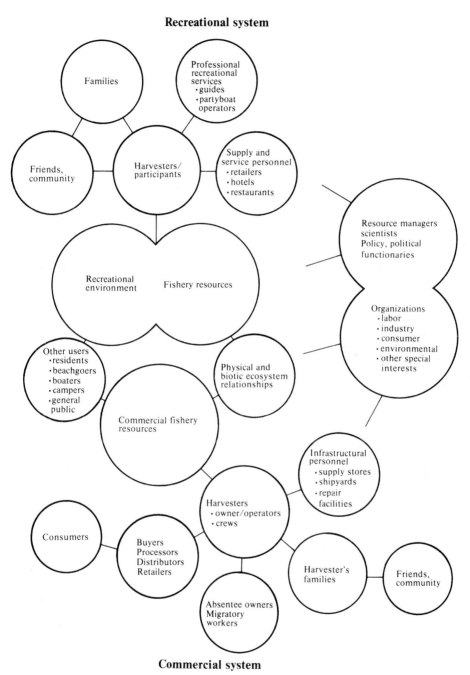

Fig. 6.1. Model human components of fisheries systems. *Note:* While every particular system will differ both in its component parts and in their structural arrangement, these diagrams present a general idea of the kinds of linkages which tie the people in a resource exploitation system together.

that there are three kinds: those who eat the resource; those who sell it; and those who play with it. This is a concise statement of the fact that there are fishermen whose purpose is to provide direct nourishment to their families from the fish which they catch; others whose purpose is to harvest a resource which they can sell, trade, or barter into a market economy; and still others who fish primarily for the pleasure or challenge of the experience, for the recreation which it provides. People in these various categories often look and act very differently from one another. A Palauan fisherman fishes out of a sailing canoe near a coral atoll in the Western Pacific. He has built his boat himself, and he will use the fish he catches to feed his family and village not only because they have few other food resources, but also because there is no market in which he may sell his catch. The Palauan fishes barefoot, in shorts and a thin shirt, and catches the fish with a hand line. A fisherman on a Polish factory trawler catches fish which are processed at sea solely for export in bulk from his country to foreign markets. He dresses in woolens and slickers against the freezing cold of the North Sea where he fishes, and the dangers of working around heavy equipment and machinery add to the disadvantages of the physical environment. A third fisherman may fish for the thrill of hooking a marlin off Mazatlan, Mexico, from a plush cabin cruiser based in that resort city. He wears the sophisticated but casual dress of one who is able to spend large sums of money for his amusement and uses equipment manufactured solely for the pleasure of the vacationer.

We must be careful, however, not to overemphasize the differences in use or benefit among these harvesters. A subsistence fisherman often trades or barters fish products for other needs even though these transactions are outside of a cash economy. A commercial fisherman may derive a significant amount of pleasure from the aesthetics of his surroundings. The satisfaction he derives from the independent nature of his activity may be in fact a major part of the total benefit he derives from his occupation. The Mexican marlin fisherman is a commercial necessity to the charter captain who has taken him on his outing, and his catch may provide needed food to local residents after the sportsman's prize has been recorded in a record book or a photograph. Each harvester of a fishery resource derives or produces a mix of benefits which transcend any simple categorization as subsistence, commerce, or recreation.

Even though the exact benefits which specific individuals derive from fishing defy broad generalization, there are general statements to be made about categories of activity and technology among fishing peoples around the world. Total subsistence fishermen—in the sense of those who have a primary nutritional dependence on the consumption of fisheries products, without commercial processing or involvement in a market system—are found today in very few places throughout the world; in certain circumpolar regions such as

Alaska, Greenland, or Lapland, in the lesser developed islands of Oceana and the Caribbean, and in remote locations on rivers, lakes, and oceanic shorelines. These fishermen typically use very elementary technologies, even in cases where the quarry is very large such as in the subsistence harvest of bowhead whales by Eskimos. Although subsistence fisheries provide essential nutrition, very few people depend on this consumption for the main portion of their nutritional intake.

Perhaps the largest number of fishermen throughout the world are those for whom the harvest is important as both a nutritional contribution and as an element in some form of market economy. In South America, Africa, the Middle East, Pakistan, India, Southeast Asia including Indonesia and the Philippines, portions of the harvest may be consumed directly but the majority is usually sold or traded, either fresh or after simple processing such as drying or smoking, in a localized market. These fishermen use a wide variety of gears—nets, hook-line, spears, traps—from the beach, riverbank or shore, or from small boats powered by hand, sail, or small gasoline engines. Their boats and other tools are made of indigenous materials, usually wood, supplemented by items obtained at the end of long marketing chains originating in industrialized countries. In terms of strict numbers most of the fishermen in the world fall into this category. They are not the yellow-slickered fisherman of Gloucester, Massachusetts, the English trawlerman of Hull or Grimsby, or the Russian factory ship worker, but rather the villager in Sri Lanka, dhow fishermen in the Middle East, handliners in Hong Kong, shark fishermen on the west coast of Mexico, or beach-seiners in small towns in Africa. The payment for their efforts is as likely to be an in-kind share of the catch as it is to be a piece of monetary currency or an instrument of commercial credit, and their involvement in the fishing is probably in accordance with familial or local community patterns or obligations. Fishing may be only part of a monthly or yearly round of activity which includes farming, indigenous crafts, or periods of wage labor in other endeavors. Formal education is rare, and mobility in residence or occupation is usually restricted.

The majority of the total tonnage of fisheries products, however, is harvested by fishermen from industrialized nations; Japan, Poland, the Soviet Union, the United States, East Germany, and others. These fishermen produce fish products in mass quantities, and for mass markets. Their market economies and their technologies are symbiotic. The demand which is created by the market fosters the development of fishing technology to satisfy that demand, and the technology produces more and more products for which markets and consumers must be found. More importantly for the human dimension of fisheries management, these markets and technologies are imbedded in complex social and political systems. Many fishermen operate from or around metropolitan areas such as San Diego, Bremen, or Hong Kong with high population densities

Fig. 6.2. Primary and secondary employment in fisheries in selected countries. These figures below are estimates from Food and Agriculture Organization of the United Nations Fishery Country Profiles. The primary sector employment figures given here represent full-time, part-time, and subsistence fishermen. They are presented to give a rough comparative picture of the numbers of people who make their living from fishing worldwide. The year in which the figures were estimated is indicated in parentheses, and the numbers below the country name indicate primary/secondary sector employment.

1 Argentina (73)
4,500/15,000
2 Chile (75)
19,860/10,200
3 Columbia (72)
130,000/650–700
4 Brazil (75)
325,000/6–14,000
5 Peru (70)
51,700/10,800
6 Mexico (69)
44,662/NA
7 United States (68)
128,449/88,742
8 Canada (73)
51,000/19,000
9 Cuba (74)
14,500/16,000
10 Jamaica (73)
9,000/NA
11 Iceland (74)
4,900/6,400

12 United Kingdom (77)
16,435/50,000
13 Norway (77)
33,241/200,000
14 Poland (77)
17,576/21,200
15 Finland (69)
9,233/NA
16 West Germany (74)
5,850/15,000
17 France (69)
37,163/NA
18 Spain (70)
124,027/NA
19 Portugal (69)
38,186/NA
20 Italy (74)
65,000/12,000
21 Greece (68)
16,435/NA
22 Morocco (73)
15,000/20,000

23 Mali (70)
60,000/40,000
24 Ghana (73)
80,000/NA
25 Kenya (72)
25–30,000/NA
26 Malawi (69)
19,000/26,000
27 Dahomey (70)
30,000/24,000
28 Benin (76)
8,000/24,000
29 Egypt (70)
100,000/NA
30 India (75)
3–400,000/NA
31 Pakistan (71)
163,998/NA
32 Bangladesh (73)
500,000/NA
33 Nepal (78)
18,000/NA

34 Burma (71)
342,000/NA
35 Indonesia (75)
854,000/NA
36 Hong Kong (70)
45,000/NA
37 Philippines (70)
687,877/30,000
38 USSR (70)
500,000 (combined)
39 Japan (69)
570,200/NA
40 Korea (73)
979,206/NA
41 Australia (73)
20,900/NA
42 Fiji (76)
2,409/NA
43 Western Samoa (72)
7,000/NA

and diverse alternatives in employment, lifestyle, and ideology. They are participants in political economies with pervasive national philosophies concerning the proper and acceptable use of capital, labor, and resources ranging from the state-owned fishing fleets of Poland, to the strong individual or corporate ownership patterns of the United States, to the mixed public and private sector fishing industry of Mexico. They live under the influence of cultural patterns which subvert traditional or religious tastes and preferences with pervasive media advertising and conflicting socialization and resocialization, often in situations of burgeoning rural-urban migration and rapid technological development such as we find in Alaska, Southeast Asia, or in many parts of Africa. These characteristics are the earmarks of the milieu within which many fishermen from industrialized or rapidly-developing nations perform their activities.

Many of these generalizations are born out in Fig. 6.2. The large numbers of fishermen in the primary sector (i.e., directly involved in fishing) and the absence of information on the secondary sector (i.e., processing, marketing, etc.) employment in countries such as Ghana, Hong Kong, India, Jamaica, Indonesia, Burma, the Philippines, Kenya, Nepal, Pakistan, Fiji, Finland, Columbia, and Bangladesh, indicate that the fishing activity is pervasive and may emphasize either subsistence uses, family industries, or localized processing or consumption patterns. In countries such as Iceland, Norway, Morocco, Poland, and the United Kingdom, all of which have large secondary segments, a significant portion of the fisheries activity is directed toward exports. In others, such as West Germany, Malawi, or Argentina, the large secondary effort is directed largely at domestic commercial distribution and sales. In yet others, such as Dahomey or Benin, the large secondary numbers result from the extensive inland aquaculture activity. In each of these cases the stream of uses and benefits have widely differing implications for the management of the fisheries.

Perhaps the narrowest and most specialized category of fish harvester is the recreational fisherman. It is primarily in the United States, Canada, and secondarily in Australia and New Zealand that fishermen in large numbers are found who devote significant time and energy to fishing as a part of their recreational activity. People all around the world may enjoy fishing in some sense, but only in the above locations is the concept of the *fishing sportsman* common, and only in these places does activity surrounding recreational fishing occupy a significant place in the economy. Local economies in many places may be dependent upon tourism as a source of income, but only in the above locations does recreational fishing constitute big business. As in the commercial cases, urbanization and industrialization also create quandries for recreational fishermen. They may no longer have access to nature in a virgin state. Fishing piers, artificial reefs and lakes, and aquarium environments may eventually rival natural settings in the provision of recreational fishing opportunity.

6.3   WHY CONSIDER THE HUMAN COMPONENT?

The first reason for examining the character of the fisherman is our need to be aware of and understand the impact of potential management decisions on the people involved in fisheries activity. What will be the economic impact of a catch quota on individual fishermen in the fleet? How much will their income be reduced, and what alternatives are available to them to make up that deficit? What is the impact of an area closure which may induce labor migration to fishing bases in other areas? What effect does such migration have on the families of the fishermen involved? We also must consider the impact of events and behaviors which occur outside of the local fishery system, but which will some-how affect people in that system. For example, what are the effects of competition from fishing activities of other countries, either in terms of spatial competition on the fishing grounds or of market competition from imported fishery products? It has been the case in the past that fish taken from the North Atlantic stocks by European factory trawlers could be exported to the United States, for example in frozen fillets, and reach the dinner table of a family in Indiana or Idaho at a much lower price than if the same fish were caught by United States fishermen in New England and processed and distributed domestically. In these cases, is it more important to supply a product more cheaply to the consumer, or to maintain the employment, industry, and community of fishermen in New England? To properly assess situations such as this, we need knowledge of the place of fisheries activity in the communities where the fishing takes place—the role of the occupation in the economic and social life of the community and the alternate local employment sources, for example.

We must also consider the traditions and historical values which surround maritime activity as a total lifestyle and as an adaptation to the physical and economic environment. To fishermen in the outports of Newfoundland, Canada, for example, fishing is a part of their culture as reflected in songs, oral traditions and dress; as a context for educating the young and socializing them into the values and social mores of the community; and as an integral part of a flexible occupational pattern which allows residents to adapt to the meager natural resources of their environment on land. What would happen if their fishing were curtailed? It is a part of their lives which they have historically resisted abandoning for an urban existence as wage laborers.

The second reason to investigate the human component is that management is often the business of allocating scarce resources among competing user groups. The scarce resource may be fish, it may be credit for boat construction loans, or it may be a physical space which will not support commercial and recreational uses at the same time. This allocation process demands a record of the mix of benefits, uses, and needs which are derived or satisfied by the fisherman's activity.

For example, in the early seventies violence broke out between inshore and offshore fishermen in Malaysia over, among other things, the availability of fish stocks. Inshore fishermen demanded preference because they are less mobile than the offshore fishermen, and thus can not take advantage of a wide range of alternatives. For these inshore fishermen the boat serves not only as a business, but also as a residence. Their children grow up aboard the boat amidst fishing and other marine-related activity. This floating family unit is part of a community which literally lives on the water, a very common phenomenon in Southeast Asia. The offshore fishermen demanded preference because they are more important to the economic infrastructure of the shoreside communities. Offshore fishermen purchase more capital equipment; their economic activity— dockage, unloading and processing, purchases of fuel and supplies—is more centralized. They are, in the short run, more efficient and visible targets of large-scale foreign aid and government assistance programs. Their presence in Malaysia, although less pervasive, is somewhat more powerful.

Japanese long-liners fishing for tuna in the United States fisheries conservation zone also capture species of billfish which they sell for food in Japan. These same billfish—marlin, for example—are the targets of extensive and highly capitalized sport fisheries based in the southeastern United States, Hawaii, and California. Taking Florida as a case in point, in many counties the age and income profiles of the population are both high, making recreational fishing of this kind an important activity. Which is more important, the opportunity for sport and recreation for Florida residents, recognizing that the provision of this recreational opportunity also supports a wide range of business and industry, or the provision of foodfish to the Japanese public to satisfy demands for nutrition and culinary pleasure?

In the 1970s the International Whaling Commission set quotas on commercial whaling which had severe impacts on the people involved in industries in several countries. The unemployment in Japan alone resulting from these quotas ran into the thousands of workers. The healthy and diverse economy of Japan presumably provides opportunity for alternative employment of these workers. Yet, when quotas were suggested for bowhead whales which are taken primarily by Eskimos in Alaska, the place of the whale harvest in the nutrition, culture, and social structures of these native groups—who take the whales solely for subsistence and limited bartering purposes—was seen as a crucially important set of variables. In the isolated, harsh environment of the Eskimos, fishing and hunting activity forms not only the most viable source of nutrition, but the structure of the hunting activities themselves reflects and in many cases determines the relationships of political authority and social connection in their villages. Without a knowledge of the characteristics, behaviors, and values of the people involved in these systems it is impossible to make rational decisions in allocating such scarce and valuable resources.

The third reason behind our need to become familiar with the human components of fisheries is the simple principle of understanding. Only by knowing how and why the people in these systems act as they do can we understand how the system works. All of the interactions in a fishery—among harvesters, managers, owners, processers, scientists, consumers, and so on—will take place more smoothly and with a better chance of satisfactory results if each participant understands the point of view, motives, goals, and constraints of the other participants.

An illustrative case is the tuna-porpoise controversy involving the high-seas tuna fleet which fishes out of San Diego, California. Because tunas follow the porpoise during parts of their migratory or feeding patterns, visual sighting of porpoise is used to locate schools of tuna. As a result of problems with the purse-seining method which the tunamen used, porpoise would often become entangled in the nets before they could swim free and would drown. The tunamen were aware of this problem from the early days of their adoption of the seining method and had made several attempts to correct the problem through gear changes, new net retrieval procedures, and so on. However, the porpoise is a mammal with which the American public had a great amount of emotional attachment, and the porpoise mortality caused by the tunamen's activity created a furor. The problem was compounded by the fact that the tunamen's trips lasted from two to three months, and the fishing was done in the central eastern Pacific. Thus the tunamen, year after year, spent eight or nine months of the year thousands of miles out at sea, effectively isolated from the furor their industry was creating ashore and the public emotions and attitudes which lay beneath the controversy. Conversely, the public was unaware of the tunamen's constructive efforts, and an impression of the fishermen as unconcerned, ill-intentioned individuals who do not value the lives of the creatures around them was an easy impression to develop. Much of the resulting controversy in this and other cases could have been avoided, or minimized, if all concerned had a proper understanding of the activities, perceptions, motives, goals, and constraints of others.

6.4   INVESTIGATING THE HUMAN DIMENSION

The kind of factors and considerations which have been mentioned in the preceeding pages have always been significant, if not primary components of decision-making processes in fisheries management. Like biological and physical aspects, these factors and their place in the management process can be assessed and recorded in a scientific manner. This task, the social scientific analysis of the human dimension of fisheries, may be approached through several avenues.

Large amounts of data and information on the human component of fisheries are available through secondary sources. Monographs, articles, and documents written by social scientists and journalists often contain pieces of information or insight about fishermen, their industries, and their communities. Because these items are often produced with a specific application in mind, their documentation, completeness with respect to the fishery, level of accuracy, or format may render them of less use than if the fishery system itself would have been the focus of study. If an economist studies the costs and returns of a certain fishing vessel class; if a psychologist studies the effects of father absence on a distant-water fisherman's family; if a journalist writes an article on preservation of marine mammals; if a fisheries advisor reports the problems of raising pondfish on idle agricultural land; if a sociologist studies the aesthetic benefits derived by a trout fisherman; if a legal scholar traces the effects of extended fisheries jurisdiction; each of these will constitute only a small part of the total set of information necessary to an investigation of the human dimension in a given fishery. Each piece of information must be combined with others, and the results or implications of their work translated into objectives, goals, management measures, or some other form which fits into the process of evaluation or management (see Chapter 8). In addition, in the form in which they are available these secondary sources generally do not give us the dynamic picture of the fishery necessary to assess social or economic changes or impacts.

There are several ways to actively and directly investigate the human dimension of fisheries. One is through *census surveys*. This type of survey will yield information such as age, education, residence, location of fishing activity, income, family size, ethnicity, migration patterns, and other demographic information. Just as statistical sampling methods are used in biological research, variants of these methods can be used to scientifically sample human populations and expand the resulting data into meaningful statements. Even though a great deal of this census-type information is collected by organizations such as the Bureau of the Census or state employment agencies in the United States or by the Fisheries and Agriculture Organization internationally, it is often either missing key variables or aggregated in such a way as to make assignment of the information to specific occupational or economic systems impossible or impractical. Some information of this type has been collected by fisheries managers and used for descriptive purposes in management regimes which basically are biologically-oriented.

With more and better information, we may in many cases be able to bring our management regimes more in line with a definition of a fishery which includes as integral parameters not only the biological species and their habitats, but also the patterns and characteristics of the user groups. Thus, instead of defining a fishery as an 'albacore', 'salmon', or 'crab' fishery, as mentioned in Chapter 1, we may want to define a 'west coast troll and pot fishery' based on the fact that the same fishermen take all three species from the same boats but with different gear throughout the year. It may be more reasonable and efficient to license fishermen or boats *once* as multi-purpose units and include catch limits on individual species, rather than license the same fisherman or boat three or more times for each separate species. Census surveys designed to gather specific data about fisheries can facilitate such decisions, leading to management tactics more likely to be accepted by the fishermen.

Census-type information may be supplemented by *attitude* or *opinion surveys*. Such surveys attempt to measure what people like or dislike and how they would respond to specific management decisions. For example, people's perceptions of their alternatives or preferences in residence or employment can be used to assess probable impacts of alternative management options on the local fishing community. Opinions concerning the management options themselves— or of the desirability or necessity for any management at all—are extremely valuable pieces of information. Surveys can help avoid unnecessary conflicts between managers and resource users or among user groups themselves by making areas of potential conflict explicit and resolving conflicts before rather than after effort is expended in attempts at management. As with census surveys, attitude or opinion surveys can be performed with rigorous methods of sampling and analysis. Non-parametric statistics, formal elicitation procedures, various forms of cluster analysis, and other specialized techniques are useful in cases where exact values cannot be assigned to the variables. For example, if a fisherman is asked to rank various aspects of his activity in order of importance to his economic well-being, to his role in his family, or to some other specific end, these data can be analyzed using non-parametric techniques without asking the individual to actually assign values to those aspects of his activity.

The most in-depth and complete information on human populations is obtained through combinations of *ethnographic techniques*: participant observation, a method favored by anthropologists whereby the researcher lives and works with his or her subjects for extended periods of time; series interviews, such as a social psychologist might use to investigate personal histories or familial relationships; group discussions, where the responses of each individual are supplemented, qualified, or expanded on by others involved in the fishery. Ethnographic research is useful in situations where an understanding of social and cultural patterns is important, such as in assessing the role of subsistence fishing in village political relationships, the role of ethnicity in recruitment to an industry or occupation, or the ways in which recreational fishing activity contributes to mental health or childhood socialization. Ethnographic research is also a necessary adjunct to survey research, because without an understanding of the people and activities under consideration it is often difficult to produce relevant and meaningful questions for the survey. Asking a respondent which management option is preferred when the best course may be no management at all is a misdirected question. Asking a businessman who uses charter boats for business entertainment what the role of this recreational activity is for him and his family is meaningless. Proper ethnographic groundwork enables one to ask the right questions.

For any of these methodologies, it is often helpful and sometimes essential to engage fishermen and others in the fishery to carry out the work. This helps assure that surveys are not influenced by the presence of an outsider who may not be trusted or who may not properly interpret the words or actions of the fishermen.

6.5   CONCLUSION

All of these techniques and approaches produce information which can be combined with biological, ecological, and other information to create a picture of the fishery in its totality. This in turn enables both manager and fisherman to properly assess the state of the system and the potential impacts of any activity which might affect resource populations, habitats, people, businesses, industry relationships, or any other part of the system.

Although the human dimension of fisheries management is complex, it is also subject to rigorous observation and description. All of the situations and issues which have been raised in this chapter can be assessed, recorded, and documented as can virtually all of the situations and issues which will arise in any given fishery. Once again, we rarely manage fish—we usually manage people. To help insure that we begin with the right objectives, expend effort only where it is necessary, and avoid subverting or simply not attaining our management objectives, we need this clear, documented picture of the human dimension in our fisheries.

6.6   REFERENCES

Acheson J.M. (1975) Fisheries management and social context: The case of the Maine lobster fishery. *Trans. Am. Fish. Soc.* **104**, 653–668.

Alexander P. (1976) The modernization of peasant fisheries in Sri Lanka. In D.M.Johnston (ed.), *Marine Policy and the Coastal Community*. London: Croom Helm.

American Friends Service Committee (1970) *Uncommon Controversy. Fishing Rights of the Muckleshoot, Puyallup, and Nisqually Indians.* Seattle: Univ. Washington Press.

Andersen P. and C.Wadel (eds.) (1972) *North Atlantic Fishermen; Anthropological Essays on Modern Fishing.* Newfoundland: Memorial Univ. of Newfoundland.

Firth R. (1966) *Malay Fishermen; Their Peasant Economy.* 2nd Ed. London: Routledge and Kegan Paul.

Forman S. (1970) *The Raft Fishermen: Tradition and Change in the Brazilian Peasant Economy.* Bloomington: Indiana Univ. Press.

Moreda Orosa G. (1966) Fishermen's guilds in Spain. *Internat. Labor Rev.* **94**, 465–476.

Norr K.F. (1975) The organization of coastal fishing in Tamilnadu. *Ethnology,* **14**, 351–371.

Orbach M.K. (1977) *Hunters, Seamen, and Entrepreneurs; the Tuna Seinermen of San Diego.* Berkeley: Univ. California Press.

Poggie J.J.Jr. and C.Gersumy (1974) *Fishermen of Galilee; the human ecology of a New England Coastal Community.* Kingston: Univ. of Rhode Island.

Polanska A. (1965) *The Profession of the Deep-Sea Fisherman in Poland.* Gdynia: Wydawnictwo Morskie.

Smith C. (1976) Intracultural variation: Decline and diversity in North Pacific Fisheries. *Human Organization,* **35**, 55–64.

Smith C.L. (1977) The failure of success in fisheries management. *Environ. Manage.* **1**, 239–247.

Smith M.E. (ed.) (1977) *Those who live from the Sea.* Minneapolis: West Publ.

Spaulding I.A. (1976) Socio-cultural values of marine recreational fishing. *In* H.Clepper (ed.), *Marine Recreational Fisheries.* Washington: Sport Fishing Institute.

Tunstall J. (1962) *The Fishermen.* London: MacGibbon and Kee.

Van Stone J. (1962) *Point Hope; American Eskimo Village in Transition.* Seattle: Univ. Washington Press.

Yoshida T., K.Maruyama, and E.Namihira (1974) Technological and social changes in a Japanese fishing village. *J. Asian and African Studies,* **9**, 1–16.

PRINCIPLES OF
MANAGEMENT

Chapter 7
Systems Principles in Fisheries Management

CARL J. WALTERS

7.1 INTRODUCTION

Throughout this text it is emphasized that most fisheries problems are complex and contain human as well as biological dimensions. Too frequently we see the consequences of trying to deal with complexity in a fragmentary or narrow way. Management plans based on the soundest of biological information fail when it is discovered that fishing pressure cannot be controlled because of unforeseen political or economic constraints. Economic policies fail when unforeseen biological limits are exceeded. In short, fisheries represent dynamic (time varying) systems with interacting components, not a series of static building blocks that can be manipulated or studied independently.

In recent years there has been much thinking about how to deal with complex, interacting systems, in fields ranging from space exploration through economics and ecology. The result has been a set of ideas and techniques that cut across the various disciplines, and a general approach known as *systems analysis*. The common denominator in this new field has been the notion of trying to capture how a system works by expressing the interactions in terms of a *mathematical model*.

A model is a precise set of statements about how the components of a system affect one another; to solve the model equations is to use it as a *deductive engine* to predict consequences that might otherwise be overlooked. Unfortunately for many biologists and managers who do not like to use mathematics, no better way of thinking about systems has been discovered. Unfortunately for even the practitioners of systems analysis, mathematical formulations permit an extreme form of shorthand description (symbols instead of words, equations instead of sentences) that can be frighteningly illegible.

This chapter will try to remove some of the mystery from systems analysis, by discussing some of its principles and limitations in nonmathematical terms. The following section looks at the anatomy of mathematical systems models, viewing them simply as sets of rules for predicting changes. Next is a discussion of the unresolved problem of how to formulate models that are broad enough (consider a wide enough range of factors) and detailed enough not to be deceptive, yet are not so complex that they cannot be solved or understood. Two particularly useful sets of systems methods are then discussed—identification and optimization; system identification refers to the estimation of unobserved constants and variables from time series data, while optimization is concerned with finding best policies given a model and a set of alternative management actions. The final section identifies some misapplications of the system modeling approach.

A word of caution is necessary before proceeding. The tools of systems analysis usually look quite sensible, but no fishery has yet been managed by using them for a long period of time. Thus we cannot be sure that they will work, in the sense of producing better average harvests or fewer serious mistakes than traditional approaches.

7.2   ANATOMY OF SYSTEM MODELS

The systems analyst begins his examination of a problem by presuming that he can represent system behavior in terms of a finite list of *state variables* such as population sizes and levels of fishing effort. However, it is not assumed that all the state variables can be directly observed or measured in the field. Some kinds of information, such as the rich visual content of a topographic map, cannot be easily represented by a list of numbers, and this information will not appear directly in the model calculations. The problem of how state variables are chosen is a crucial one that will be discussed in the next section; for now, let us assume that it is possible to find a good list.

The state variables can be used in at least three kinds of mathematical expressions. The first is the *system dynamic model*, which is a set of rules for predicting how the state variables will change, usually over time, in relation to one another and to management actions. The second is the *observation model*, which hypothesizes how the state variables are related to observable quantities such as catch and catch per effort. The third is the *objectives model*, which hypothesizes how to measure performance of the system (in human terms such as dollars) in relation to the states that might occur.

Usually the system dynamic model is the first and most complex part of the analysis. Observation models are often ignored entirely or stated in very simple terms like 'It is assumed that population size has been correctly estimated.' However, the complex field of fisheries statistics and stock assessment is based on various elaborate observation models. Objectives models try to represent human desires in a simplified, quantitative way, and so will always be a point of considerable debate.

The essential features of system dynamic models can be illustrated with a simple, one variable example:

$$\begin{array}{c}\text{population}\\\text{next year}\end{array} = \begin{array}{c}\text{population}\\\text{this year}\end{array} + \begin{array}{c}\text{net}\\\text{production}\end{array} - \begin{array}{c}\text{catch}\\\text{this year}\end{array}$$

where net production is calculated from 'population this year' by using the simple *functional relationship* shown in Fig. 7.1, and catch is to be established outside the model and put into each year's calculations as a *driving variable*. Three points can be made from this example. The first and most critical is that the equation makes only short, incremental predictions; a variant on the same theme would have been to model rate of population change as net production minus catch, in the form of a *differential equation*. Suppose instead that we had demanded a single equation to predict population 10 years hence directly from this year's population. We would then have faced a nearly impossible task of constructing a 10-year functional relationship for net biological production; the population base for production could change radically, and in many different ways, within that 10-year step. In short, we could not have captured the *feedback* between population changes and subsequent production.

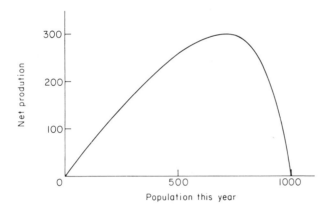

Fig. 7.1. Net production as a function of population size for a simple population model.

The second point to note about the example is that it divides the prediction problem into two components of change, production and catch. If we are careful to define 'net production' as recruits plus growth minus natural deaths plus immigration minus emigration, then the basic balance equation of the previous paragraph is a *tautology* (definition that is true by the way the words in it are used). Tautologies are essentially trivial statements about a problem, but they help to define it more precisely for further study: in this case, by dividing it into components about which we can then make empirical assertions in the form of functional relationships.

The third point is that the dynamic model contains constants, or *parameters*, as well as variables. These parameters are the numbers used to scale relationships like Fig. 7.1; in this example three parameters are particularly critical: the maximum net production (300), the natural equilibrium population if catch is zero (1000), and the population giving maximum production (about 800 in Fig. 7.1). Much discussion about whether a model is useful can be boiled down to two specific questions about parameters: (1) do the parameters give the correct shape of functional relationship (i.e., where should peak of Fig. 7.1 be?); and (2) are the parameters really constant over time? In essence, parameters represent the consequences of unmodeled forces operating within or upon the system; for example, to assume that the population equilibrium remains at 1000 is to pretend that there are no changes in biological limiting factors such as habitat and natural mortality agents. Often, a good way to proceed in building a model is to propose a simple option like Fig. 7.1, and then ask the two questions above. If response to the second is negative, then more variables and relationships are added until each relationship does appear to represent a stable, *repeatable* feature of the system.

An important class of models arises when relationships like Fig. 7.1 are assumed to represent average responses, and random variability is explicitly added to the calculations. The result is called a *stochastic model*, as opposed to a *deterministic model*. By including stochastic (random) effects, we can indirectly reflect the operation of factors that change rapidly in the system but do not have persistent, long term consequences in terms of changing parameters.

If dynamic models are generally formulated in terms of short term changes like the previous population example, how then are longer term predictions generated? The answer is simple if we view the short term equations as rules that can be applied repeatedly to build a long term *trajectory* of states:

$$\longrightarrow \text{state now} \longrightarrow \text{rules for change} \longrightarrow \text{new state} \longrightarrow$$

We plug the 'state now' into the equations (rules) to get 'new state', which in turn becomes the 'state now' for predicting a second 'new state', and so forth. The crucial requirement for carrying out this *recursive* calculation over many time steps is that the rules be expressed in terms of system state rather than time directly. In the simple example, we will not know at year one whether the population at year five will be 200 or 300 or 1000; that will depend on the stream of catches we plug in as well as the production relationship. To get from year five to year six we must have the production rule (Fig. 7.1) to accommodate any population state.

The idea of predicting how a system will change by repeatedly solving a set of component rules is almost ridiculously simple, yet it is the basis for most system modeling. When the 'state now' is described by many variables, and the rules are complicated, it is often necessary to have a computer do the repeated calculations; this is called *computer simulation.* One very large computer simulation model of North Pacific fisheries in relation to oceanographic factors has several thousand state variables (Levastu and Favorite 1977); most fisheries simulations involve 10–100 variables.

A point is worth stressing here about dynamic models and the people who build and use them. The modeling process involves breaking problems down into sets of specific variables and relationships that can be examined one at a time, then fitted back together. This means that several individuals, usually of different disciplines (biologists, economists, etc.), can work in parallel on a problem. The model provides clear tasks for each person (particular rules he is responsible for formulating) while connecting these tasks together in a well-defined package. Thus the modeling process can be an extremely powerful tool for promoting interdisciplinary cooperation and communication (for procedures and examples, Holling 1978; Walters 1974).

7.3   BREADTH AND DEPTH IN MODEL FORMULATION

System modeling requires the analyst to choose a series of quantitative state variables; the perceptive student will have realized that this choice is the most difficult and potentially misleading step in the analysis (and for that matter, in the design of good empirical studies). Choices of variables inevitably place arbitrary bounds on the study system, in terms of breadth and detail of factors to be considered. Should the model deal only with biology, or should it look more broadly at economic factors as well? Should economic measures like fish prices be considered constant, or should there be a broader look at how these prices are determined in a larger economic setting? In the other direction, should biological populations be represented only by gross measures such as total biomass, or is it necessary to spell out the details of age structure, spatial distributions, and the like? There are no absolute standards for answering questions like these.

A fundamental concept in systems theory is that most problems can be viewed in a hierarchic manner, just as ecology was discussed in a hierarchy from individual to community levels in Chapter 2. It is particularly important to note that any fishery is embedded in (forms part of) a larger biological ecosystem, and a larger set of economic or recreational opportunities. Often the most powerful analysis is one that steps back, to a higher hierarchical level, and looks at the fishery as just a small piece in the larger whole. An example of looking at Pacific salmon problems this way is shown in Fig. 7.2; the manager who concerns himself only with the seasonal details of regulating catches will miss larger, regional economic influences on salmon habitat, but the manager who looks only at the broadest influences will also err by failing

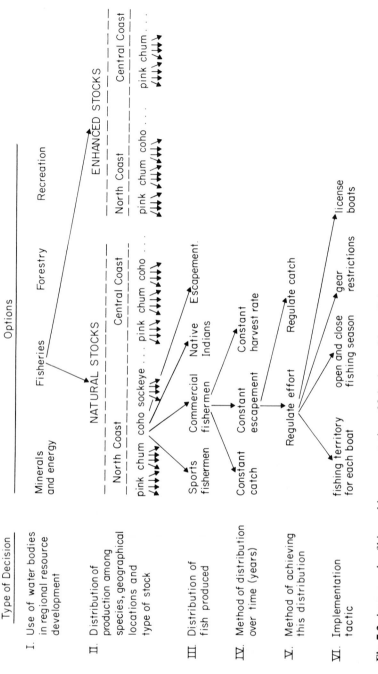

Fig. 7.2. An example of hierarchic representation of fisheries problems. Salmon management on a British Columbia River involves the range of concerns shown. From Holling (1978).

to consider the practical, day-to-day decisions that can have serious cumulative effects.

The lesson here is that the fisheries manager should deliberately sketch out a hierarchic representation like Fig. 7.2 of his problems, so that he can see how various models and analyses fit into the broader picture. Some have argued further that model variables can most efficiently be identified by first working out a hierarchic, qualitative description of a system in terms of 'linked subsystems'; this description may make the appropriate quantitative variables obvious (Klir 1969; Overton 1977).

A key choice in most fisheries modeling is whether to consider fishing effort (man the predator) as a dynamic variable or a parameter. Usually fishing cannot be controlled completely, and it may show some surprising changes when other regulations are introduced. For example, in a recent analysis of the sport and commercial troll fishery for chinook and coho salmon in the Gulf of Georgia, British Columbia, a main objective was to quantify how regulatory measures such as minimum size limits might reduce daily catches, and how seasonal closures would affect the escapement of spawning fish. We tried to treat fishing efforts as given constants based on historical patterns, but we noticed by chance that past efforts (both sport and troll) have often been higher in periods of high fish abundance. This relationship is certainly not surprising, so we decided to incorporate it into the calculations as a bit of realistic window-dressing. We simply made effort proportional to fish abundance. The simulated result was dramatic: assuming fixed fishing effort, we had predicted large benefits (increased escapement) from increased size limits; with responding effort, these benefits were almost completely cancelled. The size limits saved some small fish, but this resulted later in increased abundance of larger fish. Fishing effort then responded to the increase, and most of the saved fish were cropped before they could escape. Now, the predictions might be wrong and regulations like size limits might really work, but we had identified a key uncertainty that deserves careful study before or as new regulations are implemented.

Many biologists argue that the critical need in fisheries is for more study of fish and their ecosystems. This is a *reductionist* philosophy: the workings of a system are revealed by taking it apart and studying the pieces ever more carefully. An alternative view is the *holist*: the system is best treated as a whole, with properties (measurable behaviors) that would not be predicted by looking at all the parts. In terms of practical systems analysis, both these points of view are naive.

A reductionist might have great hope for systems modeling as a way to bring all the pieces together, but he will soon encounter three difficulties. First, without a model to identify all the pieces in the first place, some are bound to be missed; scientists tend to follow one another into similar research areas, and to study a few, experimentally tractable pieces over and over. Second, if the pieces are arranged in a hierarchy, so each system component has, say, two sub-components, each of these has two sub-sub-components, etc., then the number of pieces that have to be studied is 2^N where N is the number of levels (degree of detail) through which the increasingly detailed breakdown is carried; the number of details soon becomes practically unmanageable, and some are bound to be missed. Third, the sensitivity of overall system predictions to each detail considered does not decrease as the number of details increases; each detail does *not* become progressively less important as more are added to the analysis (effects are usually not additive).

Detailed studies can thus fail as miserably as simple and comprehensive ones. One possible way around this dilemma for the reductionist is to couple modeling and empirical studies in an iterative process: use modeling to identify critical data which are then collected to test and improve the model; the tests will reveal model weaknesses requiring further data, and so forth. Unfortunately, no one knows whether such a process will eventually converge to some acceptable level of understanding. In a way the holist faces a similar difficulty: it may be necessary to observe the system for a very long time before its emergent properties are fully exhibited.

Good systems analysis involves a mixture of reductionist and holistic approaches. Historical observations of system behavior (holist properties), especially in response to accidental or deliberate perturbations, can provide valuable clues about which component processes deserve more careful (reductionist) attention. For example, sudden large changes in fish population size may indicate the presence of depensatory mortality agents, such as natural predators, that the analysis might otherwise have overlooked.

There has been a tantalizing suggestion from some fisheries modeling that detailed analysis can be a waste of time. Often, detailed models exhibit simple dynamic behavior (or holistic properties) reminiscent of logistic growth, predator-prey oscillations, and the like. After seeing this simple behavior, it is often possible for the analyst to *compress* the model by sorting out irrelevant details and by aggregating some variables and relationships. For example, several survival rates, s_1, s_2, etc. that operate in series may be compressed to a single rate, $\bar{s} = s_1 \cdot s_2 \cdot s_3 \dots$.

A special bonus from simplification is that the dynamic behavior of the compressed model often can be expressed in terms of highly informative graphs such as *catastrophe manifolds* (for example see Holling 1978). Also, many different complicated models may give rise to the same simple model (imagine how many different birth–death hypotheses will give rise to logistic population growth); in other words, the simple model may be *robust* to many (possibly unresolvable) uncertainties about the detailed organization of the system. Thus some analysts claim it is best to start with a simple model in the first place, and just assume the details will not be important; you may wish these people luck, but be especially careful about applying their recommendations.

7.4 VALIDATION AND SYSTEM IDENTIFICATION

Systems analysis involves arbitrary choices of model variables, and the previous section has discussed how difficult it is to make good choices. There will always be a need to carefully compare models to field data, and usually it will be necessary to estimate at least some unknown parameters from that data. The determination of whether a model is consistent with observation is called *validation*, and the estimation of unknown parameters from time series data is called *system identification*. These ideas are obviously interrelated; system identification may be used to 'tune' a model so as to make it as valid as possible within the limits of its mathematical structure.

7.4.1 Validation

Before discussing system identification as a tool in fisheries, it is necessary to be more precise about the meaning of model validation. This concept is widely misunderstood; it is often believed that a model which has been validated (i.e., found or made to be consistent with observation) is likely to be functionally correct: to be based on true assumptions. This belief is more than naive: it is dangerously wrong. A perfectly valid model can be based on quite incorrect assumptions, and this point can be illustrated with a simple example from fish physiology.

Suppose it is desired to measure the respiration rate of a fish by placing it in a closed, well-mixed flow chamber (Fig. 7.3), then following oxygen concentration in the chamber as measured at the outflow pipe. Suppose two experimen-

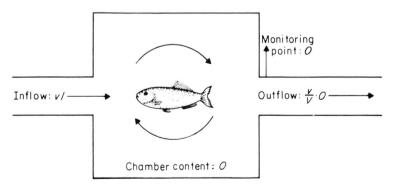

Fig. 7.3. A simple but potentially deceptive system for studying fish respiration. Explanation in text.

ters in different laboratories are using this procedure. Both will note that oxygen dynamics in the chamber can be described by the rate equation (tautology):

$$\frac{dO}{dt} = \text{inflow} - \text{respiration} - \text{outflow}$$

where O is total oxygen in the chamber. If the water inflow rate is v liters per minute at a constant oxygen concentration I per liter, and the chamber volume is V liters, both researchers will elaborate the above model as

$$\frac{dO}{dt} = vI - \text{respiration} - (v/V)O$$

Now comes the difficulty. Suppose the first researcher believes that respiration rate is just proportional to oxygen concentration, i.e., he believes the fish has no way to maintain steady respiration in spite of lowered oxygen concentration. He will express this belief with the relationship respiration $= r\,O$, and his chamber system model will then be

$$\frac{dO}{dt} = vI - rO - (v/V)O$$
$$= vI - (r + v/V)O$$

He will note that this model is of the general form $dO/dt = A - BO$, and his freshman calculus book will tell him that models of this form have a solution equation that can be easily fitted to a time series of O values; the curve fitting or system identification procedure results in estimates of A and B which the experimenter can then interpret as $A = vI$ and $B = r + v/V$.

Suppose the second researcher also has a pet conviction about respiration, namely that the fish regulates the flow of water past his gills so as to maintain a constant respiration rate over a wide range of oxygen concentrations. Our second researcher will then assume respiration $= R$, where R is a constant per time, and will model the chamber as

$$\frac{dO}{dt} = vI - R - (v/V)O$$
$$= (vI - R) - (v/V)O$$

He will note that his model is of the form $dO/dt = A - BO$ and his calculus will tell him the same thing it told the first researcher. Researcher two will go through the same curve fitting exercise as number one, but he will interpret the estimated constants as $A = vI - R$, $B = v/V$.

Both models will appear equally valid though they have almost diametrically opposed assumptions about the fish! The researchers can resolve which functional hypothesis is correct, but only if they (1) get together and compare the alternative models, and (2) add an extra piece of information to their test information, namely the volume flow rate v. We can be easily deceived by building only one model, yet by deliberately comparing alternative hypotheses using models, it may be possible to find simple criteria for testing which is correct.

7.4.2   System identification

System identification is a type of optimization. To find 'best' parameter estimates for a model in relation to a set of data, it is necessary to define a quantitative *objective function* that measures the goodness of any estimate relative to others. Often this function is taken to be a simple sum of squared deviations between predicted and observed data. The best estimate is then found by varying the parameters until a combination giving the minimum sum of squares is found. There are various mathematical procedures to make the search for a minimum as efficient as possible. If the model is 'linear in its parameters', i.e., has the form

$$\hat{y} = \sum_j b_j f_j (x)$$

(where $\hat{y}$ is the predicted observation, the b_j are parameters, and the f_j are functions of other observed variables, x), then the b's can be estimated by standard regression equations.

In system identification it is usually necessary to postulate an observation model (see Section 7.2) as well as a dynamic model. An example from fisheries is the widely used surplus production model (see Section 5.2). Here, population biomass growth is assumed to follow the logistic equation

$$B_{t+1} = B_t + k B_t (1 - B_t/B_\infty) - C_t$$

where B_t is biomass in year t, k is intrinsic rate of population increase, B_∞ is the equilibrium biomass in absence of fishing, and C_t is the catch in year t. Usually B_t is not observed directly; instead, we must *reconstruct* population changes from some abundance index, q_t, such as catch per unit effort, that is assumed proportional to biomass,

$$q_t = cB_t$$

This simple observation model is the basis for much practical stock assessment and decision-making in fisheries, in spite of its obvious weaknesses. If we express it as $B_t = q_t/c$, we can substitute this version of the observation model back into the logistic equation. Rearranging the terms a bit, the result is

$$q_{t+1} - q_t = kq_t - \frac{k}{cB_\infty} q_t^2 - cC_t$$

This equation has the linear regression form mentioned above if we take $y = q_{t+1} - q_t$, $b_1 = k$, $f_1 = q_t$, $b_2 = k/cB_\infty$, $f_2 = q_t^2$, $b_3 = c$, and $f_3 = C_t$. Thus if we can estimate b_1, b_2, and b_3, we can 'reconstruct' the unknown parameters k and B_∞ of the system dynamic model, as well as the unknown c of the observation model. One basic criterion for being able to do the estimation is that there are strong contrasts over time in the values of q_t and C_t; that is, the system must have been disturbed considerably. This is known as an *identifiability* criterion.

Simple examples like the surplus production model suggest exciting possibilities for using system identification procedures to gain insight about more complex system and observation processes. An important area of systems research is the question of how much information can be extracted from situations where there are more system variables (i.e., biomasses) than observation variables. More specifically, what observation set is needed to uniquely determine system parameters (what should be monitored)? How 'noisy' can the data be? How should different observations on the same variable (i.e., abundance indices) be weighted against one another?

System identification provides a quantitative basis for studying fisheries management activities as *adaptive*, or learning processes. As a fishery develops, information accumulates on abundance, catches, and fishing effort. This information is filtered through observation and dynamic models (explicit or intuitive), and predictions are made about the future of the system. Errors in the predictions are used to revise the models. With luck, adaptive revisions lead to adequate predictive models before the system is irreversibly damaged.

Adaptive management approaches using only monitoring data and system identification techniques are not a complete substitute for 'component research' on important ecological and human processes. As the fish respiration example above points out, there will remain unresolved functional questions that warrant direct experimental study. Also, no statistical procedure can be trusted when it is desired to extrapolate beyond system states that have already been observed. Extrapolation is a necessary part of fisheries management and involves intuition and guesswork; modeling may help to refine intuition and expose guesswork for constructive criticism, but the key need is for experience and functional understanding.

7.5   OPTIMIZATION AND FEEDBACK CONTROL

A common practice in fisheries management has been to base harvesting policy only on trend data, with decision rules like 'abundance seems to be decreasing so we should reduce catches.' Management by trends tends to lock systems into unproductive and uninformative equilibria, which in turn have a nasty habit of turning into disastrous collapses for which the manager is totally unprepared. Supposedly, a better approach is to use the data to construct models for estimating maximum or optimum sustained yield and the system equilibrium that will produce it, and then to manage so as to move toward this target. Unfortunately, targets keep moving as new information becomes available, management objectives change, and nature introduces large disturbances in the form of weak year classes, changes in fishing technology, and so on. Thus it is much more important to have an optimal rule for deciding what to do when the system is not at equilibrium than it is to know where that equilibrium is.

In the jargon of systems analysis, situations with uncertain parameters and unpredictable inputs are called *stochastic optimal control* problems. A large literature is beginning to appear on how to deal with them (Walters and Hilborn, 1976, 1978), using optimization techniques such as *dynamic programming*. So far it has only been possible to study small models (1–3 variables) with these techniques, but a few general principles have emerged from the examples that have been analyzed.

The most important principle from stochastic optimization studies to date is that management actions should be varied with system state according to a *feedback control policy*. When there is a single *control variable*, e.g., harvest rate, and a single state variable, e.g., population abundance, the feedback control policy is simply a graph relating best control to state. An example from salmon management is compared to actual management practice in Fig. 7.4; as the salmon run size varies along the *x*-axis from year to year, due to environmental effects, the manager can determine best action for each year by reading a harvest rate value from the curve marked *max H*. In this example, following the optimal policy happens to result in a fixed escapement of salmon to the spawning grounds most years, with no harvest and lower escapement in occasional very bad years.

When the management objective is to maximize catch or some other simple linear function of stock size, optimal harvest policies generally involve leaving a fixed escapement of fish after each harvesting period. When harvest rates cannot be changed rapidly because of considerations of social or economic hardship, then the best incremental changes are those which will move the stock most rapidly toward the best escapement level.

Optimum harvest policies in the presence of random environmental effects are generally more conservative than would be predicted from deterministic (no random effects) calculations of MSY, OSY, or MEY. Harvest rates should be lower, and escapement (stock after harvest) levels higher than the models of Chapters 5 and 8 would predict.

A final principle is that harvests should be deliberately varied to provide information on uncertain parameters. Policies that experimentally 'probe' a system's response capabilities are known as *actively adaptive* or *dual control* policies. The term dual control comes from the observation that management actions (if their effects are monitored!) have the dual effect of producing benefits

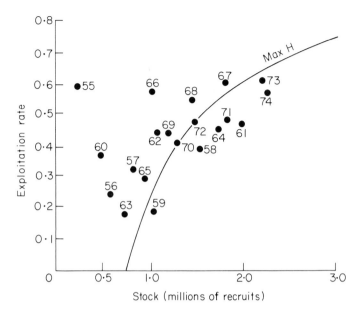

Fig. 7.4. A feedback control policy for harvesting sockeye salmon from the Skeena River, British Columbia. Numbered points are actual management 'choices' in recent years. The curve marked Max H gives optimum harvest rates when the management objective is to maximize expected long term catch. From Holling (1978).

to system users and experience for the manager. Unfortunately, the goals of gaining information and keeping users happy are often in conflict; fishermen who prefer steady, predictable opportunities do not look favorably at variable harvest experiments, at least not when these involve short term reductions in catch. The trick is to find an optimum balance between the conflicting interests.

7.6   MISAPPLICATIONS OF SYSTEMS ANALYSIS

As a fisheries manager, the reader is likely to see some misuses of systems analysis even if he never becomes closely involved with the techniques. This final section identifies a few of the more glaring abuses; two of these have already been mentioned.

Modeling is often used as a justification for expensive and poorly conceived field programs; it is claimed that modeling will be used at the end of the program to 'synthesize and interpret' the masses of data to be collected. This is nonsense; unless the model is at least partly developed at the start of the study and used to set some field priorities, how can its eventual requirements be anticipated? It must be remembered that models are sets of functional relationships, not collections of numerical data; having the numbers is not a prior requirement for identifying the relationships.

Sometimes, bad models will fit available data very well. Consider the fish respiration experiment again. The experimenter who believes respiration to be independent of chamber oxygen concentration could use his good fits to data as justification for allowing oxygen depleting pollutants into some field environment. No one would be foolish enough to make this particular mistake, but the dangers are much greater with complicated models where various bad assumptions can be hidden away or go unnoticed. The lesson is simple: beware even those models that seem to fit very well.

Simplistic thinking and misleading assumptions can be hidden by elaborate mathematical equations. Who would think that the complicated algebra of Beverton–Holt yield equations (Chapter 5) is based on only three very simple assumptions (constant recruitment, constant natural mortality rate, density-independent growth according to the simple Von Bertallanfy curve)? The surplus production model analysis in this chapter is another good example.

An apparent application of system thinking might be called the 'boxes-and-arrows syndrome' (Fig. 7.5). It is trivially easy to develop 'picture models' of ecological components (boxes) and interactions (arrows), as a way to identify and display relationships that will be further pursued through more careful functional thinking. The trouble is, picture models are often where analysis *ends*, rather than where it starts; for example, they are often used in field research programs and environmental assessment synthesis reports as evidence of having pursued a systems approach. One can see how meaningless this is just by taking a simple box and arrow diagram from some ecology text, then imagining how many distinctively different quantitative models and types of system response can arise from the system organization as displayed.

Finally, there is a tendency to apply myopic optimization results without giving careful thought to broader management objectives. We see strong advocates of maximum biological yield, maximum net economic yield, maximum social benefits, and so forth. But how often has it been advocated that fisheries should be maintained just so that people can have the freedom to choose whether or not to fish? Diversity of opportunities and options is obviously an important social objective to which fisheries have much to contribute, but to do so requires that the fisheries manager view his problems and actions from a much broader perspective than has traditionally been encouraged.

7.7　REFERENCES

Astrom K.J. and P.Eykhoff (1971) System identification—a survey. *Automatica*, **7**, 123–162.

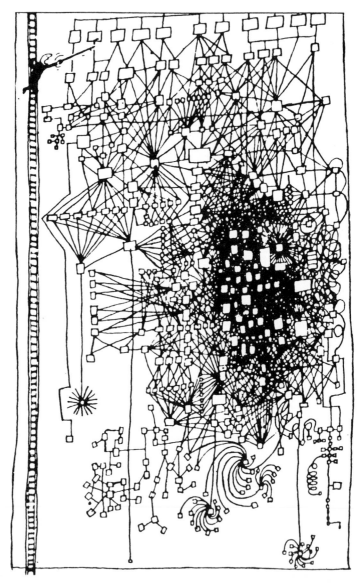

Fig. 7.5. Dr Ralph Keeney used this diagram to chide his colleagues at the International Institute for Applied Systems Analysis about the triviality of 'picture models'.

Bard Y. (1974) *Nonlinear Parameter Estimation*. New York: Academic.

Buckley W. (ed.) (1968) *Modern Systems Research for the Behavioral Scientist: A Source Book*. Chicago: Aldine Publ.

Churchman C.W. (1971) *The Design of Inquiring Systems*. New York: Basic Books.

Clark C.W. (1976) *Mathematical Bioeconomics: The Optimal Management of Renewable Resources*. New York: Wiley.

Forrester J.W. (1969) *Principles of Systems*. Cambridge, Mass.: Wright–Allen Press.

Holling C.S. (ed.) (1978) *Adaptive Environmental Assessment and Management*. London: Wiley.

Klir C.J. (1969) *An Approach to General Systems Theory*. Princeton, New Jersey: Van Nostrand-Reinhold.

Levastu T. and F.Favorite (1977) Preliminary report on dynamical numerical marine ecosystem model (DYNUMES II) for eastern Bering Sea. *U.S. Dept. Commerce, NOAA*, Northwest Fisheries Centre, Seattle, Washington, 81 pp.

Linstone H.A. and W.H.C.Simmonds (1977) *Futures Research: New Directions*. Reading, Mass.: Addison-Wesley.

Overton W.S. (1977) A strategy of model construction. *In* C.A.S.Hall and J.W.Day (eds.), *Ecosystem Modelling in Theory and Practice: An Introduction With Case Histories*. New York: Wiley.

Patten B.C. (ed.) (1975) *Systems Analysis and Simulation in Ecology*, vols. I–III. New York: Academic.

Regier H.A. (1978) *A Balanced Science of Renewable Resources*. Seattle, Washington: Univ. Washington Press.

Walters C.J. (1974) An interdisciplinary approach to development of watershed simulation models. *Technol. Forecast. Soc. Change*, **6**, 299–323.

Walters C.J. and R.Hilborn (1976) Adaptive control of fishing systems. *J. Fish. Res. Board Can.* **33**, 145–159.

Walters C.J. and R.Hilborn (1978) Ecological optimization and adaptive management. *Ann. Rev. Ecol. Syst.* **9**, 157–188.

Watt K.E.F. (1968) *Ecology and Resource Management: A Quantitative Approach*. New York: McGraw-Hill.

Chapter 8
Planning and Policy Analysis

A. A. SOKOLOSKI

8.1   INTRODUCTION

Planning is an often used, often abused, almost magical word that means many things to many people. In order for the term 'planning' to be useful, however, we must define it more precisely. The dictionary defines planning as '... preparing a scheme of action ...' A more explicit definition relevant to fisheries is that planning is a management process done prior to implementation and devoted to clearly identifying, defining, and determining the best alternative courses of action necessary to achieve determined goals and objectives or to refine existing goals and objectives. I will return to elements of this definition many times, as I develop the critical components of *goals*, *objectives*, and *alternatives*. The term 'policy' suffers from the same abuses as 'planning' and likewise requires careful definition. 'Policy' connotes authority, rules, constraints. A suggested definition is that policy is a definite course or method of action selected from among alternatives and in light of given conditions to guide and determine present and future decisions.

The common reference in both definitions to 'course of action' suggests that planning and policy are closely connected. Planning begets policies at one level. These policies provide guidance and boundaries to planning at subsequent levels. Planning cannot be done in a vacuum and policies cannot spring forth from a vacuum. Thus, the critical differentiation is partly the *level* of the planning or policy formulation activity. To a 'technician,' planning can only legitimately embrace technical matters. To the 'policy maker,' these technical inputs to planning are critical, but his planning or policy formulation must embrace a full range of legal, technical, social, and political considerations.

The key linkage is that some elements of most plans can be extracted and communicated as generalized rules (policies) which hold under certain sets of conditions and assumptions. They reflect alternatives that have already been evaluated at a certain level, they are to provide guidance within a certain framework and to a certain level of specificity. When an employee at a hatchery discusses alternatives with the supervisor relative to the maintenance 'plan' for pumps and generators at the raceways, he is not formulating policy. He is guided by a policy statement from a central office which states that there shall be certain backup capacity for both the pumps and the generators at all times. This 'policy' is one among many other policies governing safety factors, overtime, competitive bidding procedures for acquiring pumps and generators, etc. At higher levels in the agency, managers responsible for all hatcheries had previously developed management and operation plans specifying that the best alternative is that the specified pump and generator capacity be available at all times, that overtime be governed in a specified way, and that pumps and generators be acquired by a certain procedure, etc. Clearly, 'policy' becomes a concept of growing import at progressively higher levels in the management hierarchy.

The processes of planning and policy analysis are dynamic, demanding the constant review of new laws, rules, regulations, and guidelines and the use of all knowledge and skill available *to assess and reassess alternative ways of achieving identified goals and objectives.* In what follows I will avoid focusing on discrete products of these activities to be labelled either as plans or policies. Rather, I will focus on the process, with the objective of illustrating the integral nature of planning and policy analysis to the management of fisheries resources.

8.2   PLANNING IN THE FISHERIES MANAGEMENT PROCESS

For management to succeed, the planning process *and* the resultant plans must be orderly. That order must come from the primary orientation and principal responsibilities of the management agency, usually characterized initially in the legal codes under which that agency operates. Some codes may specify the responsible agency by name; other codes may be less specific, but responsibility has been assigned to an agency by some higher level in the management hierarchy. Original legislation and accompanying clarification are followed by regulations, executive orders or their equivalent, memoranda of agreement, and judicial interpretation which, through time, combine to form the continually evolving body of those codes.

Whereas legal codes are the ultimate source of authority for management, they usually fail to provide guidance for the operation of management agencies. The legal codes must be translated into a useable plan. This should take place in the first hierarchy of planning, beginning with the stating of the principal missions of the agency (Fig. 8.1). From this mission would come broadly stated *goals* covering the main areas of responsibility. For each goal, a set of *objectives* would be written. Each objective would define a desired type and quantity of output relevant to the appropriate goal and a time schedule for achieving those outputs. For each major objective there would be a series of operational sub-

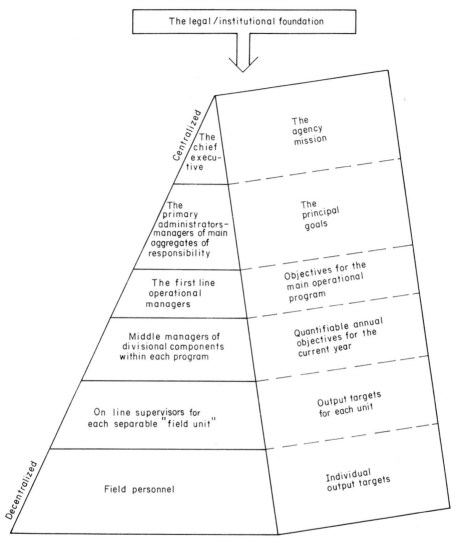

Fig. 8.1. The generic organization/management/planning hierarchy.

objectives identifying discrete products to be produced by each program activity, each region, or each function, depending on the organizational structure of the agency.

A plan developed in this way identifies the outputs, or targets, of management and is a document for achievement, a 'how to' prescription. To be effective, however, the plan must be a truly dynamic document. In subsequent planning periods it will reflect changes in legislation, implementing rules, legal interpretations, and the evolution of technology and skills that may be applied to attaining the targets.

The formulation of goals, objectives, subobjectives, and the like, all imply that outputs of some kind are desired in some quantity. While obvious that management programs need to be framed and evaluated in terms of outputs, the concept of measureable outputs and their use in planning and directing fisheries management programs is a complex question. Controversy arises because while most people can agree about vague notions of desired outputs, the kind of specific outputs desired depends upon many variables.

The issue of defining appropriate kinds of output can be examined by considering the outputs of a mechanic in a garage specializing in auto repair. For profit purposes the garage manager needs some indication of the productivity of all of his mechanics. His motivation could be to dismiss some of the least productive, to ascertain clues to increased productivity associated with shop layout, or to determine the need for purchasing certain types of ancillary equipment. Focusing on a single mechanic, let us assume that he has just replaced a carburetor in a customer's car. Which of the following might be the most appropriate measure of his productivity?

1. Time spent replacing the carburetor;
2. Time spent correcting a combustion problem in the car;
3. Time spent providing a satisfactory solution to the problem identified rather superficially by the customer as 'the car just doesn't have the old pep, and it is really guzzling the gas.'

Many people would agree that each of these three qualify as measures of output. How do we differentiate? The key is to return to the motivation of the manager. If his goal is to replace carburetors as rapidly as possible, then that is the type of mechanic that he wants. Perhaps all customers are referred to this establishment by another firm which operates a diagnostic center. If so, the mechanic who replaces carburetors the fastest would have the highest output. If, however, the goal is to correct combustion problems, then finding the cause of problems needs to be a part of the objective and the measureable output. Or, even further, if the objective is to provide solutions for customers whose complaint is less specific, then complete analysis and customer satisfaction is a major part of the output, with diagnosis being a premium.

Now consider the potential goals and outputs of an agency operating fish hatcheries. Is the goal to run hatcheries and produce fish? If so, then a rather simple and straightforward set of objectives would relate to quantifying fish production at even the highest levels of planning. Or is the goal of the fisheries management effort to attain some particular status for the fisheries in the nation (or the state, county, river, lake, estuary, basin, drainage area, etc.)? Even more than in an auto-repair garage, there are many possible choices for measureable outputs, ranging from success–effort ratios for both commercial harvest and recreational use to the pounds of production of 10-cm rainbow trout fingerlings at hatcheries (see Table 8.1). Any of these may be appropriate at the right level of management, but can be horribly inappropriate at the wrong level of management.

Table 8.1. A few of the possible outputs from a public fish hatchery.

Number of fish produced
Weight of fish produced
Size of fish produced
Mortality rates
Food conversion ratios
Absence of disease
Return of stocked fish to anglers
Area of water stocked
Cost per fish produced
Cost per unit weight of fish produced
Number of visitors
Cooperation with other agencies
Punctuality of report submission
Frequency of accidents
Employment of minorities
Number of complaints to elected officials

If the head of a fisheries agency has a clear mandate to improve the opportunity for recreational fishing experiences in his state, but all of his objectives, the ones around which all plans are built, speak only to the kilograms of fish produced at his hatcheries, he may accidentally contribute to improvements sometimes, but he still has no objectives or measureable outputs related directly to his principal goal. Ideally goals and objectives should be directed toward those parameters which prior multi-disciplinary research has defined as critical to attaining the broad goals and outputs of the agency. Note that each member of the agency would extract goals from the resulting master plan and design a personal subplan with subobjectives that related to habitat, water quality, genetics, diseases, or perhaps the *productivity* of the hatchery. It is important to note that most fisheries and wildlife professionals carry with them their early career attachment to goals and objectives from early responsibilities in an agency and attempt to retain these same measures at the higher levels of management. If the above agency head retains a fixed attachment to the hatchery output goals of his earlier career, he is likely to substantially misdirect the budgetary resources of his current job.

As a final element of this section on planning, let us consider *strategy* formulation. Like the other concepts in this chapter, strategy can be an ambiguous term. In one sense it can be a beginning point, a broad, non-specific statement of an approach to accomplishing desired goals and objectives. It can also be the result of all elements of planning, that summary statement of all of the resultant key pieces of the 'plan.' A plan which is truly multi-disciplinary—including all elements of the physical and social sciences, all based on an analytical foundation—leaves no additional consideration which would differentiate it from a strategy. However, such a plan may never have been designed. Usually, one or more of the various disciplines comprising the plan cannot be subjected to analysis. Other discipline areas, even if they can be critically analyzed, are developed only in general terms because an agency may wish to avoid exposing its internal constraints. In these circumstances, the substitute for a sufficient level of analysis is a series of determinations by senior level officials, perhaps labelled as policy proclamations, of what the strategy will be.

For example, an Executive Order may state quite clearly and unequivocally that, in certain types of program actions, local citizens, industries, or other advocacy groups will play a specific role in certain management actions. Henceforth, these will become fixed elements of the plan and part of the strategy mandated by the Chief Executive. There may also be mandates regarding the timing, such as 'this will be accomplished by December 2, 1991.' The resources required to perform the action may be defined further, such as, 'These functions will be completed by restricting agency employment to 5000 employees and supplementing these skills with consultants as necessary and subject to budget constraints.'

One may call these elements of strategy edicts, mandates, policy statements, Executive Orders, or whatever. The appellation is irrelevant. Planning can range from being completely self-sufficient if the topic is subject to analysis in all of its components or planning may be only a component itself if part of a larger strategy which includes other components representing the expert judgment of senior managers as a substitute for analysis in each case. The important point is that a plan which is incomplete by some component must be corrected, either by analysis or a seemingly more arbitrary process. The combination of all of these factors may be more commonly referred to as the strategy.

8.3 POLICY ANALYSIS IN FISHERIES MANAGEMENT

In preceeding pages planning and policy analysis were shown as being closely related. I will not diverge from that perception here, although the word 'policy' will be used almost exclusively, and I will emphasize analysis. I shall use the word evaluation interchangably with analysis, actually preferring the former, because it implies a less formal process. This is not necessarily desirable, but it is closer to reality and actual practice.

Policy is an all-encompassing process which can be divided into the four stages of development, communication, implementation, and evaluation. In the absence of any of these four dimensions, policy is incomplete at best and contains the risk of being incompletely understood and improperly implemented. Figures 8.2 and 8.3 illustrate policy analysis as performed within the U.S. Fish and Wildlife Service. They easily translate into equivalent actions at equivalent levels in any public agency.

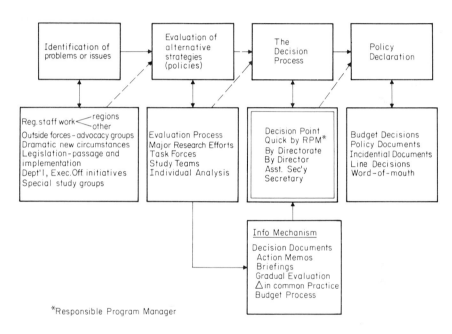

Fig. 8.2. Stages in the development of agency policy.

As the first stage in the process, policy development closely parallels the planning process (Fig. 8.2). The process begins by identifying new or emerging problems or issues. Professionals within an agency may identify issues through research, development, and evolving management practices, and many issues surface in the public arena, with input from public interest or industrial lobby groups and others. These issues may take several forms, which generally may be characterized as some diversion from the present practices or at least a signal that diversion should be considered. The fundamental process of identifying alternative responses comes next. Identification of alternatives must be accompanied by evaluation of the pros and cons of each. As the historical environment surrounding the conduct of programs continues to absorb new dimensions without shedding many of the old dimensions, the complexity of the evaluation of alternatives continues to grow. This complexity continues in the decision process, where through time more people from every part of society, both public and private, are becoming involved in making fisheries and wildlife management decisions. The evaluation process is necessarily enlarged to incorporate these varying forces. Task forces, study teams, and permanently-staffed research and analytical teams are likely to be a regular part of the policy formulation process. The final step of policy development is the declaration of the policy statement and its associated mandates.

Policy communication must be performed in a manner which assures that relevant information reaches all affected persons (Fig. 8.3). Part of the system should include documents produced regularly throughout the year which summarize the current state of all principal policies of the agency, highlight changes that have been made recently, and indicate where changes should be

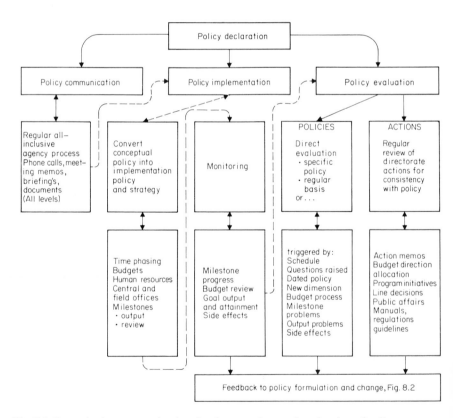

Fig. 8.3. Stages in the communication, implementation, and evaluation of policy.

anticipated. This is facilitated by requiring that every policy proposal, especially the decision or action document on which the final policy determination would be made, include a plan for communication of the ultimate decisions. By this device, the decision maker can evaluate the policy change itself and the adequacy of the suggested communication plan. Many policy decisions have foundered because they lacked a practical method for informing people about the policy and convincing them that the policy was necessary, desirable, and efficient.

Policy implementation includes two major components—the conversion of thought to action, and the monitoring of those actions. At this stage, the use of dollars and human resources must be fully considered and priorities must be set. One may say that a policy is a policy, but there must be a hierarchy of policy priorities. Budgets and manpower are never unlimited, and only high priority policies are likely to approach full implementation. The need for considering priorities is justification for including an implementation plan within the policy process, for naive policy which is totally out of the realm of reality is bad policy, no matter how well-intentioned. As dollars and manpower become scarcer and as our technical and scientific ability outstrips our ability to implement, a greater emphasis is needed on the rational choice of implementation strategies which are likely to resolve the issue or problem at which the policy is directed.

The implemented policy is monitored by tracking progress past planned milestones, identified again as output attainment. Within the scope of this stage of policy analysis, monitoring implies the objective reporting of progress, not the evaluation of whether the policy has been successful. Whereas failure to reach planned objectives is a certain signal that all is not going well, achievement of those objectives indicates only that the implementation plan has been successful, not that the policy accomplished the desired goals. Monitoring provides inputs to the last stage of policy analysis, the evaluation of policy.

Policy evaluation can be a regular part of all policies or keyed to individual policy needs as they arise. A regularized mode requires an established staff which operates under a predetermined schedule and plan for policy evaluation. All policies, including communication and implementation, would be scheduled for periodic review and analysis. Presumably, no policy could become outdated by neglect or actually die of old age. For those needs which arise at random, evaluation probably would be performed by task forces or study teams composed of staff temporarily shifted from their regular responsibilities. In a large agency these random needs are sufficiently predictable, in terms of the volume and man-months of analysis required annually, that permanent staff may be assigned for this function also.

8.4   POLICY AND PLANNING IN FUTURE FISHERIES MANAGEMENT

Planning and policy analysis have been depicted here as being closely allied, multidisciplinary efforts directed toward the attainment of resource quantified outputs. While certain activities can be differentiated as plans, policies or strategies depending upon the level and location within an organization, these different labels are of minor importance. The guiding principle for all activities is the necessity for including certain basic steps relating to formulation, implementation, communication and evaluation of plan and policy components throughout the management process.

There is no likelihood of a successful retreat to a simpler life when only one discipline, fisheries biology, sufficed to resolve those few management problems that existed in a situation of plentiful resources. The physical sciences continue to advance, bringing science and technology to new frontiers that just as rapidly become old. With this march comes a distressing mix of both solutions to problems and the creation of multitudes of new problems. This becomes increasingly relevant to the fisheries manager because this march appears to focus more on both the direct (planned) and the indirect (unplanned) impacts on natural resources and the resulting trade-offs. Elements of resource use that were scarcely perceived a few decades ago, such as endangered species and energy supply, are now becoming regular parts of decision-making.

We also find that despite a long history of experience with democracy, we are really just beginning the process of incorporating public participation in public sector decision-making. Present experiments with mechanisms of involvement and the eventual institutionalization of certain mechanisms will bring whole new dimensions of complexity to the management process. While there may be some extraordinary persons who can manage that complexity via intuitive judgment in decision-making, such will represent ever diminishing examples of the exception to the rule. For most managers, the need for broader application of the expertise of a multitude of disciplines in a rigorous framework of planning, policy formulation, and policy analysis is certain. Readers of this text who will be a part of the future must expect to function within that reality.

8.5   REFERENCES

Ackoff R.L. (1970) *A Concept of Corporate Planning.* New York: Wiley-Interscience.
Beneviste G. (1972) *The Politics of Expertise.* Berkeley: Glendessary.
Burkhead J. and P.J.Henningan (1978) Productivity analysis: A search for definition and order. *Public Admin. Rev.* **38**, 34–40.

Dye T.R. (1972) *Understanding Public Policy.* Englewood Cliffs, N.J.: Prentice Hall.

Ewing D.W. (1968) *The Practice of Planning.* New York: Harper & Row.

Ewing D.W. (ed.) (1972) *Long Range Planning for Management.* New York: Harper & Row.

Hahn W.A. and K.F.Gordon (eds.) (1973) *Assessing the Future and Policy Planning.* New York: Gordon and Breach Science Publ.

Hatry H.P. (1976) The status of productivity management in the public sector. *Public Admin. Rev.* **38**, 544–550.

McDonough A.M. and L.J.Garrett (1965) *Management Systems: Working Concepts and Practices.* Homewood, Il: Richard D. Irwin.

Rosenthal A.H. (ed.) (1973) *Public Science Policy and Administration.* Albuquerque: Univ. New Mexico Press.

Chapter 9
Fisheries Economics

FREDERICK W. BELL

9.1 ECONOMIC TRENDS IN WORLD FISHERIES

The estimated world production of fish has more than tripled over the 1948–1970 period, from less than 20 million metric tons in 1948 to more than 70 million metric tons in 1970. This growth, which is considerably faster than that of either human population or the overall production of food, means that up until 1970 fish were making an increasingly important contribution to world food supplies. After 1970, world catch showed no appreciable trend averaging about 70 million metric tons from 1971 through 1976. Although much of this turnaround in world fish production has been due to a collapse in the world's largest fishery, the Peruvian anchovetta, this new trend may be indicative of economic factors that greatly threaten increased fish production so critical to an ever growing world population. The world's fisheries should be viewed within the context of a world population/food imbalance that according to Meadows *et al.* (1972), portends a collapse in per capita food production early in the twenty-first century. A proper understanding of fisheries economics is essential to understanding both what has happened to world fisheries and how they may be managed more wisely in the future to insure increasing production under economically efficient conditions.

To properly analyze recent trends in world fisheries, we must not only look at catch—an aggregation of metric tons of products ranging from high priced lobsters to relatively lower priced menhaden—but dollar values and dollar values per metric ton. According to Bell (1978), over the 1963–1973 period, the total real value or dollar value adjusted for inflation of the world's fisheries catch increased by 6.8 percent annually while the real value per metric ton increased by 3.9 percent annually.[1] In 1973, the total estimated value of the world's catch was over $18 billion at dockside and the value per metric tons was approximately $288. In contrast to trends in total value, fisheries catch increased by only 3.7 percent annually over the 1963–1973 period. The increase in real value

[1] All rates of growth in this chapter are compound rates of growth rather than average rates of growth.

per metric ton indicates that the cost of catching fish and shellfish is increasing faster than it is for other commodities. Fishermen experience rising real value, while consumers find that fish are becoming relatively more expensive. Economists call this trend *increasing* resource scarcity or where the economic factors of production, capital, and labor, experience falling productivity or output (catch) per unit of inputs (fishermen and vessels) declines. Of the 29 major fishing nations, 25 experienced rising real value per metric ton over the 1963–1973 period. These trends are indeed ominous since resource scarcity in agriculture was predicted by Malthus over 200 years ago and was the basis for the conclusion that economic growth would cease and most of mankind would exist at a mere subsistence level. Although these predictions have not materialized in agriculture for most of the advanced societies, the last decade and a half have seen the Malthusian spectre emerge in world fisheries.[2]

Table 9.1 shows the trend in catch, value (unadjusted for inflation), and value per metric ton for the major fishing powers. Japan, China, and U.S.S.R. increased total fish catch over the 1963–1973 period while the U.S. production was stable and Peru's catch fell off dramatically by 1973 despite a rapid growth over the 1963–1970 period. Notice that although Peru's catch was almost five times as great as the U.S catch in 1970, the value of the Peruvian landings was only one third of those in the U.S. because of the low value per metric ton received for anchovetta used for fish meal, a food for chicken and other poultry. The U.S. and other fishing powers shown in Table 9.1 concentrate on a variety of species mix from low to high value per metric ton. Thus, it is not always the volume of catch that is important, but the kinds of species and their corresponding per unit value that determines total value from the sea.

Up to this point, we have been discussing commercial fisheries. No data are available on a world basis for recreational fisheries. However, a U.S. survey reports that expenditures by fresh and salt water anglers increased from $1.9 billion in 1955 to $15.2 billion in 1975 (unadjusted for inflation). Although these surveys are not directly comparable because of change in the definition of the universe to be sampled, they do indicate the enormous growth and size of recreational fisheries. In 1975, nearly 54 million Americans participated in recreational fisheries. The demand for recreational fishing has increased even more rapidly than the demand for fish as food; therefore, the recreational fishing industry, which is based upon a limited resource, suffers from many of the problems of commercial fisheries. We shall discuss the economics of recreational fisheries below. However, we shall first discuss the economics of commercial fishing.

---

[2] In the developing countries, the Malthusian theory is quite operative. Per capita food production in less developed countries has not increased over the 1961–1973 period.

Table 9.1. Total fisheries production, value, and value per metric ton for selected countries, 1963–1973*

	1963	1966	1970	1973
Total Production (mil Metric Tons)	46.6	57.3	70.0	65.7
Selected Countries				
Japan	6.7	7.1	9.4	10.7
China**	6.7	6.0	6.9	7.6
U.S.S.R.	4.0	5.3	7.3	8.9
U.S.A.	2.8	2.5	2.8	2.7
Peru	6.9	8.8	12.6	2.3
Total Value (Ex Vessel) (Billion of dollars)	$6.3	9.2	13.1	18.0
Selected Countries				
Japan	1.3	1.7	2.7	2.5
China	0.8	1.4	2.0	3.1
U.S.S.R.	0.6	1.0	1.5	2.6
U.S.A.	0.4	0.5	0.6	0.9
Peru	0.07	0.1	0.2	0.09
Value Per Metric Ton (world)	$131	159	190	288
Selected Countries				
Japan	194	245	290	235
China	119	227	292	404
U.S.S.R.	156	178	210	307
U.S.A.	136	188	221	340
Peru	10	13	15	37

 * Dollar values are *not* adjusted for inflation.
** Includes both People's Republic of China (mainland) and the Republic of China (Taiwan).
    Source: Bell (1978).

9.2   THE DEMAND FOR FISHERIES PRODUCTS

Fish is demanded for consumption throughout the world, supplying some 10 percent of the world's protein. The demand influence of consumers varies considerably from 'market economies' to 'centrally planned economies.' A market economy is strongly associated with the term capitalism, a form of economic organization that allows producers and consumers wide latitude in

their economic decision-making. A centrally planned economy is less individualistic and more centralized among a few government leaders who
determine what is produced. Thus, what is produced in market economies is
more responsive to consumer demand. In a centralized economy such as the
Soviet Union, the state determines what is produced and consumers must
make choices among a reduced variety of products. This distinction is important since Robinson and Crispolde (1971) estimate that centrally planned
economies consume about one-third of the world's fisheries catch. Whatever
type of economic organization, there are certain so-called laws of consumer
demand that must be understood.

Let us look at a typical market economy such as the U.S. For example, if
one looks at the per capita consumption for the average U.S. consumer of
canned tuna since 1947, one generally sees a persistent rise in this series (from
0.4 kg in 1947 to 1.3 kg in 1971). What determines how fast the per
capita consumption of canned tuna has risen? We can attempt to answer these
questions by hypothesizing what is called a demand function or relationship
for canned tuna as

$$q_T = f\left(P_T, \frac{Y}{N}, P_1 \ldots P_n\right) \tag{9.1}$$

where

$\quad q_T =$ per capita consumption of tuna (canned)

$\quad P_T =$ price per unit (usually weight)

$\quad \dfrac{Y}{N} =$ real per capita income (i.e., real aggregate consumer income,

$\qquad$ Y, divided by population, N)

$\quad P_1, \ldots, P_n =$ price per unit for goods that are substitutes for canned tuna

It is important in economic analysis that we be able to tell the separate influence
of each demand determinant *when other demand determinants are held constant.*
Consider Fig. 9.1. *DD* shows an inverse relation between price and quantity
of tuna and is referred to as a demand curve. This is a law of demand based
upon the fact that as prices increase consumers will substitute relatively less
expensive foods for tuna such as eggs, vegetables, or hamburger. Real income
per capita is another factor that affects the consumer's choices. As income
increases, a consumer may increase his consumption of tuna. Thus, at a given
price $\bar{P}_T$ the demand curve will shift upward and to the right increasing consumption from q_1 to q_2 or $D' D'$. Per capita income in the U.S. has been trending
upward through history. Increased earnings (income) are not necessarily spent to
merely increase purchases of the same item that was formerly consumed. For
example, per capita consumption of all food fish (excluding fish used for

industrial purchases) has increased little in the U.S. since 1909 averaging about 5 kg; however, per capita consumption of canned tuna, shrimp and crabs have increased while that of salmon, sardines, and oysters have not changed appreciably over the 1948–1971 period. Finally, one may theorize that tuna, for example, may have many *close* substitutes such as canned salmon, chicken and beef. Thus if the price of canned salmon were to increase, this would make it *relatively* more expensive compared to canned tuna; thus, the demand curve would shift from DD to $D'D'$, increasing the consumption of tuna from q_1 to q_2 due to the relative price changes as shown in Fig. 9.1. Thus, a relative price change of close substitutes may have the same impact on consumption as income changes.

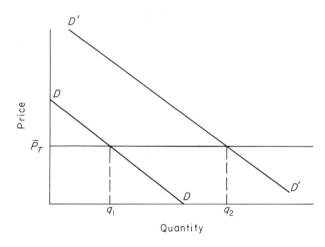

Fig. 9.1. Demand analysis.

We have so far been talking about the average consumer. Statistically, the per capita consumption would be derived by dividing *total* consumption, say for the U.S. or any other country, by population. Thus, once we have determined per capita consumption, it must be multiplied by population to derive the *total market demand*. At the beginning of this chapter we stressed the 'population bomb' or the population/resource imbalance. Thus, the growth in population is an extremely important demand determinant. For example, in 1977 there were 215 million Americans with a per capita consumption of lobsters of 0.43 kg or aggregate consumption of about 93 million kg. In 1967, the per capita consumption of lobsters was 0.37 kg with a population of 195 million or aggregate consumption of 73 million kg. The increase in population *alone* accounted for a 7.4 million kg increase in aggregate lobster consumption over the 1967–1977 period. Thus even if per capita consumption is constant or even slightly declining, population growth will contribute to the growth in the total market for lobsters and hence demand pressure upon the lobster resource.

9.3   BIOECONOMIC SUPPLY CURVES

A hundred years ago most people, including leading scientists, believed that the living resources of the sea were essentially inexhaustible. This assumption has been greatly invalidated by the subsequent intensive exploitation of many of the world's most valuable and vulnerable species. The first stocks to show depletion were those close to the ports of the industrial nations. A stock that has been so reduced through overfishing that a *substantial* reduction in fishing effort must be achieved so that the stock can replenish itself is said to be *depleted.* For example, the blue whale is nearly extinct while haddock, herring, yellowtail flounder, and sea scallops are depleted off the coast of New England. This has led to severe economic problems in coastal cities dependent on these fisheries stocks.

To understand why the economic system functions in a manner that results in depletion of valuable fisheries stocks, we must first develop the concept of a bioeconomic supply curve. This curve is composed of two elements: (1) *the biological relation between catch and fishing effort* and (2) *the economic relation between total cost and fishing effort.* Let us consider the catch-effort relation first or what is generally called population dynamics.

The effect of fishing on a stock of fish has been described by a range of models of varying mathematical complexity. In this chapter we consider the surplus production model developed in Section 5.2. An alternate way of expressing this relation is the parabolic form where C = total catch; f = total fishing effort and A and B are parameters:

$$C = Af - Bf^2 \qquad\qquad (9.2)$$

Next, we must introduce the *cost* of fishing effort. All cost may be divided into explicit and implicit cost. *Explicit cost* is actual money outlays of the fishing firm to purchase fuel, food, and gears as well as expenditures to hire labor, acquire fire insurance, and handle interest payments on the vessel's mortgage. *Implicit cost* is sometimes called opportunity costs associated with the firm's use of resources which it owns. These costs will not involve a direct monetary payment. The returns or profit to vessel and management must be included in cost. If the vessel owner invested $100,000 of his own money in a vessel, he must try to recover an adequate return on this investment, say, 12 percent. The vessel owner could have invested his money in a gasoline station or the stock market—other opportunities. Thus, the owner would expect 12,000 in profit (before taxes) to remain in fishing. If total cost (explicit and implicit) are, for example, $500,000 for a 30-m tuna vessel per year and that vessel renders or applies 5000 units of fishing effort then the cost per unit of fishing effort is $100. Thus, *total* industry cost (TC) is a linear function of *total* fishing effort (E) by the fleet:

$$TC = cf \tag{9.3}$$

where $c = \$100$ in our example. Two additional concepts are necessary: (1) average (AC) and (2) marginal (MC) cost. As fishing effort expands, average cost per kg (or metric ton, etc.) rises. Define

$$AC = \frac{TC}{C} \tag{9.4}$$

AC can be computed for any level of catch (C_1) by first finding the necessary level of total fishing effort using eq. (9.2) for f_1. Next, f_1 can be substituted into 9.3 to obtain TC_1 and AC_1 follows from 9.4. Average cost rises as catch increases because average catch per unit of effort falls as effort is increased while the cost per unit of effort remains the same. The AC curve will continue to rise, but bend backwards at MSY since increases in f result in declining catches. This is shown in Fig. 9.2.

Another extremely important cost concept is the notion of marginal or incremental cost (MC) or

$$MC = \frac{\Delta TC}{\Delta C} \tag{9.5}$$

Marginal cost is the incremental cost associated with incremental changes in the quantity of fish harvested by the fleet. MC is above AC or rises more rapidly because marginal catch per unit of effort drops off faster than average catch per unit of effort. Positive MC is asymptotic to MSY; but would be negative if catches started to decline since the change in C would be negative in eq. 9.5. Consider Fig. 9.2. if C_1 is caught with f_1 units of fishing effort, point c on the AC curve is associated with AC_1; however, if overfishing takes place and fishing effort expands to f_2 at a, AC rises to d or AC_2. Thus overfishing is economically inefficient since the same catch, C_1, can be caught with a smaller TC at c than at d ($TC_c = AC_1C_1 < TC_d = AC_2C_1$). AC and MC are both bio-economic supply curves since they show the costs at varying levels of supply from the fishery in question.

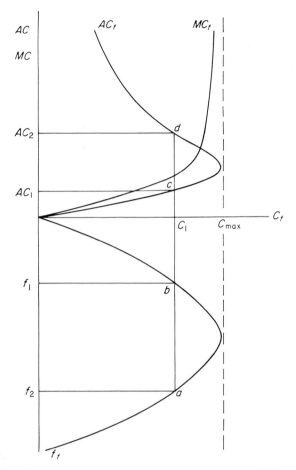

Fig. 9.2. Relation between average and marginal cost and quantity of fish harvested and fishing effort.

9.4   FREE ENTERPRISE AND THE MANAGEMENT OF FISHERIES

9.4.1   Commercial fisheries

A basic characteristic of all wild fisheries is that they are a common property natural resource. Like many common property resources, such as water and air, they can be used without cost by economic enterprises. That is, no single user has to pay for the right to use the resource, nor does he have exclusive rights to the resource or the right to prevent others from sharing in its exploitation. The commercial users of fisheries resources are in competition with one another to get a larger share of the resource for themselves. By allowing the resource to be common property, a vessel owner's productivity is directly influenced by how many firms are exploiting the resource. The entrance of more firms increases the fishing effort on the common property resource. Each fishing firm is influenced by what economists call a technological externality—technological because each fishing firm's productivity or catch per unit of effort declines as fishing pressure increases and external because the firm has no control over this aspect of its productivity.

In market oriented or free enterprise economies, the fishing industry generally operates under a market form that approaches what is called pure competition. This market form is characterized by the following: (1) homogeneous products; (2) large number of buyers and sellers; and (3) easy entry and exit from the industry. These conditions characterize the majority of fisheries in the countries of the world that have market-oriented economies. That is, fish are homogeneous; there are numerous vessels and buyers of fish and investment required to enter most fisheries is relatively small. We are finally ready to bring together the demand and supply for fisheries products developed above. Consider Fig. 9.3.

The intersection of DD, the demand curve, and MC, the marginal cost curve, is of particular interest (point f). At that point, price is P_1 and the corresponding total revenue (TR) is $P_1 C_1$. But, what is the total cost? At that level of harvest, average cost is AC_1 and $TC_1 = AC_1 C_1$. Thus, $TR > TC$. It should be

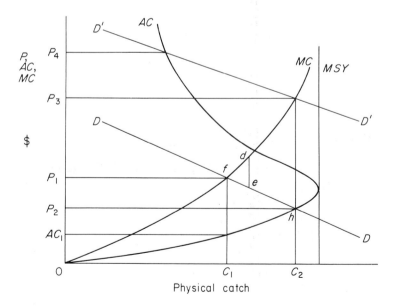

Fig. 9.3. Market equilibrium for a fisheries resource.

remembered that total cost includes a normal return on invested capital. Hence, point f reveals that fishermen would be earning an economic profit defined as a return well above normal. Remember this important point! Where marginal cost equals price, excessive returns accrue to those engaged in fishing. However, at point f, society through individual action (i.e., free enterprise) has allocated so many units of effort where the marginal cost of producing that last fish is equal to *what consumers are willing to pay*. Any production to the right of f is sub-optimal since the marginal cost of producing the last fish exceeds the price consumers are willing to pay ($d > e$). Point f is called *MEY* or maximum economic yield.[3]

However, *MEY* is unstable with a common property resource since $TR > TC$ and fishermen are earning what is called an economic profit defined as a *return well above normal*. Since entry to the fishery is relatively easy, the above normal returns will encourage more fishermen to enter the fishery. Catch will increase thereby depressing prices and raising *MC* and *AC*. Entry will continue until $AC = P_2$ at point h. This is called market equilibrium since $TR = TC$ and no entry or exit of fishermen will take place and economic profits will be zero. Thus, the free enterprise system will lead in the case of common property resource to overproduction and consequently too many fishermen and vessels in the fishery.

Why do we overfish? Although our argument above showed production at less than *MSY*, we can easily show using the general analysis why overfishing takes place. Suppose population and per capita income increase and shift the demand curve for the catch upward and to the right to $D'D'$. Initially, C_2 will be produced with f_2 units of effort since the fleet cannot expand instantly. This increases *TR* considerably to $P_3 C_2$, but not *TC* (at the initial market equilibrium associated with the *lower* level of demand). Thus, economic profits increase considerably inducing more fishing effort into the fishery. Individual economic behavior leads to a reduction in catch with consumers paying higher prices. At P_4, the fishery will be in equilibrium, but overfished. Hence, it is the pressure of demand, a finite maximum sustainable yield from any resource and most importantly the structure of property rights (common property) that lead to overfishing and depletion as discussed above.

9.4.2   The economics of fisheries management

Historically, the problem of common property fisheries resources leading to overfishing had been attacked by two basic methods: (1) regulated inefficiency and (2) overall quotas. Regulated inefficiency attempts, through the force of the

---

[3] At point f, the difference between *TR* and *TC* is also called *economic rent* since this is the sum that could be charged to users (fishermen) if the resource were owned by someone (e.g., the government). Since the resource is common property, the 'rents' go to the fishermen in the short run.

law, to prohibit technological improvements in the fisheries. This is done through prohibiting, for example, spotter planes or helicopters in the salmon fishery. In this way, fishing effort does not increase as rapidly and proponents argue that conservation is encouraged. This is not a solution with a favorable economic impact since the fishery becomes technically backward in comparison to other countries and fundamentally treats symptoms, not causes. The *overall* quota system attempts to control fishing effort indirectly by setting catch quotas, usually on an annual basis. When the quota is taken, the fishery is closed and vessels are either forced into less profitable uses or remain idle. This technique is used in the tuna, halibut, various groundfish, mackerel, herring, and king crab fisheries in U.S. as well as other countries.

Recently, there has been an attempt to abandon the overall quota system and regulated inefficiency in favor of direct control over fishing effort itself, i.e., limiting the number of fishermen, vessels, tonnage of vessels, etc. This is a form of *limited entry*. For example, direct controls have come in the form of Canadian and Alaskan limited entry legislation. Usually, an entry permit or license is issued to a limited number of individuals based upon some criteria such as economic need, years in the fishery, etc. The results indicate that this scheme suffers from a constant battle between fishermen and the regulatory authority to outwit each other as fishermen with these permits improve their fishing power (i.e., effort) while authorities place more restrictions on the limited number of permits. These permits are usually transferable and once entry to the fishery is limited, they immediately acquire a value caused by increasing expectations of rising demand with a finite supply in the hands of a relatively few individuals; hence the prospect of rising economic profits.

A second method of limiting entry is called the fishermen or boat quota. Under this system, quotas are allocated to individual fishermen on the basis of past catch or other criteria. In a developed fishery, the summation of the *individual* catch quotas may be equal to *MSY*. The chief advantage of this approach is that the fishermen would be free to take their shares using whatever technology they wish, subject only to regulatory constraints of a non-economic nature. These quotas, sometimes called stock certificates, may be transferable and thus have economic value. Both licensing and boat quotas involve the transfer of a potentially valuable property right from the public in general to a limited number of individuals. Thus, economic profits (or rents) would accrue to the fishermen unless taxes were levied on these excess profits.

A third limited entry scheme would place a tax on gear or catch. Since fishing beyond *MEY* or *MSY* is produced, as discussed above, by the appearance of economic profits and more entry into the fishery, the idea is to siphon off these rents in the form of a tax which increases the cost of fishing (i.e., it requires that $TR > TC$). The tax would be calculated to inhibit entry at any point perhaps so that either the objective of *MEY* or *MSY* could be obtained. Alternatively, this procedure could be left to an auction of catch whereby fishermen would bid for a right to fish a certain fixed quantity of fish. In theory, the tax collected would be equal to the amount collected by auctioning a given catch, say *MSY*. This plan differs considerably from the first two in that economic rent is collected by the general public for the use in the management of the fishery. As with offshore oil or federal forest lands, private interest would be charged for the use of public fisheries resources. Also, if the objective was *MEY*, economic efficiency would be maximized (i.e., $MC = P$).

One thing is quite clear from the above discussion. The world economic pressure for food makes open access or common property fisheries doomed to overfishing, overcapitalization, and ultimate depletion. Some form of limited entry seems to be the answer, but not overall quotas or regulated inefficiency.

9.4.3   An empirical bioeconomic model

The U.S. American lobster fishery ranges from Maine to Delaware. The fishery is largely based upon fishing wooden traps or pots in inshore areas. Bell (1970) has estimated the market equilibrium under open access (i.e., common property) conditions for this fishery. Thus, we have an actual empirical illustration of the theory embodied in Fig. 9.3 and this will introduce much more realism into our previous discussion.

First, the demand equation for all lobsters (American, spiny, imports) was estimated. It has the same form as eq. 9.1 in our demand analysis section; that is, per capita consumption of lobsters is inversely related to lobster prices, P_L, and positively related to real per capita income. Close substitutes were not statistically important. The 1966 population, real per capita income, and other sources of lobsters (i.e., imports, Florida, etc.) were 'plugged into' eq. 9.1 in its statistical form not shown here, but discussed in detail in Bell (1970). The final *demand curve (see above) for 1966 was the following*:

$$P_L = 2.053 - .0264 \ C_A \qquad (9.6)$$

where

$$C_A = \text{landings of American lobsters (mil Kg)}$$

Next, the bioeconomic supply curve was derived. As stated in eq. 9.2, the simplest catch-effort function is parabolic; therefore, the American lobster catch was related to traps fished or f (i.e., fishing effort). For American lobsters, seawater temperature is positively related to catch and the catch-effort equation was adjusted to the 1966 seawater temperature which is close to the average of that over the past 60 years. The following catch-effort equation was computed:

$$C_A = 22.46f - .0000109f^2 \qquad (9.7)$$

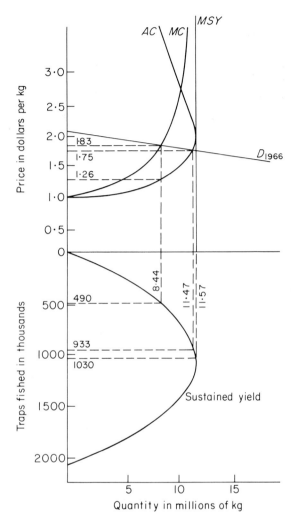

Fig. 9.4. Bioeconomic model: American lobster fishery, 1966. MC = Marginal Cost; AC = Average Cost; MSY = Maximum sustainable yield.

The maximum sustainable yield (C_{MAX}—see Fig. 9.4) is 11.57 million kg taken with 1,030,000 traps for the American lobster fishery.

By taking a representative sample of lobster boats in 1966 from this fishery with respect to fishing cost (i.e., implicit and explicit cost) and effort, the average cost per unit of fishing effort (i.e., per trap) was $21.43; therefore total industry (fleet) cost is as defined by eq. 9.3:

$$TC = \$21.43f \qquad (9.8)$$

As discussed in our theoretical section on bioeconomic supply curves, eq. 9.7 and 9.8 may be used to derive the average and marginal cost per kg of lobsters as the fishery expands to *MSY* and then contracts with further fishing effort.

The demand curve or equation, together with the marginal cost curves, is plotted in the top half of Fig. 9.4; the lower half displays the catch-effort curve as expressed in eq. 9.7, the number of traps being measured in a downward direction. The American lobster fishery is common property; therefore, open access equilibrium will occur where the demand curve intersects the average cost curve ($AC = P$). The empirical model *predicts* that the intersection will be reached at an ex vessel price of \$1.75 per kg of lobsters and a total catch of 11.47 million kg obtained by using 933,000 traps. The *actual* figures for 1966 were \$1.68, 11.57 million kg and 947,113 traps.

MEY occurs where $MC = P$ or an ex vessel price of \$1.83/kg and total yield of 8.44 million kg obtained by using 490,000 traps. The average cost per kg of American lobsters at this output level is \$1.26. Thus, the total economic rent earned at this point is the difference between the ex vessel price and average cost multiplied by catch or \$4.8 million. Why is *MEY* optimal? A reduction in catch will reduce total costs by approximately \$9.5 million. This means that goods worth that amount can now be produced in other parts of the economy. The reduction in catch, even though price will increase, will cause the total value of American lobsters to fall from about \$20.1 million to \$15.5 million, a loss of \$4.6 million. Thus, the loss in value of lobster production will be more than offset by production of other goods; the net benefit to society of *\$4.8* million (i.e., \$9.4 − \$4.6 million) accrues.

MEY may be accomplished in the following manner. At an ex vessel price of \$1.83 ($MC = P$), economic rent will be \$4.8 million ($TR − TC$) and if proper management schemes were instituted, the government, for example, could impose a tax on catch equal to the difference between price and average cost so that *MEY* can be obtained. The tax revenue of \$4.8 million (i.e., \$0.57 × 8.44 million kg) could be used for enforcement as well as research to guide management. The fishery could become self-financing rather than a drain on the federal treasury. Also, a boat quota set at a catch of 8.44 million kg would allow those selected for the quota to earn positive economic profits ($TR > TC$). Open access dissipates profits and creates overfishing and overcapitalization. An *overall* quota system set at the present catch of 11.47 million kg will result in excess capacity since the common property problem is *not* addressed. Since 1966, the demand for lobsters has expanded creating temporary economic profits that have been dissipated through an increase in fishing effort. Because of management failures, the American lobster fishery stands today *overfished* and vastly *overcapitalized*. One final point. A return to *MEY* may create too much social upheaval in terms of unemployment. However, it may be desirable to at least prevent the fishery from expanding in terms of fishing effort to such an extent that the fishery stock is depleted and massive amounts of capital and labor become redundant (i.e., overcapitalization). New management plans of limited entry are being instituted in Canada, Norway, Japan, Alaska, Washington, and even inland states such as Michigan and Ohio.

9.5 RECREATIONAL FISHERIES

Too little attention has been given to fisheries resources used for recreational purposes. In 1975, gross expenditures on U.S. saltwater recreational fishing were $3.4 billion, and approximately $11.8 billion was spent on freshwater recreational fishing, for a total of $15.2 billion. In contrast, the 1975 commercial domestic landings are estimated to have a retail value of $3–$4 billion. Of the total food fish consumed in the U.S., it is estimated that the recreational catch may be as much as 28 percent of total consumption. From an economic standpoint, recreational fisheries are extremely important and they should be given more recognition by both the state and federal governments. The saltwater recreational fishing sector expanded rapidly over the 1955–1970 period, but future projections are less optimistic owing to limitations on resource supply. There are indications that the resource productivity of marine recreational fisheries, as in commercial fishing, is declining.

One principal problem with recreational fishing is that it is a nonmarket activity, where no 'price' or user's fee is charged for the use of the resource. The most pragmatic way of approximating a unit of recreation is by defining the experience in terms of time, or more specifically, a unit day measure. But, what price per day should be used? Most studies compute gross expenditures on transportation, food, lodgings, tackle, bait, etc., as a measure of total value and divide this by user days. In 1975, salt and freshwater fishermen spent $16.65 and $10.43 per recreational day, respectively. However, economists argue that such values are not the real value of the recreational experience. First, if there were no fishery resource, such gross expenditures would undoubtedly be reallocated to other forms of recreation. Second, it is argued that the real value of the resource is what people would consume (i.e., recreational days) if a user charge of varying degrees were placed upon the resource. Thus, we would trace out a demand curve or the 'willingness to pay' as user charges are increased. For example, using the travel cost approach (see Clawson, 1959) Singh (1965) found that a user charge of $6 per angler day would maximize revenue from the salmon-steelhead trout recreational resource. The state, in this case, would collect about $2.5 million in revenue. This would be the *value* of recreational fishing. However, if no price is charged, consumers or recreationalists do derive a value called consumer surplus or a dollar measure of the utility they derive by using the resource, paying nothing. Individuals are usually asked how much the cost (i.e., user charge) would have to rise until they would discontinue recreating. Reported values run from $20–$30 per day. Unless proper valuation studies are carried out, it is difficult to compare the economic trade-offs between resource use whether it is between recreational vs. commercial fisheries or other forms of land use (e.g., housing developments in coastal areas).

9.6   SPECIAL ECONOMIC ASPECTS OF THE FISHERIES

9.6.1   Environmental deterioration
The visual effects of water pollution are obvious enough to us all (see Chapter 10). Shellfish areas are closed to fishing because of a danger to public health. Fish kills are numerous and are directly attributable to various pollutants. Pollution of the water exists because water, like fisheries resources, is common property. Industrial dischargers of pollutants do not have to pay for the use of the water. In essence, chemical, pulp and paper, and fertilizer plants, for example, produce at a reduced cost since the true social cost of producing chemicals, paper, and fertilizer is not reflected in their cost of production. Polluters introduce external diseconomies or 'spillovers' that increase costs for other industries such as commercial fishing. If pollution reduces the fish stock, catch per unit of effort will drop and cost will increase. Indirectly, the fishing industry is forced by the present circumstances to bear the cost of producing chemicals. From society's standpoint, the fishing industry is forced to bear part of the cost of producing chemicals, paper, and fertilizer, for example. Economists would say that polluters should be made to 'internalize' the full cost (direct plus those they escape by transference to pollutees) of production.

What has been the economic impact of pollution on the fishery? The impact has been three fold: (1) reduced marketability; (2) reduced biological productivity; and (3) reduced opportunities for protein production such as fish farming or aquaculture. Commercial shellfish areas are most often closed because of bacterial pollution. Total shellfish areas closed because of bacterial pollution increased in the U.S. from approximately 0.9 million hectares in 1966 to 1.5 million hectares in 1971, an increase of 65 percent. According to a separate survey by Bell and Canterbery (1976a), closed areas did not increase between 1971 and 1975. They estimate that approximately $38 million at dockside of shellfish was lost in 1975 due to bacterial pollution. Because of mercury and DDT contamination in swordfish, tuna, mackerel, etc. the U.S. Food and Drug Administration has either seized these fisheries products or banned their sale. The lost revenue has been estimated from $3.5 million in recall of tuna to $4.25 million for swordfish, tuna, and jack mackerel during the early 1970s.

Fish kills are almost a daily experience and fluctuate greatly from year to year, averaging 34 million fish killed per year due to pollution over the 1968–1972 period. If two thirds of fish kills in 1972 had commercial value, it is estimated that the loss would be $5 million. Sublethal effects have not been quantified in terms of economic damage, but are thought to be a major factor in the decline of some fisheries such as the short-nosed sturgeon, Atlantic salmon, and herring.

Economists generally favor taxing polluters up to the point where the marginal cost of abatement is equal to the marginal reduction in damages. In this way, polluters are forced to pay for the free resource, water, and the adverse effects of pollution would be significantly reduced. By and large, this is a theoretical solution, one that would be difficult to implement. The United States and other governments of the world have chosen regulations that apply increasingly stringent standards to the discharging of pollutants. The object of the U.S. Federal Water Pollution Control Act is to end discharge by 1985. Although this objective may not be attained, it does force polluters in a direction which will relieve the commercial and recreational fisherman of bearing the social cost of pollution. If the goals of the Act are met many closed shellfish areas can be reopened, and improved water quality will increase the productivity of existing fisheries.

9.6.2   Aquaculture feasibility

Aquaculture is now flourishing in Asia and the Far East, particularly in China (see Chapter 16). Aquaculture currently provides an estimated 10 percent of the world's water-derived protein with a harvest of between 5 and 6 million metric tons valued in excess of $2.5 billion. The growing demand for protein is likely to spur the growth of cultured fish species, especially in light of the widespread depletion of the ocean's wild stocks discussed earlier.

Some countries now get more of their finfish from aquaculture than from wild stocks. For example, Hungary and Czechoslovakia—both landlocked countries—derive over 75 percent of their fisheries protein (catch) from aquaculture. Indonesia and Israel derive 10 and 36 percent, respectively, of their catch from aquaculture. Although the United States derives only 1.5 percent of its catch from finfish aquaculture, indications are that the potential is great, especially for high-value species such as shrimp and lobsters.

Throughout the world, the most commonly cultured species are carp, tilapia, shrimp oysters, milkfish, mullet, clams, and mussels. So far, the U.S. major commercial aquaculture enterprises are bait minnows, oysters, catfish, trout, and crayfish, which at the retail level constituted a $219 million industry in 1969. These five major aquaculture sectors may account for approximately 14 percent of the retail value of U.S. commercial landings from wild stock fisheries.

The land requirements for aquaculture often permit the use of low-value land—ravines, swampland, saltwater marsh, or mangrove areas—that is not well suited to other uses. Generally, a kg of fish can be produced more cheaply than a kg of red meat. That is, fish are better feed converters than land-based animals, a fact that makes aquaculture particularly attractive.

In attempting to establish an aquaculture enterprise, an economic feasibility study should be made to determine the least-cost size farm. This is generally dependent upon the relation between harvest and economic inputs plus environmental factors. Pillay (1973) has found fairly high rates of return to capital on aquaculture enterprises throughout the world. Ultimately, aquaculture must be competitive with wild stock species as well as with other protein-producing sectors.

The world potential for aquaculture development lies in a combination of (1) expansion of areas already under cultivation, (2) increases in productivity through research and development, and (3) transfer of existing technologies to countries where aquaculture is not now extensive. The last possibility would significantly aid in increasing protein production through aquaculture. A study by Bell and Canterbery (1976b) showed great potential for technology transfer, especially to developing countries where the food/population imbalances are most pronounced. Given the demand for protein it would seem that the FAO projection of 20–25 million metric tons per year from aquaculture is certainly conservative. The Bell–Canterbery study indicates that the Bardach and Ryther (1968) projection of 40–50 million metric tons annually by the year 2000 is realistic. Aquacultured species are usually for direct food consumption. If the food fish catch from the wild stocks does not increase it is quite possible that more fish will be produced through aquaculture than through commercial fishing in all the oceans of the world. The major limitations to aquaculture development are both technical and social. Conflict over land use, especially in developed countries, and pollution may severely hamper the expansion of aquaculture.

9.6.3   Underutilized resources

The projection of the world's marine resources indicates that there is an unquantified and virtually untapped reserve of lesser-known fisheries stocks, which, pending the development of efficient and economical harvesting methods, may prove to be a significant addition to the world's sources of fish and protein (see Chapter 15). To a large extent such stocks are located off the shores of the developing countries in the Southern Hemisphere. Excluding crustaceans, Suda (1973) has estimated a total potential increase of unconventional pelagic, demersal, and cephalopod species of 43.1–55.3 million metric tons. Gulland (1971) is much more optimistic when he indicates that lanternfish and cephalopods may provide up to 200 million metric tons.

Although Alverson (1975) is rather optimistic about our ability to make slight alterations in existing technology to catch unconventional species, Suda feels that such apparently vast resources as the lanternfish may require new technology, especially in mid-ocean areas. Consumer acceptance of unknown species depends to a considerable degree upon industry's ingenuity in packaging its product. The krill has had some success in the Soviet Union as a *food* fish and there is great potential for squid and other cephalopods as direct fish food. However, lanternfish and other pelagic fish, such as the Falkland herring and thread herring, are most likely to be used for industrial purposes.

What countries are likely to gain from these unconventional and untapped fisheries resources? The demersal and pelagic resources are mostly tropical and are found at some time in their life cycle primarily off developing countries. Thus, the developing countries may gain greatly from extended fisheries jurisdiction. The Antarctic krill may become the exclusive property of the Soviets and Japanese since they are the countries best equipped to exploit such species. Lanternfish, squid, and seaweed are ubiquitous throughout the world. For example, a rapidly growing squid fishery is developing off the U.S. East Coast. Within 200 miles of the United States (including Alaska and Hawaii), there are many underutilized fisheries resources, such as alewives, mackerels, flying fishes, squid, sharks, grenadiers, ocean quahog, skipjack tuna, mullet, snappers, bonito, and tanner crabs. The low level of utilization can easily be explained by our supply and demand analysis. In summary, considerable food resources from the ocean remain to be exploited. These resources may eventually be used, but if past trends continue, the distribution among the countries of the world will probably be uneven, and 'have-nots' may not share to any great extent in this abundance unless extended fisheries jurisdiction becomes a major factor. The benefits will go to enterprising nations such as the U.S.S.R. and Japan. The United States is likely to play a minor role in exploiting unconventional species such as krill, lanternfish, and cephalopods, species that require deep-sea fleets. The U.S. tuna fleet is now the only U.S. fleet with such long-range capabilities.

9.7 THE ROLE OF GOVERNMENT IN THE FISHERIES

Although there is considerable debate over the role and extent of government intervention in the private sector among market-oriented, as opposed to centrally planned, economies, most would agree that the recreational and commercial fishing industries are subject to at least four market failures. That is, the working of the private and decentralized marketplace in the fisheries does not deliver products to the consumer at the lowest possible price—in other words, it does not make the most efficient use of capital, labor, and the fisheries resources. The four distinct market failures are (1) overcapitalization and overfishing attributable to the common property nature of the resource; (2) environmental deterioration of fisheries habitats resulting from the indiscriminant use of water—a common property resource—by polluters; (3) misallocation of fisheries resources among commercial and recreational users since the latter have no market price or user charge imposed; and (4) anticompetitive practices such as massive subsidies granted to one country's industry, but not to others—where each has to compete in a world market. Of course, governmental intervention is sometimes predicated on slogans such as 'save our fishing industry,' which has an emotional appeal—the public might perceive that 'our' resources are being lost to foreigners.

Because of the principle of freedom of the seas, fisheries resources have been common property to all countries. Countries have attempted to deal with the problem of 'conserving fish' through international fisheries commissions. Generally little if any economic analysis has been applied to the problem. Unfortunately, the problem has been viewed almost exclusively within a biological context. Thus, market failures of overfishing and allocation of fish between commercial and recreational use have not been adequately dealt with. The historical experience of international fisheries commissions is one economic disaster after another (e.g., whaling commission, see Chapter 15). Some of these overfishing disasters stem from acting too late; others are rooted in a failure to adopt limited entry in favor of overall quota systems or regulated inefficiency as 'conservation measures'. These conclusions are shared by many governments involved in fishing and certainly have had much to do with the current wave of extended fisheries jurisdiction.

Government has greatly extended its involvement in fisheries well beyond these four market failures. For example, the United States has provided financial assistance in the form of construction subsidies, loans, loan guarantees, and deferrals of federal income tax to those wishing to construct, acquire, or reconstruct fishing vessels. In the main, this assistance has gone to the groundfish, shrimp, tuna, crab, and salmon fisheries, all of which are greatly overcapitalized. Such market-oriented economies as Canada, Japan, the United Kingdom, and West Germany actually offer more financial support to their respective fishing industries than the United States does. For example, the recent formation of alliances among oil-producing nations and the accompanying higher prices for fuel have prompted many governments to offer fuel subsidies to their fishing industries. Price supports and construction loans are substantial for foreign fishing fleets. The motives for this financial assistance are diverse. Some countries desire to maintain a fishing industry for food, foreign exchange, and employment. However, these relatively large subsidies are in violation of U.S. trade laws (i.e., a market failure due to reduction in competition), and the U.S. government has rarely imposed sanctions against offending governments.

9.8   REFERENCES

Alverson D. (1975) Opportunities to increase food production from the world's oceans. *Marine Technology Society*, **9**, 33–40.

Anderson L.G. (1977) *The Economics of Fisheries Management*. Baltimore and London: Johns Hopkins.

Bardach J.E. and J.H.Ryther (1968) *The Status and Potential of Aquaculture*. Clearinghouse for Federal Scientific and Technical Information, P.B. 177, 768. Springfield, Virginia.

Bell F.W. (1970) Estimation of the Economic Benefits to Fishermen, Vessels and Society from Limited Entry to the Inshore U.S. Lobster Fishery. Working Paper 36, National Marine Fisheries Service.

Bell F.W. (1978) *Food from the Sea: The Economics and Politics of Ocean Fisheries*. Boulder, Colorado: Westview.

Bell F.W. and E.R.Canterbery (1976a) *An Assessment of the Economic Benefits Which Will Accrue to Commercial and Recreational Fisheries from Incremental Improvements in the Quality of Coastal Water*. National Commission on Water Quality, NTIS. U.S., Department of Commerce PB-252172-01-12.

Bell F.W. and E.R.Canterbery (1976b) *Aquaculture for the Developing Countries: a Feasibility Study*. Cambridge, Mass.: Ballinger Publishing.

Christy F.T. and A.Scott (1972) *The Common Wealth in Ocean Fisheries*. Baltimore: Johns Hopkins.

Clawson M. (1959) *Methods of Measuring the Demand for and Value of Outdoor Recreation*. D.C.: Resources for the Future.

Crutchfield J.E. and A.Zellner (1962) *Economic Aspects of the Pacific Halibut Fishery*. *Fishery Industrial Research*, U.S. Department of Interior.

Gordon H.S. (1954) The economic theory of a common property resource: the fishery. *Jour. Pol. Econ.* **69**, 124–142.

Gulland J.A. (1971) *The Fish Resources of the Ocean*. London: Fishing News (Books).

Meadows D.H., D.L.Meadows, J.E.Randers, and W.W.Behrens (1972) *The Limits to Growth*. New York: Universe Books.

Pillay T.V.R. (1973) The role of aquaculture in fishery development and management. *J. Fish. Res. Board Can.* **30**, 2202–2217.

Robinson M.A. and A.Crispoldi (1971) *The Demand for Fish to 1980*. FAO fisheries circular no. 131.

Schaefer M.B. (1965) The potential harvest of the sea. *Trans. Am. Fish. Soc.* **94**, 123–128.

Scott A.D. (1955) The fishery: the objective of sole ownership. *Jour. Pol. Econ.* **63**, 116–124.

Singh A. (1965) *An Economic Evaluation of the Salmon-Steelhead Sport Fishery in Oregon*. Ph.D. dissertation, Oregon State University.

Suda A. (1973) Development of fisheries for nonconventional species. *J. Fish. Res. Board Can.* **30**, 2121–2158.

Chapter 10
Environmental Analysis

JOHN CAIRNS, JR.

10.1 INTRODUCTION

In an article entitled 'Chemicals: How Many Are There?' Thomas H. Maugh II (1978) gives a one page summary of the overall problem. The American Chemical Society's computer registry contains 4,039,907 distinct entities. In addition, the number of chemicals in the register has been increasing at about 6000 per week. Of these, the American Chemical Society has given the U.S. Environmental Protection Agency (EPA) a preliminary list of approximately 33,000 chemicals that are thought to be in common use. EPA believes there may be as many as 50,000 chemicals in daily use not including pesticides, pharmaceuticals, or food additives. The U.S. Pure Food and Drug Administration estimates approximately 4000 active ingredients in drugs and 2000 more used as excipients as well as 2500 additives for nutritive and 3000 more to promote product life. Taking all of these together, Maugh estimates about 63,000 chemicals in common use. When one considers that many of these chemicals may end up in water and that relatively few have had extensive determinations made of their toxicity to aquatic life, the magnitude of the problem is evident. In addition, the rate of production of new chemicals is such that the available laboratory facilities, budgets, and trained personnel are presently inadequate to keep up with the testing of new materials, let alone making determinations on those already in use for which little or no information is available.

Before World War II relatively few publications were produced on environmental analysis for pollution assessment; in the last 30 years an astonishing number of species have been proposed as bioassay organisms and an impressive variety of responses have been used to evaluate reactions. One can obtain an idea of the magnitude of this effort by examining the annual literature summaries published by the *Journal of the Water Pollution Control Federation*. Persons unfamiliar with the field of environmental analysis should keep in mind that these lists are selected citations (by no means exhaustive) and certainly more oriented toward English language publications, particularly American and Canadian, with much less emphasis on publications in other industrialized countries. The biologists who produce these methods frequently forget that the results will be meaningless unless understood by policy makers and decision makers, frequently engineers, lawyers, economists, and so forth who are relatively unfamiliar with biology. The nonbiologists are understandably confused by the overwhelming amount of information that could be gotten if they utilized every one of the now available methods. They are appalled by the cost of getting all this information and uncertain as to the best means of selecting that information most useful in resolving the particular problem at hand. Fisheries biologists must remember to state why the information is useful and how it should be used.

Environmental analysis may be divided into three basic categories: (a) *base line surveys*, which might be likened to the annual physical checkup for a human; (b) *bioassays or dose response analyses*; and (c) *biological monitoring*, the continual or routine systematic assessment of biological condition as a quality control measure. The baseline surveys may be used in an attenuated form as a part of the site selection process for a dam, steam electric power plant, manufacturing plant, or some other development, or in a more detailed and comprehensive form as a preoperational baseline assessment of ecological conditions before the proposed development, such as construction and operation of a steam electric power plant, begins. This is probably the single most important use of a baseline survey, although it also may be done on a regular and continuous basis to determine long-term ecological cycling or the long-term subtle effects of exposure to a particular pollutant.

The most commonly used bioassays are short-term tests using fish and generally do not exceed 96 hours duration with lethality as an endpoint. However, at greater cost it is also possible to carry out long-term chronic exposures using much more subtle endpoints, although the subtle endpoints also can be used for short-term tests. Customarily, bioassays involve a single age class of a single species and occasionally different life history stages of a single species. However, it is possible to use a mixture of trophic levels (e.g., primary producers, detritus processors, primary consumers, secondary consumers) and at the same time determine the effect on interactions among the trophic levels.

Biological monitoring usually is subdivided into early warning systems and receiving system methods. The former generally are used on waste discharges or effluents before they reach the receiving system and typically involve the exposure of a single species to a mixture of receiving water and effluent. The receiving system methods generally involve analysis of community structure (such as diversity indices) but may also involve analysis of a community function such as primary production.

10.2   BASELINE SURVEYS

Baseline surveys should include biological, chemical, and physical evidence. These surveys should show the kinds of organisms present, relative abundances, characteristics of the chemical–physical environment (including descriptions of significant habitats) and, when possible, the variability in the system caused by seasonal change. Ideally, a baseline survey should also include the type of variability normally occurring within the system from year to year, but rarely do circumstances permit such long-term advance studies.

The baseline survey should be as comprehensive as possible and include an inventory of all major trophic levels and all major taxonomic groups. Whenever this is not possible, careful judgment should be used to pick the most useful organisms to assess change. Unfortunately, the background of the investigator often determines which group of organisms will receive primary attention since scientists frequently assume that the group of organisms they study is the most important. As a consequence, we often see baseline studies of photosynthetic organisms in streams where the major energy is detritus from leaves and other organic materials originating outside of the stream system itself rather than photosynthetically produced energy from within the system. Even the 'best' and 'most sensitive' organisms are not appropriate in every situation, and persons who are known for using only a single group of organisms should be avoided when determining the best group to use in a limited survey. It is also well to preserve representatives of each species collected, both for verification of identification at some later date as well as providing tangible evidence that the specimens were actually there. In certain circumstances, it might also be well to preserve a limited number of specimens for residue analysis of bioaccumulative chemicals (Buzas *et al.* 1976).

10.2.1   Selection of sampling stations

Stations are selected most easily when one is trying to cope with a point source discharge (i.e., waste discharged through a pipe, as opposed to non-point or run-off from a land mass). Basically, one needs one or more reference stations (the word *control* is not used because it implies that one can keep certain variables constant, which one cannot do in a natural system). A reference station should show the degree of change that is occurring within the system due to natural phenomena. A second station should be selected to show the direct toxicological or inhibitory response of the natural system to the discharge either at the point at which it is first thoroughly mixed with the receiving water or at some point between there and the discharge outlet itself. The selection of this area will depend on the extent of the mixing zone authorized by the regulatory agency. A third station should be selected to show the effects of the maximum oxygen sag (i.e., lowering of dissolved oxygen concentration) due to the presence of degradable organics in the waste discharge. This can be estimated through use of various incubation techniques of the wastes that will enable one to predict the time required to produce the maximum oxygen utilization, and then one can use the reaeration and reoxygenation equations for the stream to estimate the location of this station. A fourth station should be a delimiting station so that if damage does occur at either stations 2 or 3 one can show that the damages only occurred in a certain stretch of the river. Of course, one has to be realistic in picking this delimiting station because a major spill of toxic materials could severely affect it also if it is too close to the other stations. Therefore, one has to assume 'worst possible case' circumstances in locating the fourth station or use more than one station for this purpose. Since fines may be assessed on the total amount of ecosystem affected, the delimiting station may be the most important one of the group.

(*a*) *Streams and Rivers.* Streams and rivers are probably the easiest of water ecosystems on which to set up stations for comparison and will be used as an example of general design. Nevertheless, there are difficulties. Since the study of all point source pollution must necessarily focus on the discharge area, the station just below the discharge pipe should be examined first. Among the general ecological conditions which one wishes to standardize among the stations are: (a) structure of the river bed, (b) current, (c) contour, stability, and composition of the substrate, (d) sedimentation, (e) vegetation in the surrounding drainage area, (f) quality and quantity of debris, and (g) collectibility of the study area. Since one wishes to select stations elsewhere for comparison, it is important to make detailed notes of these and other ecological conditions so that one can match them as closely as possible or eliminate evidence from dissimilar areas when collecting at other stations. Once the station nearest the point source discharge has been selected, one should then go upstream to find the most ecologically comparable area unaffected by the discharge being considered. Ideally, this should not be under the influence of any other waste discharge, but, in heavily industrialized areas, this desirable condition may not be available. Under such circumstances, it might be well to have two reference stations: one in a comparable situation (even though many miles away) that is unaffected by waste discharges or reasonably so, and one just above the discharge in question so that if deterioration is the result of upstream discharges this can be segregated from the discharge being evaluated. It is important to emphasize that ecological differences may exist between stations 1 and 2. These must be given careful consideration since not only is each station its own reference through time but also the stations must be compared with each other at a single point in time. Station 3 should be placed in the oxygen 'sag' zone (only used if the industry has a degradable discharge). Since this is a zone rather than a fixed point, one can make considerable adjustments in location in order to best match the ecological quality of the two previous stations. The same is true of delimiting station 4, although this station's location may be strongly influenced by the location of additional downstream point source discharges that would complicate analysis.

(*b*) *Lakes and Reservoirs.* Both lakes and reservoirs have currents, with the latter being closest to a riverine system (see Chapter 12). Nevertheless, they pose difficulty because a point source discharge in a lake or reservoir, particularly one which undergoes periodic thermal stratification, may follow a temperature gradient and not be mixed as well as it might be in a river. A heated waste-water discharge, such as one from a steam electric power plant, is likely to fan out over the surface and have its greatest effect on planktonic organisms which are more difficult to assess than benthic organisms associated with substrate.

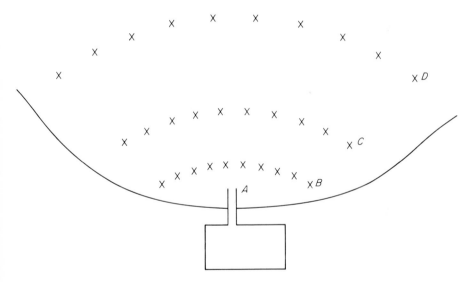

Fig. 10.1. Waste discharge through pipe *A* into a lake or reservoir can be assessed using stations *B*, *C*, and *D*. If the relationship is constant, there is no effect, but a severe effect primarily on *B* (or part of *B*) would indicate deleterious conditions near the discharge. Deterioration in both *B* and *C* relative to *D* would indicate even worse conditions.

However, artificial substrates can be strategically anchored for the collection of diatoms, etc., at points around the discharge and establish a toxicant or nutrient gradient (see Fig. 10.1). Although somewhat more complicated, the problem is essentially the same as for rivers and streams, but one must take into consideration the limnological characteristics and morphometry of the lake in locating the sampling stations (e.g., see Fig. 6 in Goldman and Kimmel, 1978).

(*c*) *Wetlands.* Recent literature indicates that certain wetlands are extraordinarily efficient decomposers and transformers of organic wastes (e.g., Odum *et al.* 1978). There is no question that nutrient processing is phenomenal in many wetlands, and proposals are being made to use them more extensively as biological filters for sewage and other societal wastes. Probably the most important single difference of wetlands from other ecological systems is the importance of anaerobic processes in their function. These should be given particular attention in both selection of stations and criteria for the assessment process.

10.2.2   Selection of parameters

Almost any biological parameter, that is anything that can be measured, might be appropriate under some circumstances. The key in selecting the parameter is to recognize that each ecosystem has unique properties and each situation where pollution assessment is called for also has its unique attributes. In short, one should not *a priori* select a group of parameters such as diversity, chlorophyll, or a particular group of organisms such as fish before evaluating both the type of pollution being assessed and the ecological characteristics of the aquatic system to be studied. Once the local situation has been analyzed from both ecological and waste characteristics standpoints, there are some questions one might ask to help in the selection of the parameters to be used:

1. Which parameters are most likely to predict effects on the entire system?
2. Are the methods to be used accepted by professionals in this particular discipline?
3. Is quantification and statistical analysis possible given the particular sampling problems for the parameters selected?
4. Will any of the information to be obtained duplicate any other information in the list of parameters to be analyzed?
5. Can this quality be measured on a regular basis with sufficient sample size to assure statistical reliability without the collecting itself influencing the parameter being measured?
6. Are the parameters selected likely to be affected adversely by the potential source of pollution?
7. Are the parameters selected subject to wide and erratic fluctuations likely to make sampling and analysis difficult?

(*a*) *Cursory Surveys.* The parameters selected for cursory surveys should have somewhat different characteristics than those selected for the long-range or comprehensive surveys. Cursory surveys are generally carried out either to determine where one might place a more permanent station in the future or to verify that unexpected changes are not occurring between comprehensive surveys. In both cases, it is desirable that the methods used be reasonably rapid, preferably generating relatively inexpensive 'on the spot' data. It is also desirable that they be sufficiently standardized to be carried out by persons with less training than those involved with the comprehensive survey.

(*b*) *Comprehensive Surveys.* Only the best methods and the most highly skilled personnel should be used for the comprehensive surveys since these are designed to detect more subtle differences than cursory surveys. Since comprehensive surveys are generally designed for long-range use, careful planning is essential in the selection of parameters to be measured, in the acquisition and interpretation of data, and in the selection of personnel and methods. Because of the cost involved, it is well to verify by careful review process and preliminary investigations that the selection has been appropriate. A discussion of comprehensive survey design may be found in Dickson, *et al.* (1975), Dickson and Cairns (1972), and Cairns (1971). An example of the general components of a broad study is given in Fig. 10.2.

10.3   BIOASSAYS

A bioassay is merely a dose response evaluation, that is, living organisms are used to determine their response to a series of different chemical concentrations (i.e., doses). An example is given in Table 10.1. The exposure may be to a

Table 10.1. An illustrative dilution series for a fish bioassay from the Ohio River Valley Water Sanitation Commission 24-Hour Bioassay (Smith *et al.* 1974). To prepare test solutions mark a test container at the 16-liter volume, add volume of wastewater specified in right-hand column for appropriate concentration, and fill container to the mark with dilution water.

Test concentration (percentage waste)	Milliliters of wastewater to be added
100	16,000 = 16 liters
50	8000
25	4000
10	1600
5	800
2.5	400
1.0	160
0.5	80
0.25	40
0.1	16
0.05	8
0.025	4
0.01	1.6

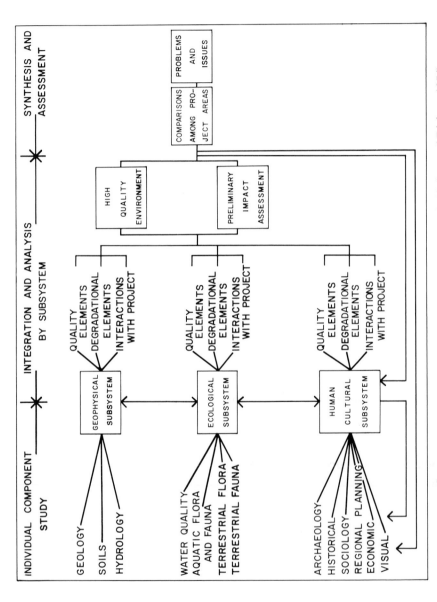

Fig. 10.2. Study process for inductive phase of a broad environmental study (Cairns and Dickson 1974).

beneficial material such as vitamin B12 or a potentially harmful material such as cyanide. These usually are given in a graded series of concentrations in which the increased dose is accompanied by an increased response. The response may be crude and brutal (death) or more subtle such as changes in growth rate, cell division, behavior, or respiration. Without such tests, one has no confidence about the concentrations that are essential to life for the beneficial materials or those detrimental to life for the harmful materials. Bioassays are particularly useful in evaluating such things as industrial waste discharges, which are usually a mixture of a rather impressive array of compounds rather than the pure chemicals frequently used in the more academic tests. There are three reasons for carrying out bioassays with industrial wastes: (a) some toxicants have biological effects at levels below chemical analytical capabilities; (b) some compounds interact synergistically (the combined effect is more than the individual additive effects) or antagonistically (the combined effect is less than the individual effects); and (c) the toxic response may be strongly mediated by water quality (e.g., zinc toxicity to bluegills is greater in soft water than in hard; see Table 10.2).

Table 10.2. The effects of water hardness and temperature upon zinc toxicity to bluegill.

Temp °C	Dilution water	Concentration of Z^{++} allowing		
		100% Survival	50% Survival	No Survival
18	Soft	1.59–2.25 ppm	2.86– 3.78 ppm	4.80– 5.80 ppm
30	Soft	0.90–2.10 ppm	1.93– 3.63 ppm	4.00– 5.81 ppm
18	Hard	6.60–9.47 ppm	10.13–12.50 ppm	12.60–14.10 ppm
30	Hard	6.18–9.50 ppm	10.15–12.30 ppm	13.50–14.10 ppm

However, the biological tests alone do not provide all the information for without the chemical analysis one would not know the precise concentration that elicited a specific response. Therefore, these tests should be done in concert and are not mutually exclusive. Those wishing more information on fish bioassays should consult the superb papers of Sprague (1969, 1970, and 1971) or the more condensed version (Sprague 1972).

10.3.1 The selection of test organisms

Out of the millions of organisms living in the world, how does one select the best test species? One cannot, of course, use rare and endangered species since it is illegal in some areas of the world and ecologically undesirable in all. In addition, species that are scarce should not be used not only because of the threat to their existence but because it is difficult to get sufficient organisms for a valid bioassay. One might consider using the most sensitive species, but the sensitivity of species relative to each other will vary among compounds (Patrick *et al.* 1968) or life history stages (Cairns *et al.* 1965). However, the greatest difficulty in selecting the most sensitive species is the paucity of information available to make this determination. Because most species have not been kept alive or cultured in the laboratory, it is impossible to run a reliable bioassay with them. Whatever species are selected, it is necessary to extrapolate from a small number of individuals to an enormous number and from a small number of species to an enormous number, and the more reliable the data base from which the extrapolation can be made, the more likely that the results will be valid. As a consequence, it is usually preferable to select the species commonly used by other investigators for which the requirements for laboratory rearing and the means of keeping healthy controls are both well understood. In addition, where legal considerations are involved, it is best to use a species for which standard methods have been developed because these are more likely to be accepted in court.

(*a*) *Trophic Levels.* For years it was felt that only fish bioassays were necessary when estimating the probability of harm to aquatic life from industrial wastes. After all, the fish were the top of the food chain and, if they were protected, all other organisms would inadvertently be protected. Unfortunately, there is compelling evidence now that fish are not invariably the most sensitive group of species in the aquatic system (e.g., Patrick *et al.* 1968). While it is true that effects upon the lower organisms would ultimately be reflected *in the field* on fish and, therefore, detected, it is now considered unwise to let ecosystems deteriorate to this degree in order to get answers which can be obtained by adding a few additional trophic levels to the toxicity testing series. Therefore, increased attention in toxicity protocols (a means of estimating probability of harm to aquatic life) are designed to include algae, invertebrates, and fish toxicity tests together (e.g., Cairns and Dickson, 1978).

(*b*) *Standard Reference Organisms.* The field of aquatic toxicology has not yet sufficiently matured to have available standard reference strains comparable to the white rats and other species of mammalian toxicity testing. However, among the fish, the fathead minnow, bluegill, rainbow trout, and a few other species are relatively commonly used in the United States and abroad and serve a somewhat similar purpose. There is a substantial body of literature devoted to their response to various toxicants and a comparison of life history stages in terms of their sensitivity to various toxicants. One might wish, when testing a new compound, to run a few tests on well known species to see the relationship between these and a less commonly used species.

10.3.2   Dilution water

The dilution water is the material used to dilute the test substance and, most important, is the material in which the test species must survive and, preferably, even thrive. The quality of the dilution water can influence the outcome of the toxicity test as much as the selection of the test species, although it is rarely given equal consideration in the selection process. In selecting a dilution water, one has the choice of picking a natural water and carrying out appropriate analyses to determine its quality or producing 'synthetic' water to reach a predetermined set of characteristics.

Natural waters are problematic for many reasons. In most aquatic habitats, water quality is subject to natural seasonal, daily, and even hourly fluctuations which may be quite enormous, particularly for such things as suspended solids, pH, and dissolved oxygen concentrations. Waste discharges in a heavily industrialized and thickly settled area may markedly influence the tests and be subject to substantial variability themselves. A grab sample taken at a specific time and place may represent the best conditions of that system in terms of the toxic response (i.e., those in which the substance produces the least toxicity) rather than the worst, which is the most desirable set of conditions for such testing. Water quality does not remain constant when the water is removed for transport to the laboratory and may alter further on storage. Finally, a test run in mid-winter may not simulate the worst conditions the system is likely to experience if those conditions occurred in August or September. Obviously, carrying out bioassays in natural water requires an extensive reading of the literature as a necessary precondition to significant testing.

Natural water (a) is abundant and usually easily obtained, (b) contains all of the trace elements, many of which are in concentrations below present analytical capabilities, (c) is capable of supporting the reproduction and survival of indigenous species, and (d) contains contaminants from other sources which should be considered when being used to develop waste control management practices. However, natural water changes and is impossible to replicate precisely. One can never get exactly the same water twice and others cannot replicate the experiments. Additionally, the quality available at the time of the test must be accepted as it is and the range of conditions cannot be easily explored as is possible with synthetic water.

Synthetic waters are usually produced by using distilled or deionized water and adding appropriate chemicals. The advantages of 'synthetic' water are: (a) it is completely reproducible in other laboratories and the tests, therefore, can be replicated by other investigators, (b) it can be made to mimic any desired set of qualities, and (c) it contains no toxic materials other than those deliberately introduced. The disadvantages of 'synthetic' dilution water are: (a) it is expensive, (b) it may lack trace materials critical to the test being carried out, and (c) test species may require an acclimation period before testing.

10.3.3 Application factors

No matter how many bioassays one does, it is not possible to obtain information on all species likely to be exposed at all life history stages under all conditions of water quality and with all other deleterious substances to which exposure may occur simultaneously. Therefore, once one has established, with a limited series of tests, a response range under laboratory or even field conditions, one must assume that more extreme conditions will occasionally occur or that there has been some error in carrying out the test, such as an important factor omitted or not adequately considered. As a consequence, a safety or application factor must be used to lower the concentration below the experimental no response point to allow for a reasonable number of these additional contingencies.

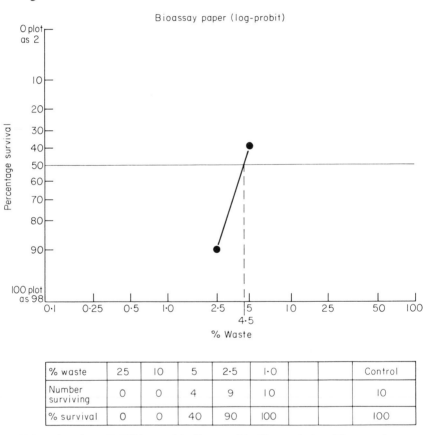

% waste	25	10	5	2·5	1·0			Control
Number surviving	0	0	4	9	10			10
% survival	0	0	40	90	100			100

Fig. 10.3. Estimation of LC50 by straight-line graphical interpolation. The waste is hypothetical. (Smith *et al.* 1974).

The most common number to which the application factor is applied is the LC50 (median lethal concentration) for which an example is given in Fig. 10.3. The LC50 is a more reliable number than either 100 percent survival or 100 percent death because at the extremes one very sensitive or one very tolerant organism can significantly alter the threshold whereas the median concentration will probably be less altered by one specimen. Even if more sensitive parameters than death are used, the same comments apply. As an illustration, however, one might develop an application factor based on the difference in response concentration from an acute short-term lethality bioassay and a long-term bioassay using fecundity as the endpoint. As this chapter is being written, there is an extensive discussion in progress in the U.S. Environmental Protection Agency and the industrial and academic communities on this very subject. It is unlikely that a single all purpose application factor such as 1/10 times the 96-hour LC50, a commonly used application factor in the recent past, will persist. The means of deriving new application factors have not yet been resolved, but it is highly probable that they will be developed on a chemical by chemical basis rather than a single number for all.

10.4   BIOLOGICAL MONITORING

Hellawell (1978) uses the following definitions:

Survey: an exercise in which a set of standardised observations (or replicate samples) is taken from a station (or stations) within a short period of time to furnish qualitative or quantitative descriptive data.
Surveillance: a continued programme of surveys systematically undertaken to provide a series of observations in time.
Monitoring: surveillance undertaken to ensure that previously formulated standards are being met.

The king's wine taster and the canary in the mine are examples of early biological monitoring systems so the concept is not new, although application of biological monitoring to problems of pollution is a newly developing field. Literature on this subject may be found in Cairns *et al.* (1977) and Cairns and van der Schalie (in press). The words *biological monitoring* appear in legislation (e.g., U.S. Public Law 92–500) and are appearing very frequently, although often used incorrectly, in the literature.

The basic assumptions made for supplementing traditional bioassays and field surveys (Cairns 1975; Cairns *et al.* 1970a, b) with a rapid biological monitoring system were (a) that the characteristics of the natural environment and industrial waste discharges changed regularly, sometimes hourly, (b) that in the absence of information about the condition of the receiving system, one would either under or over react in attempts to protect it, and (c) that adequate quality control would be possible when one had a continuous flow of information about the impact of industrial waste discharges and other stressors upon living organisms. The assumption was made that an environmental quality

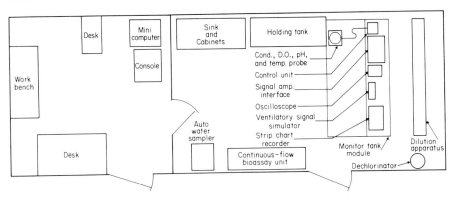

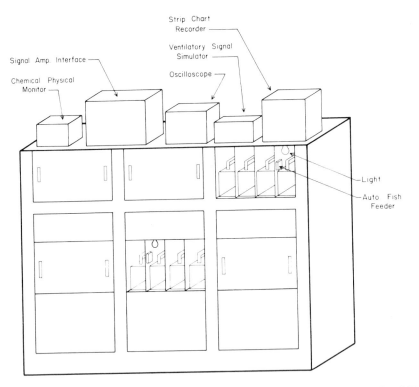

Fig. 10.4. (a) A mobile monitoring unit housed in a 9.75 × 3.66 × 2.44 m (32 × 12 × 8 ft) trailer. (b) Monitoring tank module and auxiliary components. (c) The basic monitor 'tank' (housed in groups of four in (b). BS represents the barrier strip to which the electrodes E are coupled in the electrical impulse from the opercular movement fed to an amplifier. FW are the walls preventing the fish from touching the electrodes. SF is a sloping floor to enable easy removal of waste products. SP is the adjustable standing pipe, and MC is the maintenance chamber.

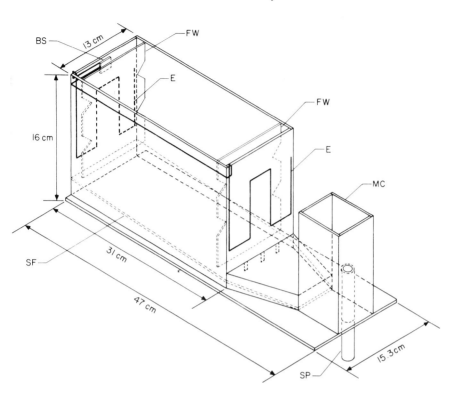

control system was no different than an industrial quality control system in that it required: (a) sensors at critical spots within the system, (b) rapid generation of information immediately relayed to a central decision-making area, and (c) a management group capable of taking immediate corrective action when deleterious conditions develop.

Although monitoring systems were developed for protecting the environment from the insults of an industrial society, the most common present use is to protect human health from contaminated drinking water supplies. Three units are now being set up by the Anglian Water Authority in England to evaluate the use of monitors as operational tools and six biological monitoring units using rainbow trout already have been developed by the Water Research Centre, Stevenage Laboratory, England. Figures 10.4 and 10.5 are of the fish monitoring unit developed by the Center for Environmental Studies, Virginia Polytechnic Institute and State University, for the U.S. Army Armament Research and Development Command, Chemical Systems Laboratory, Aberdeen Proving Grounds, Maryland.

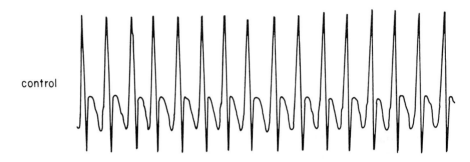

control

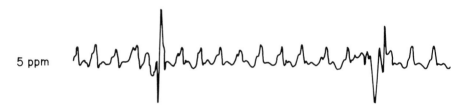

5 ppm

Fig. 10.5. The initial ventilatory response of a bluegill exposed to dilution water only (control) and one exposed to 5 ppm total chlorine (from Gruber *et al.* 1978). This is the type of early warning signal generated by a monitoring system.

10.4.1   Future research needs

For biological monitoring for inplant systems the following questions need to be answered (reproduced from Cairns *et al.* 1977):

1. Will the system detect spills of lethal materials before they reach the receiving waters?
2. If only one organism is used as a sensor (for example, the bluegill sunfish), will this organism be so much more tolerant to the particular toxicant in question that it will pass undetected and harm other members of the aquatic community in the receiving system (for example, algae and invertebrates)?
3. Is it possible to monitor chemical–physical parameters and achieve the same results at lower cost and greater efficiency?
4. Since the biological response alone will not identify the particular toxicant causing the response but only indicate that some deleterious material is present, is it possible to couple a chemical–physical monitoring system

with a biological monitoring system that will expedite the identification of the particular deleterious component causing the warning response?

5. Will a false signal cause an expensive shutdown of the plant or an undue expenditure of time and effort by the waste control personnel?

6. Should an organism indigenous to each receiving system be used, which would require a long site-specific developmental period for each new drainage basin, or can some 'all-purpose' organism, such as the bluegill, be used for all types of systems (or perhaps one organism for a warm-water and one for a cold-water system)?

7. Is it possible to use in-plant biological monitoring systems to detect the presence of spills of materials having either acute lethality or long-term effects or only the former?

8. Are the in-plant monitoring systems only for very large industries with sizable waste control staffs, or is it possible to develop compact miniaturized reliable in-plant monitoring without undue expenditure of time, etc.? (Reprinted with permission of American Society for Testing and Materials, 1916 Race Street, Philadelphia, Pa. 19103, Copyright).

For the 'in-stream' biological monitoring systems very similar questions apply. These might be summarized in the following way: (a) out of the array of possible parameters, have the right ones been selected? (b) is there a possibility of damage occurring that will not be detected at once? and (c) are there cost-effective practical measures that can be applied by persons who are not well trained scientists?

10.5   HAZARD EVALUATION AND RISK ANALYSIS

10.5.1   Origin of need

Fortunately, the vast majority of compounds mentioned in the introduction are either of low toxicity or are present in such small concentrations that toxic effects are not noticed. However, there are enough major catastrophes that we cannot continue with the casual approach toward these deadly materials characteristic of the past. The lower James River, Virginia, was closed to fishing in 1975 and for substantial parts of recent years because of kepone pollution. It is estimated that seven billion dollars are needed for removing and treating the kepone in the lower James River or that it would take over 100 years for the material to degrade naturally. If the projections are reasonably accurate and the standards now used are not altered or become more stringent, it is likely that the James will be closed to commercial fishing for many years to come. Kepone is by no means the only case but, rather, the most recent to come to widespread public attention. Mercury contamination, contamination by

PCB's, and a variety of other situations have arisen throughout the world. These are clear and unmistakable warning signals that failure to properly estimate the hazard of a chemical before it is manufactured and used on a large scale will exact a heavy price from society.

10.5.2 The basic hazard evaluation process

A hazard assessment designed to estimate risk of introducing a chemical substance into an aquatic ecosystem requires evidence to make a scientific judgment on: (a) toxicity—the inherent property of the chemical that will produce harmful effects to an organism or community of organisms after exposure of a particular duration at a specific concentration; and (b) environmental concentration—those actual or predicted concentrations resulting from all point and non-point sources as modified by the biological, chemical, and physical processes acting on the chemical or on its byproducts in the environment. One may depict the process of hazard evaluation graphically. In Fig. 10.6(a), the 'no adverse biological effects' level and the concentration that will result from introducing the chemical into the environment (e.g., Mill *et al.* 1977) are well apart. In fact, only estimates of these concentrations are known and the degree of uncertainty is indicated by the spread of the dotted lines that envelope the solid concentration lines. This is because tier 1 testing consists of comparatively crude short-term tests. Tier 2 is more sophisticated and, therefore, more expensive (for example, for fish tests it may call for continuous flow rather than batch testing), and tier 3 far more so. Frequently tier 1 testing will sufficiently improve the estimates to show that the concentrations are indeed different and that 'no adverse biological effects' concentration is well above the environmental concentration. In this case, there may be justification for terminating testing in tier 1 and concluding that *at a certain risk level* introduction of the chemical will probably not cause an environmental hazard.

Generally, the determination to use a specific chemical (that is, the determination of its 'safety') is a societal decision. Ideally, these decisions are made after weighing the probable benefits accruing from use against the probable unpleasant side effects. These side effects may occur at any stage in the production of the chemical including extractions from the ground, transportation, production, storage, use, and disposal. Even from this brief discussion, it

---

Fig. 10.6. This figure depicts the relationship of a chemical concentration that produces no adverse biological effects with the actual environmental concentration of the chemical. Tier I bioassays or toxicity tests are preliminary or screening tests; those in Tier II are of intermediate complexity; and those in Tier III are the expensive long-term sophisticated tests for sublethal responses. P indicates the point in the testing program where a decision is justified. See text for further discussion. Reproduced with permission from *Fisheries* (Cairns 1978).

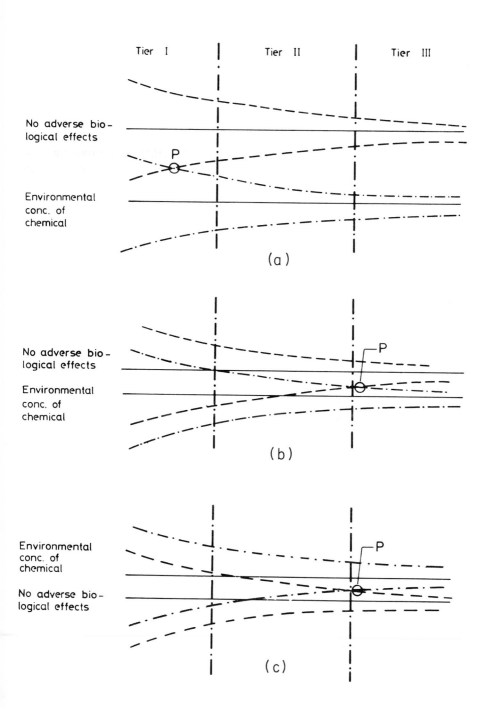

Tier I Tier II Tier III

No adverse bio-
logical effects

Environmental
conc. of
chemical

P

(a)

No adverse bio-
logical effects

Environmental
conc. of
chemical

P

(b)

Environmental
conc. of
chemical

No adverse bio-
logical effects

P

(c)

is clear that there never will be sufficient information to make an absolute prediction of the outcome of the manufacture and the use of a chemical. However, it is also abundantly clear that tests now available, properly used and interpreted, could have prevented or reduced the severity of many, if not all, of our current environmental disasters resulting from misuse of chemicals.

To properly understand cost-benefit judgments, a simple example might be worthwhile. If one were terminally ill with cancer and a new chemical appeared that provided a one in three chance of curing the ill person accompanied by a one in ten chance of killing that person because of use, the benefits would clearly outweigh the risks. However, a lipstick or after-shave lotion with a one in ten chance of killing the user is clearly unacceptable. Regretably, the determination of cost-benefit ratios is rarely that simple, but it must be done and fisheries managers must develop this capability.

10.5.3   The use of tiered protocols in the hazard evaluation process

Even today, most fisheries managers have not encountered the use of the word *protocol* in connection with their profession. However, in recent years the use of protocols has become increasingly common, and it will not be long before everyone actively engaged in the field will at least be aware of testing protocols. Just as there are large numbers of chemicals produced, there are also large numbers of biological tests which one might carry out to estimate the toxicity of these chemicals. Obviously there are neither funds, personnel, nor facilities adequate to carry out every test on every chemical. It is also clear that the relatively harmless chemicals do not require the same amount of information as the really deadly ones. Most important, a sound decision cannot be made on data generated for other purposes (e.g., basic research) and, therefore, the process for determining whether a chemical is safe to use must include the generation of an adequate data base. Protocols are merely a means of systematically generating information about the toxicity of a chemical to ensure that the data base is produced efficiently, soundly, and is adequate for the determination being made. Examples of these can be found in Cairns and Dickson (1978) and Cairns *et al.* (1978).

10.6   LEGISLATION

Most of the readers of this book will probably be from a generation that was under 30 when environmental activism reached its peak in the late 1960s and early 1970s. 'Earthday' celebrations during that period consisted of speeches, films, and slides about environmental damage, interspersed with folk songs depicting the evils of materialism and the 'man conquers environment' theme. It is a shock to students in that age bracket to hear folk songs from the

1930s extolling the virtues of taming rivers through the construction of dams and gratitude to the Tennessee Valley Authority (TVA) for making it possible for a person who has been out of work for years to again be gainfully employed and get married. Anyone interested in environmental problems would do well to get a perspective on societal attitudes toward the environment in this country by looking at the legislation directed to this end. A fine summary of the evolution of environmental legislation may be found in Bean (1979).

Protection was given first to commercially valuable fur bearing animals. This protection had a strictly economic basis and was designed to ensure that a profitable source of income was not severely damaged through loss of breeding stock. Gradually, protection was expanded to cold blooded animals, such as fish, when it was realized that these too had commercial value and supplies were finite. The protection base gradually broadened when animals for which the supply was originally thought to be limitless had greatly diminished stocks. However, the emphasis was still on commercially valuable species. Only relatively recently has legislation directed toward protection of species for aesthetic or other reasons or just for their own sakes been politically viable. Also, the protection of ecosystems is a relatively recent phenomenon, probably the most notable landmark for Americans being the establishment of the U.S. National Park System. Very recently in the Endangered Species Act, protection has been extended to species of no commercial value such as the Tennessee snail darter and the species of lousewort which may have little or no ecological significance at their present population density. The premise for this legislation is that it is now considered improper to hasten the extinction of a species and that the unique genetic makeup of a species may prove to be of practical value in the future. At the present time, such comprehensive protection has impeded several major construction projects such as the Tellico Dam, and a reverse reaction to such protection primarily based on economics is quite evident. The outcome of this battle to protect species and to ensure economic growth through construction and development is now bitter. It is clearly beyond the bounds of this chapter to discuss such legislation in detail; however, it is worth making the point that no fisheries manager can afford to ignore legislation because it impinges on practically every aspect of fisheries management. Two examples of legislation will be discussed briefly with the expectation that the necessity for professionals in the field to become reasonably well acquainted with all forms of legislation, national and state, related to one's occupation, is mandatory rather than optional.

10.6.1 National Environmental Policy Act
The National Environmental Policy Act (NEPA) of 1969 was established by the U.S. Congress to provide a national policy requiring all federal agencies to give full consideration to environmental effects in planning their programs. NEPA

prescribes specific procedures which agencies must observe as a further guarantee that they will implement this policy. Although far from perfect, this requirement has begun to bring about a substantial and fundamental reform in the decision-making process of federal agencies. One of the action-forcing procedures is the requirement that each federal agency prepare a detailed statement of environmental impact on every major federal action that might significantly affect environmental quality. As a consequence, by December 31, 1972, the various agencies had filed 3635 of these statements with the Council on Environmental Quality. As is characteristic of all pioneer activities, many of these statements conformed in a cursory way to the letter of the law but not the spirit. Some have said that the U.S. Congress made a serious error in passing this legislation and not requiring that it be accompanied by projections of types of personnel needed to implement this law. The federal and state governments as well as private industry quickly expanded their personnel assigned to cope with this law. In addition, large numbers of consulting firms sprang up to cash in on this bonanza. This expansion happened to coincide with the first serious decline in the academic job market since the late 1950s. As a consequence, the field of environmental studies was flooded with people who were educated and trained in other areas, many of whom would rather not be in the environmental arena but could not find positions in their original discipline. It would be amazing if this hastily recruited group performed with distinction. Mounds of data were gathered in attempts to conform with the law, but only a portion of this contributed in any substantive way to resolving environmental problems. The lesson for students of fisheries management should be quite clear. Work in the area of environmental management should be prepared for as assiduously as for any other profession. The field is beginning to develop professional standards, and, once in place, the quality control will undoubtedly progress as it has in all other professional fields.

10.6.2   The Toxic Substances Control Act

The Toxic Substances Control Act (TSCA), which became effective in January, 1977, provides that no person may manufacture a new chemical substance or manufacture or process a chemical substance for a new use without obtaining clearance from EPA. TSCA represents an attempt to establish a mechanism whereby the hazard of a chemical substance to human health and the environment can be assessed before it is introduced into the environment. After reviewing premarketing testing results on the potential effects of the chemical substance on human health and the environment, the administrator of EPA must judge the degree of risk associated with the extraction, manufacture, distribution, processing, use, and disposal of the chemical substance. If the chemical substance presents an unreasonable risk of injury to human health or the environment, the administrator of EPA may restrict or ban the chemical substance. According to Ritch (1978), TSCA came about because there are significant differences between the environmental contaminants that have been the focus of air and water pollution control efforts and the ones that have more recently been in the headlines—PCB's, PBB's, vinyl chloride, kepone, chlorofluorocarbons, and the like. According to Ritch, what the whole act really says can be put into two major parts: (a) find out what chemicals are already on the market and if any of them are hazardous; and (b) look at new chemicals before they get out in the marketplace to assure they pose no unreasonable risk. There is no question that some of these compounds pose what fisheries managers would consider an unreasonable risk (Martin 1978). The key question, of course, is how to identify these risks without damaging ecosystems in the process.

10.7   REHABILITATION OF DAMAGED ECOSYSTEMS

The first view of the planet earth from space was a shock to most people. It showed a small habitable unit in a vast, inhospitable universe. Probably more than any other single event, this brought home the need to protect, maintain, and restore the ecosystems that constitute our life support systems. It also brought home the fact that the frontiers are gone and that people must 'make it' where they are. A few visionaries propose space colonies as an escape mechanism or a way of reestablishing the frontier psychology. It is interesting that those who propose this are primarily physicists, engineers, and space technologists, and that the group does not include very many people thoroughly acquainted with the complexity and difficulty of managing ecosystems. There is nothing wrong with such views as long as people do not get the idea that they can continue to damage the earth because there are utopias to which they can escape. Because of the energy shortages and the likelihood that people can no longer travel vast distances for recreation as freely as they did in the past, the prospects of restoring damaged ecosystems near large metropolitan areas becomes increasingly attractive.

Fortunately, there is compelling evidence that damaged ecosystems can be rehabilitated. One of the most encouraging reports concerns the Thames River (Doxat 1977). It is perhaps not well known that Queen Victoria's Prince Albert was killed in 1861 by typhoid fever almost certainly originating from infected old drains at Windsor Castle allied to primitive local sewage into the adjacent Thames. It is an irony of history that Albert was dedicated to progress and civic improvement. Nevertheless, a little over 100 years ago both noblemen and the common citizen risked daily infection with diseases resulting from minimal or nonexistent public hygiene. From a fisheries management standpoint, the history is equally dramatic since, except for eels and a few other robust or occasional visitors, fish life was essentially extinct in the tidal Thames at the beginning of Queen Elizabeth II's reign. In startling contrast, since 1964 when the effects of the pollution abatement program first became evident, 91 species of fish have been found in the tideway. It is generally thought that the last salmon was caught upstream from London in 1833, although reported sitings and catches continued until about 1862. However, salmon fishing essentially ended between 1801 and 1820. It is encouraging that a salmon was found on an intake screen at West Thurrock in November, 1974. Although this is hardly an indication of reestablishment, the reoccurrence of other important species has led to the belief that it is practical to restore the salmon run to the Thames and funds have now been provided for this purpose. Even in the unlikely event that no further improvement occurs in the Thames fishery, it affirms three very important facts:

1. Present waste treatment and pollution abatement methodologies can be used effectively to rehabilitate a river so that a viable and productive fishery can exist where none existed before.
2. Not only did this not cause undue financial hardship upon the industries and residents of the area, but the economic benefits far outweighed the costs involved.
3. Results do not require a huge amount of time—on the contrary, the benefits are by fisheries standards relatively rapidly available.

A major opportunity now exists for fisheries professionals to participate in one of the most encouraging activities possible. It is necessary to stop the encroachment of pollution on relatively undamaged water ecosystems still remaining and to ensure their continuing safety. It is perhaps even more exciting to realize that major damaged areas of the earth's aquatic habitats may be rehabilitated to a condition more ecologically and aesthetically pleasing with a greater utility to society.

10.8   REFERENCES

Bean M.J. (1979) *In* H.Brokaw, *Wildlife in America*. Washington, D.C.: Council on Environmental Quality.

Buzas M.A., R.S.Cowan, A.L.Dahl, F.A.Fehlman, R.H.GibbsJr., C.W.HartJr., E.Jarosewich, D.W.Jenkins, P.T.Johnson, B.L.Landrum, J.F.Mello, D.L.Pawson, J.W.Pierce, and G.E.Watson (1976) *The Use of Museum Specimens for Past Pollutant Level Documentation*. Washington, D.C.: National Museum of Natural History, Smithsonian Institution.

Cairns J.Jr. (1971) Application of aquatic ecological information for water pollution control in the chemical industry. *In Industrial Process Design for Water Pollution Control*, vol. 3, pp.4–10. American Institute of Chemical Engineers.

Cairns J.Jr. (1975) Critical species, including man, within the biosphere. *Naturwissenschaften,* **62**, 193–199.

Cairns J.Jr. and K.L.Dickson (1974) *Preliminary Environmental Analysis of the Burnsville, Leading Creek and West Fork Lakes Project,* vol. 1. Washington, D.C.: U.S. Army Corps of Engineers.

Cairns J.Jr. and K.L.Dickson (1978) Field and laboratory protocols for evaluating effects of potentially toxic wastes on aquatic life. *J. Test. Eval.* **6**, 85–94.

Cairns J.Jr., K.L.Dickson, and A.Maki (eds.) (1978) *Estimating the Hazard of Chemical Substances to Aquatic Life,* STP 657. Philadelphia, Pa.: American Society for Testing and Materials.

Cairns J.Jr., K.L.Dickson, R.E.Sparks, and W.T.Waller (1970a) A preliminary report on rapid biological information systems for water pollution control. *J. Water Pollut. Control Fed.* **42**, 685–703.

Cairns J.Jr., K.L.Dickson, R.E.Sparks, and W.T.Waller (1970b) Reducing the time lag in biological information systems. *Div. Water, Air, Waste Chem., Am. Chem. Soc.* **10**, 84–98.

Cairns J.Jr., K.L.Dickson, and G.F.Westlake (eds.) (1977) *Biological Monitoring of Water and Effluent Quality,* STP 607. Philadelphia, Pa.: American Society for Testing and Materials.

Cairns J.Jr., A.Scheier, and J.J.Loos (1965) A comparison of the sensitivity to certain chemicals of adult zebra danios, *Brachydanio rerio* (Hamilton-Buchana) and zebra danio eggs with that of adult bluegill sunfish, *Lepomis macrochirus* Raf. *Not. Nat. Acad. Nat. Sci. Phila.* **381**, 1–9.

Cairns J.Jr. and W.H.van der Schalie, Biological monitoring, Part I: early warning systems. *Water Res.* (In press).

Dickson K.L. and J.CairnsJr. (1972) Ecological data for water resources management. *In* D.A.Hoffman (ed.), *Ecological Impact of Water Resource Development,* pp.1–6. Washington, D.C.: U.S. Department of the Interior, Bureau of Reclamation, REC-ERC-72-17.

Dickson, K.L., D.W.Kern, W.J.RuskaJr., and J.CairnsJr. (1975) Problems in performing environmental assessments. *J. Hydraulics Div., Am. Soc. Chem. Eng.* **101**, 965–976.

Doxat J. (1977) *The Living Thames—The Restoration of a Great Tidal River.* London: Hutchinson Banham.

Goldman C.R. and B.L.Kimmel (1978) Biological processes associated with suspended sediment and detritus in lakes and reservoirs. *In* J.CairnsJr., E.F.Benfield, and J.R.Webster (eds.) *Current Perspectives on River–Reservoir Ecosystems,* pp.19–44. Columbia, Missouri: North American Benthological Society.

Gruber D., J.CairnsJr., K.L.Dickson, A.C.Hendricks, M.A.Cavell, J.D.LandersJr., W.R.Miller III, and W.J.ShowalterJr. (1978) *The Construction, Development, and Operation of a Fish Biological Monitoring System.* Aberdeen Proving Grounds, Maryland: U.S. Armament Research and Development Command, Chemical Systems Laboratory.

Hellawell J.M. (1978) *Biological Surveillance of Rivers: A Biological Monitoring Handbook.* Stevenage, England: Water Research Centre.

Maugh T.H. II (1978) Chemicals: how many are there? *Science,* **199**, 162.

Martin R.G. (1978) Impact of aquatic contaminants on angling. *Sport Fishing Institute,* **293**, 1–3.

Mill T., J.H.Smith, W.Mabey, B.Holt, N.Bohonos, S.S.Lee, D.Bomberger, and P.W.Chou (1977) *In Environmental Exposure Assessment Using Laboratory Measurements of Environmental Processes*, pp.1–21. Corvallis, Oregon: Proceedings Ecosystem Symposium.

Odum E.P., J.S.Larson, J.CairnsJr., O.L.Loucks, W.A.Niering, J.H.Sather, and G.M.Woodwell (1978) *The National Symposium on Wetlands.* Washington, D.C.: National Wetlands Technical Council.

Patrick R., J.CairnsJr., and A.Scheier (1968) The relative sensitivity of diatoms, snails, and fish to twenty common constituents of industrial wastes. *Prog. Fish-Cult.* July, 137–140.

Ritch J.B.Jr. (1978) The toxic substances control act. *Stand. News*, **6**, 23–25.

Śmith L.L.Jr., S.J.Auerbach, J.CairnsJr., D.I.Mount, G.A.Rohlich, J.B.Sprague, and W.L.Klein (1974) *The ORSANCO 24-Hour Bioassay.* Cincinnati, Ohio: The Ohio River Valley Water Sanitation Commission.

Sprague J.B. (1969) Measurement of pollutant toxicity to fish. I: Bioassay methods for acute toxicity. *Water Res.* **3**, 793–821.

Sprague J.B. (1970) Measurement of pollutant toxicity to fish. II: Utilizing and applying bioassay results. *Water Res.* **4**, 3–32.

Sprague J.B. (1971) Measurement of pollutant toxicity to fish. III: Sublethal effects and 'safe' concentrations. *Water Res.* **5**, 245–266.

Sprague J.B. (1972) The ABC's of pollutant bioassay using fish. *In* J.CairnsJr., and K.L.Dickson (eds.) *Biological Methods for the Assessment of Water Quality*, STP 528, pp.6–30. Philadelphia, Pa.: American Society for Testing and Materials.

Chapter 11
Objectives of Management

P. A. LARKIN

11.1   INTRODUCTION

To say that one can manage anything implies that some knowledge can be applied to direct what will happen so as to benefit an identified person or group. What, then, does the fisheries manager know, and how can he use his knowledge to his advantage? A fishery comprises the fish, the other organisms, the environment in which the fish live, and the people that catch the fish. It is a *system*, and the manager manages by influencing, as he can, some or all parts of the system. It is to these themes that this chapter is addressed, first developing historically the present perceptions that guide fisheries management, and then discussing their application to present and future management practices.

11.2   A BIT OF HISTORY

To our predecessors of a thousand and more generations ago, management of fisheries, as of many other things, was left to the inscrutable whim of natural and supernatural forces. With little comprehension of the factors involved, small capacities for changing the environment, and only limited social organization, primitive man accepted what nature provided in the way of fish and other aquatic foods. Man's objective was to get something to eat for himself, his family, or his clan.

Insofar as freshwater fisheries were concerned, man had a better chance to place himself in at least partial command, and the earliest attempts at management were almost certainly made in the manipulation of lakes, ponds, and streams. Water storage and irrigation projects date from the beginning of recorded history (for example, *Genesis* 2:10), and it is easy to imagine that the diverted water was used for raising fish as well as for its prime purpose of enabling agricultural production. It was also characteristic of ancient times that royal and noble prerogatives were frequently exercised in excluding the public at large from choice hunting and fishing sites—a first step, perhaps, in the recognition that unlimited harvests had undesirable consequences, at least from the viewpoint of the privileged. As early as 1278 there were restrictive regulations on freshwater fisheries of the British Isles (Graham 1956).

Marine fisheries were, of course, something else. Fish hooks and spears, and accumulations of oyster shells, give testimony to the early use of seafoods. Certainly the ancient Greeks were accomplished fishermen. Salted and dried fish were a staple of the poor, and the rich enjoyed fresh shark meat and eels (Durant 1939). Fish were of such economic significance that they were one of Aristotle's chief interests (Pledge 1939). Nowhere in the many ancient chronicles we know is there mention of management of the fisheries of the sea. Fish, like manna, came from heaven.

This state of bliss persisted, it seems, until almost the present century. The oceans were large; fish varied in abundance from time to time; their migrations were mysterious; and understandably, it was the accepted wisdom that Man's impact was infinitesimal in the grand schemes of Nature. The fortunes of the cities of the Hanseatic League in medieval times were substantially related to the herring fisheries of the Baltic Sea and neighboring parts of the Atlantic Ocean, but there was no suggestion that the fluctuations in abundance of herring might be caused by fishing (Dollinger 1970).

The first signs that fishing might affect the abundance of fish were perhaps recognized by fishermen who have left no record of their thoughts, but the notion was not a matter for debate in the minds of scientists until the middle of the nineteenth century. (Nielsen (1976) provides an excellent account.) Salmon, of course, make their abundance conspicuous by their return to freshwater to spawn, and the decline in their numbers had long been a matter of remark. For marine fisheries, 'inexhaustibility' was the word, and was supported by Darwin's champion, Sir Thomas Huxley. Improvements in the technologies of fishing, notably trawling, led to complaints about the depletion and government Commissions. The debate came to a focus in 1883 at an International Fisheries Exhibition held in London. Ten years later, a 'British Select Committee' recognized that depletion indeed occurred, and recommended lower limits on the size of fish caught. However, it took World War I to bring the message home. The North Sea fisheries were substantially improved by the curtailment of fishing from 1914 to 1918. It became logical to ask what the best level of fishing might be.

From these roots there was a rapid development of fisheries science, particularly noted for its eventual perfection of the idea that a *stock* of fish should be harvested to provide a *maximum sustainable yield*. The various contributions which led to this notion were neither simple in their chronology, nor additive to a single theme. The combining of growth rate of individual fish with their natural mortality rate to project a peak in the biomass of a cohort was first presented by Baranov (1918). More than a decade later Russell (1931) developed the notion of an equation in which the gains from growth and recruitment were balanced by the losses from natural and fishing mortalities, all assuming that immigration and emigration were equal. On a different tack, originally suggested by Hjort, Jahn and Ottestad (1933), the logistic model of Verhulst was developed by Graham (1939) as a simple way of demonstrating that, while there was a single maximum sustainable yield, there were other less than maximum yields (Fig. 11.1). By harvesting at a low rate one could sustain a smaller yield of larger fish; and at a high rate there could be a similar total yield of smaller fish. Thus, to harvest less than the maximum was to waste fish, and to harvest more than the maximum was wasteful of effort. The understandings of the era were summarized by Russell (1942).

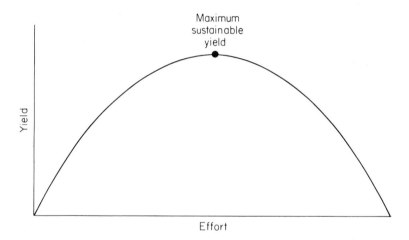

Fig. 11.1. Equilibrium yield in relation to effort for a fishery as depicted by the logistic model. The maximum equilibrium yield occurs at an intermediate level of effort.

What regulations to use to achieve maximum sustained yield was another matter. For the fisheries of the North Sea, the basic idea was to control the minimum size of fish that were caught by regulating the minimum mesh size of nets. The fish could then look after themselves. For the halibut fishery of the North American Pacific coast, Thompson and Bell (1934) advocated a 'quota' which each year fixed the permissible catch. Salmon of the Pacific coast posed a special case because they are caught on the last leg of a migration to a home stream, and they die after spawning. Rather obviously, the aim of regulation is to let that number spawn that will maximize returns in the next generation, and the maximum sustainable yield is accordingly defined as what is left over (Ricker 1954).

In the 20 years following World War II, the basic idea of maximum sustainable yield was the cornerstone of fisheries management (Larkin 1977). The idea of a single maximum became more flexible with the development of the simple Beverton and Holt (1957) model, which combined growth, natural mortality, and fishing mortality into a projection of yield. The result was an isopleth diagram depicting that for each specified smallest age of capture (determined by mesh size), there was a particular rate of fishing associated with a maximum yield, or conversely, for each rate of fishing there was a particular mesh size for maximum yield (Fig. 11.2 and Chapter 5).

From the 1920s to the 1970s, numerous refinements have been made in the methodologies of estimating the vital statistics that were needed to calculate the maximum sustainable yield. The handbooks of Ricker (1975) and Gulland (1969) became the standard texts. As might be expected, much depends on the statistics of the catch, augmented by what can be revealed by marking and tagging experiments. By contrast with wildlife management, there is little emphasis on direct observation.

Throughout the same period, complex biological aspects of fisheries management were also being explored vigorously. Among biologists there was persistent reference to the interactions of fish with each other as predator and prey, or as competitors for common resources. With the theories of Volterra (1931) as a starting point, a variety of models was constructed to depict the consequences of exploiting pairs or groups of species (for example, Larkin 1963, 1966) and to suggest how best to achieve a *joint maximum sustainable yield*. A large fund of experience was gained about the mutual compatibilities of various species of fish in pond fish culture, and there were various speculations about how best to manage the fish populations of ponds (for example, Swingle 1950). In a more holistic vein, considerable attention was given to evaluating, both theoretically and empirically, the total yield that could be harvested from whole aquatic ecosystems such as ponds and lakes, and in elucidating the characteristics of the pyramid of biomass that terminates in fish production (see Chapter 3).

As fisheries managers were pondering how to apply these concepts to the management of fisheries, another whole group of perceptions was emerging. In reviewing the history of North Sea fisheries, Michael Graham (1943) abundantly demonstrated that there was more than just biology involved. Stocks of fish are not managed to satisfy the logical arithmetic of fish production. The objective, after all, is to serve Man's needs, of which one yardstick is economic value: how then to *maximize economic return?* Harden Taylor (1951) had argued that, long before a species of fish was faced with extinction, the rationale of the market place would displace effort to other less heavily exploited species. Regulation to control fishing was accordingly unnecessary. Unfortunately for the theory, most fisheries don't work that way. People who like shad roe, like shad roe, and will pay more for it when it is scarce. Taylor was more or less ignored, and the development of fisheries economics was left to those who looked at one species at a time. Scott Gordon (1954), and subsequently Christy and Scott (1965),

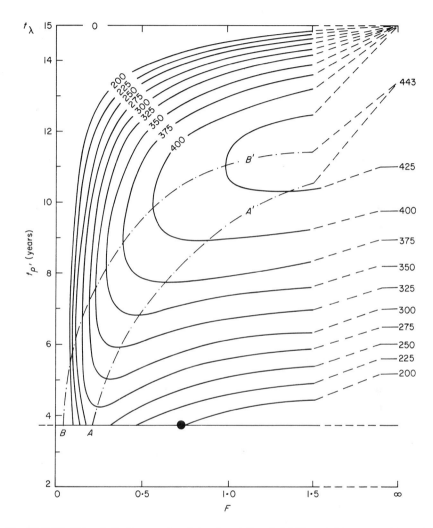

Fig. 11.2. Yield isopleth diagram for plaice. Isopleths of yield in weight per recruit are at intervals of 25 gm. The line AA′ is drawn through the maxima of yield for various rates of fishing when mesh size is 70 mm and the age of first entry (tp') is 3.72 years. The line *BB′* joins the maxima of yield for various ages of first entry associated with various mesh sizes. After Beverton and Holt (1957), Fig. 17.14.

elaborated the concept that fisheries should be managed to ensure the maximum economic net return, which would rarely and only coincidentally be the same as the management for *maximum sustainable yield* (Fig. 11.3 and Chapter 9). These perceptions, applied not only to fisheries but to other renewable resources, were part of the sustained effort on the part of economists to put an economic value on all aspects of resource management. The current culmination of these efforts is represented by Clark (1976).

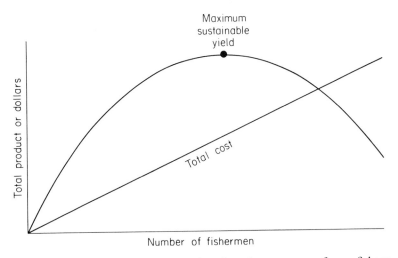

Fig. 11.3. The relation between total cost and total product or revenue from a fishery, illustrating that the achievement of maximum net economic revenue does not coincide with maximum sustainable yield. Adapted from Christy and Scott (1965).

There is more to fisheries management than biology and economics. Fishermen have their lives to lead, and governments must consider national interests. The question is not so much how to harvest fish stocks biologically, or how to promote economic return, but rather how best to proceed in the face of current social circumstances. For example, if it is apparent that too few fish are being taken by too few fishermen to achieve either a biological or economic maximum return, should the fishery be curtailed or closed regardless of whether the fishermen will no longer have a living? Should the guiding principle be drawn from the literature of welfare economics? Should social welfare be maximized? Similarly, for a sport fishery, should people be denied what they see as a 'right' to go fishing, in order that the maximum yield should not be exceeded? Should the attempt be made to manage to maximize 'aesthetic values'? Fig. 11.4, from McFadden (1969), indicates some of the alternatives for the manager of a sport fishery.

From these kinds of considerations there emerged in the United States the

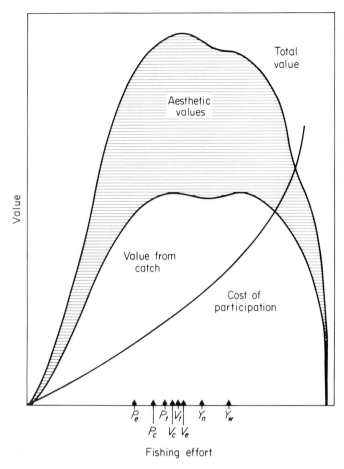

Fig. 11.4. The relation of value of catch and aesthetic value to the level of fishing effort. Arrows indicate effort associated with various maxima that a manager may set as objectives: P_e = profit from aesthetic experience; P_c = profit from catch of fish; P_t = total profit; V_c = value of catch; V_t = total value; Y_n = number of fish caught; Y_W = weight of fish caught; E = fishing effort. Adapted from McFadden (1969).

concept of *optimum sustained yield* (Roedel 1975), which was more an attempt to articulate some principles than a recipe for day-to-day decisions. A similar policy paper, produced in Canada, made a comparable effort to embrace all the considerations that might guide fisheries management (Anon. 1976). It is a major implication of their approaches that there is no single way of managing fisheries, and regional 'councils' must have a large say in making decisions (Chapter 8).

It is important to remember that the social and economic dimensions of fisheries management are international as well as national. Some fish cross national boundaries as they swim along shore, some go beyond national boundaries as they move off shore, and many species spend all their lives beyond national waters. Until recently, when national fishing zones were extended to 200 miles off shore, a large proportion of the world's fish catch could be taken (and was) by distant water fleets fishing off the shores of other nations. For many of these fisheries, management was perforce international, and accordingly beset by conflicting national goals. One country might catch fish as a means of subsidizing what was essentially a military surveillance; another might fish from the necessity to obtain a favored protein food at less expense than purchasing it from others; a third might be concerned with staking a claim to a resource that might subsequently be the subject of a treaty negotiation.

With these potential threads woven into the fabric of fisheries management, the emphasis was strongly on seeking whatever commonality of interest might bind nations together in agreement. The most common interest and first agreement usually centered on exchanges of catch statistics and relevant scientific information. With an agreed set of statistics, argument could become more orderly, and could eventually lead to negotiated understandings.

It is also important to remember that throughout the past three decades, there has been a major change in the technologies for making complex decisions. Prior to 1950, the arithmetic of fitting a model to a set of data was done with a hand calculator, or more laboriously, just by hand. Today, computers reduce this tedium to a few seconds, opening whole new doors for conjecture. While there is still elegance and concentrated comprehension in a simple analytical model, the consequences of even the most awkward conceptualization can be exploited with economy on a modern computer. The result has been a widespread proliferation of models for management, all of which rest on the various assumptions that inspire their inventors (Chapter 7).

Comparable changes have taken place in the technologies of fishing. Vessels have increased in size, ropes and nets are now stronger; new fishing techniques have been perfected, such as midwater trawling; fish detection devices are electronically sophisticated; radio-communications are almost universal; the fund of knowledge of fish migrations and concentrations is greater; and so on. Only the poorer fishermen in the lesser developed countries, and those who fish for the fun of it, have persisted in the technologies of the past.

To conclude this brief bit of history, the modern fisheries manager must have a wide range of understanding to bring to bear on his business. It is essential to know the biology of the fish, the characteristics of the aquatic environment, the social and economic setting of the fishery, and the technologies of what used to be called 'tomorrow.' Considering the variety of species of fish (about 20,000), the full spectrum of aquatic environments both freshwater and marine, the diversity of social circumstances that surround various fisheries, and the onslaught of new ways of doing things, it is difficult to generalize. What follows is a presentation, using examples, of some of the concepts that are useful over a range of fisheries management situations.

11.3   FISHERIES MANAGEMENT WITH PRIVATE OWNERSHIP

The simplest kind of fisheries management can be performed in settings in which you, as sole owner, can manipulate things as you wish. If you have an *aquarium* in your home or office, you can decide how many of which species you will attempt to grow in it; and within some obvious physical and biological limits, you can try all sorts of things. Which things you try will depend largely on your taste and interests; which things will work will depend on how much you are prepared to invest in time and money to fiddle to get what you want.

If you have a roommate, the objectives of management may immediately become a bit more complex, for there may be such things as his or her opinion or convenience to be kept in mind. Even in the simplest situations there may be various objectives of management and some constraints imposed by circumstances.

To the manager of a *fish pond* that is to make money for its owner, there is a still more limited range of ways of going about business. If the fish are to feed on natural foods but will not reproduce in the pond, and eggs or fry must be purchased elsewhere, then the objective would seem to be to achieve maximum economic return from each 'crop.' If the fish were to be sold by weight, the greatest return would be achieved by catching all the fish in the pond on the day when they reached the 'critical age', that is, when the gain in biomass from growth exactly equalled the loss in biomass from natural mortality (Chapter 5). If the market priced various sizes of fish differently, the greatest price being other than for the critical size, then the day of harvesting would come when the products of market value and the numbers of the sum of various sized fish was greatest.

Even in such simple settings there may be complications. The density at which the young fish are stocked may influence growth and mortality rates. The best density for stocking ponds of various size may be easily estimated by systematic trials, but it may subsequently be realized that there is more economic return if the pond is periodically drained and left fallow for a time to refurbish its productive capacity for fish raising. Similarly, a rotation of crops may be prudent. If the productive capacity of the pond can be increased by adding nutrients, there is the relative cost effectiveness of various kinds of fertilizers to consider. The fish may even be fed directly provided that the cost of the food is sufficiently compensated by the increase in growth and survival. The strains of fish may make a difference. The market may become oversupplied and a switch to another species may be desirable. It may be useful to grow a mixed crop of species to exploit all of the ecological potential or to hedge against fluctuations in market prices. There may be two or more quite different markets to serve. The fish farmer, like any other farmer, has lots to keep in mind in meeting the single objective of maximizing his net income.

The one consolation in the fish farming business is that annual fluctuations in the physical environment (such as temperature and water level) are usually minimal. In the jargon of economics, there are relatively few major externalities. In the fisheries of larger lakes and streams, and in the world's oceans, externalities of major proportions are common. Whatever the fish manager tries to do is complicated by naturally-induced uncertainties. If major fisheries were privately owned, it would be interesting to speculate on the management strategies that would be adopted. As it is, virtually all major fisheries of the world are publicly owned, and we accordingly turn directly from the affairs of the pond owner to the complexities of managing public resources in variable natural settings.

11.4   FISHERIES MANAGEMENT WITH PUBLIC OWNERSHIP

In theory at least, every citizen owns a piece of a public resource and is therefore entitled to a voice in management. In practice, the individual citizen can not get involved in every decision concerning all publicly-owned resources, but can only focus on matters of interest and choose some on which to express an opinion. How that opinion comes to be considered is a matter of the technique of government, and may range from direct involvement in very local matters, to helping choose a representative for 'the cause,' to delegating of prerogatives to public servants, to lobbying elected representatives—even to apathy. There is no simple recipe for describing political involvement.

Despite the complexities, it is nevertheless possible to assert that the decision concerning any fisheries resource is usually focussed on one person or group of people that may, in the abstract, be called 'the fisheries manager.' The decisions taken fall into four 'how to' categories: (1) how to *regulate* the catch of fish; (2) how to *protect* the fish and their environment from encroachment; (3) how to *rehabilitate* the resource to compensate for the inadequacies of regulation and protection; and (4) how to *enhance* the resource and augment natural production. With respect to each of these four aspects of management, it may be useful to gain knowledge by conducting investigations, investing in research and experimentation. There is also the question of how best to deploy managerial effort across the domains of regulation, protection, rehabilitation, enhancement, and the acquisition of knowledge.

11.4.1 Marine commercial fisheries

The management of commercial fisheries, particularly in the marine environment, places strongest emphasis on regulation that is more or less based on extensive statistical information, investigation, and research. Some attention has been given to the need for protection, particularly in preservation of estuaries and control of pollution in near shore waters. There have also been ventures into enhancement. For example, to increase production of oysters there are many techniques, ranging from providing a site on which the young oysters can settle, to culturing the oysters in baskets throughout their lives.

Most of the world's marine production is achieved by a hunting type of pursuit for which regulation, in the general sense, is the manager's prime method of intervention. The usual regulations are closures of areas and times of fishing, restrictions on kinds of gear, the setting of limits on catches, and the allocation of catches to various sectors of the fishery. Less obvious but no less important are the indirect techniques of regulation, such as the provision of subsidies to various sorts of fishermen, tax incentives to fishing companies, and other such means of influencing social and economic conditions.

Most of the world's marine fisheries have origins that pre-date history. The systematic evaluation of commercial opportunities and the perfection of exploratory fishing and scouting activities are relatively recent phenomena, and were brought to a fairly high degree of sophistication by the distant water fleets of such major fishing nations as Japan and the U.S.S.R. (for example, see Yudovich and Baral 1968).

The usual course of a fishery is for initial exploitation to be rapidly followed by increasing rates of harvest, which continue to increase and eventually result in a marked decline in catch per unit effort (Chapter 14). By this stage there has usually been a substantial capital investment, and depending on the fishery and the fishermen, it may be difficult to divert the effort to other enterprises (Larkin and Wilimovsky 1973). There are, then, usually demands for regulation which would: (a) ensure future harvests at some specified level; (b) protect the investments, social and economic, of those already in the fishery; and (c) accomplish some sort of catch allocation among different groups of fishermen. Even assuming that all of these objectives are compatible, it may be difficult to achieve consensus.

Suppose that a fishery for herring has expanded rapidly and the maximum sustainable yield has been exceeded. Many fishermen are not catching enough to make a living. Regulating to reduce the catch will help in the long term, but in the short term many fishermen may need some kind of social assistance if they can't find alternative occupations. Those who fish further afield may be less affected than those confined to near shore waters. What regulations should the manager apply to 'cure' this kind of situation?

In these kinds of circumstances, the manager must think more in terms of the options available than in terms of a single management objective. It may be useful to simultaneously curtail the catch to guard against a complete collapse of the fishery; to provide assistance to fishermen in diversifying their activities, both in fishing and otherwise; and to slightly modify regulations to temporarily change the allocation of catch among the different kinds of fishermen. Such compromises are typical. They reflect a judgment about what is feasible as well as what is responsible.

It is another of the facts of life of marine fisheries that conditions in the sea are not similar from year to year. Though we tend to think of these variations as capricious, random fluctuations, they are in truth the complex consequences of a great many forces, the synthesis of which is largely beyond present understanding. There are complex reasons why the survival of the young is lower some years than others, but we are far from being able to predict it some years in advance. Moreover, when the ocean is not constant from year to year, the community of fishes is not always made up of the same species in the same proportion. Ecological relations may differ from time to time. The manager obviously works in a situation of considerable uncertainty.

In many fisheries of the world it is simply not realistic to pursue the objective of maximum sustainable yield of a particular stock of a particular species. The sheer cost of obtaining the necessary information may be more than the value of the catch. If the fishery takes many species at once, and many fisheries do, some rough compromise would have to be made among the various species, even if there were no biological interactions. In a multispecies fishery where species do interact, and in most they presumably do, the only realistic, biologically-oriented objective, is to manipulate the mix in the catch with some sort of ecological prudence that will prevent the virtual extermination of some species, and so preserve the opportunity to harvest some other mix in the future. The usual pattern is to take the largest species of fish first, which commonly implies that predatory species at higher levels in the food chain are selectively removed. Subsequently, the fishery may turn to the smaller species and in the process the predators may be virtually eliminated. The total yield from harvesting lower down the food chain is theoretically larger, but the ecological association is less stable. In these circumstances, the manager must recognize that a multispecies fishery is a complex influence on a complex system, for which the consequences are at present predictable only in the most general terms. It makes sense to preserve the options for doing other things in the future by exercising restraint in harvesting now.

On the economic and social plane, the pursuit of objectives may be similarly frustrating. While it may be possible to demonstrate that a particular regulation *should* do something, it may not. For example, it has been widely believed that by limiting the number of licensed fishermen it would be possible to reduce the overall fishing effort and improve the economic return of each fisherman. What may happen, though, is that when their number is restricted, the remaining fishermen may increase their capacity to catch fish, and the stock remains at less than its productive potential. If the catch is then restricted, the fishermen lose money on their new investments, and there is a new call for restricting the number of fishermen. In an age of rapid technological developments, it is difficult to gain ground.

When questions of allocation are involved, such as among fishermen using different kinds of gear, or from different regions, or different countries, the manager can only face political realities. Instead of property rights, there are only historical precedents. There is no persuasive right or wrong concerning who gets the fish. Moreover, people's perspectives change; social and economic objectives shift. The knowledge of how to achieve maximum sustained yield or maximum economic yield is still relevant, but only as an indicator of possible consequences, not as a pointer to goals that must be attained.

The modern and highly flexible concept of marine commercial fisheries management has grown out of the fast moving events since World War II, the revolution in communication, and the capacities for computation that mark the second half of the twentieth century. By considering the many dimensions of a fishery, and by synthesizing fragmented perceptions by such means as scenarios and simulations, it is now possible to project comprehensively the consequences of a particular course of action. In that respect, then, the modern marine commercial fisheries manager has a much better chance than his predecessor of perceiving the consequences of decisions and of avoiding what could lead to misfortune.

11.4.2 Freshwater recreational fisheries

By definition, the objective of a recreational fishery is to maximize recreation. However, people define recreation in different ways (communing with nature, sightseeing, snorkelling, scuba diving, fly fishing, trolling, spearing, etc.), and many may wish to use the same lake or stream. Even if there is only one kind of recreational use it is by no means simple to decide what is to be maximized to achieve greatest recreational satisfaction. If angling is the activity, is it best to make the catch as large as possible by whatever means, or is it important to consider the size of fish that are caught? Does it matter for the recreational experience if anglers are shoulder to shoulder, or do they prefer to have some privacy? Should outboard motors be banned so that fishing is more tranquil for those who like it that way? Should fly fishing be the only kind of fishing allowed?

These are tough questions, for which a growing literature documents the necessity for careful reflection (for example, see the U.S.D.A. 1977; and McFadden 1969). The manager is faced with putting together a concoction of opportunities to comply with the aggregated taste of the users—much the same sort of challenge as that of the restaurant chef who provides a salad bar for his patrons. Fortunately for the manager, it is frequently not necessary to reconcile all possible uses in any one locality. The lakes and streams of a region may be assigned various mixes of activity.

Unlike the world's oceans, freshwater environments are vulnerable to the impact of human activities of many kinds. The protection of the resource and its rehabilitation are major concerns. Agricultural and forestry practices, road construction, industrial developments of various kinds, and indeed, almost the whole gamut of human activities (including those of the littering angler) impose a constant debilitation of the base for fisheries resources. The manager is obliged to devote much of his time to advocacy of recreational values in the resolution of conflicts of resource uses. Though his obvious objective may be to protect all of the fisheries resources, he is unrealistic to expect he can do so.

Just as freshwater environments are prone to the impact of human activities, so are they amenable to constructive efforts to compensate for past transgressions, and to ameliorate anticipated effects. Water can be stored in small reservoirs to be released when low stream flows might ordinarily restrict production. Fishways can aid migration past obstructions. Fish ponds can be created. The fish in a lake can be removed with toxicants, and the lake restocked with desirable species. A variety of favored species can be raised in hatcheries and, depending on the circumstances, planted at various sizes and densities. Streams can be 'designed' by introducing meanders to compensate for channelization, planting shade trees, and creating pools and riffles to enhance both the production of fish and the pleasure of recreational users.

The whole spectrum of activities encompassed in the words, 'protection and rehabilitation', has led to the belief that it is better to anticipate problems than to cope with them as they occur. It is now therefore the common practice to require that before any major development, an environmental impact statement should be prepared that will clearly set out what the effects could be, and how it is proposed to prevent the effects from occurring, or failing that, how to compensate for them. These kinds of requirements may demand extensive investigations which may still leave much in doubt. If the development proceeds, it is necessary to observe what actually does happen, and to take appropriate measures. The manager engaged in modern day protection work is like a fireman who must emphasize fire prevention and fire insurance at the same time he runs a fire brigade.

To conclude this brief survey of freshwater recreational fisheries, it is useful to remember that managing anything cannot be only a matter of reacting to situations. The imaginative manager will actively market his recreational wares, subtly deploying anglers by the propaganda of 'where fishing is good', or persuading them with arguments such as 'Hungarians eat carp for Christmas dinner.' The puddle in the park which might only support trout for half the year, will be stocked with fish every week, so that grandfathers may teach their grandchildren how to fish. The lake that is high in the hills that no one will walk to, will be deliberately stocked with very few fish so that the occasional angler who goes there may either come home with a 'whopper' or a relatively convincing lie about 'the one that got away.' The best managers of recreational fisheries know enough about anglers to anticipate what will bring satisfaction. For a marvellous short history of angling, and a perceptive account of what management of recreational fisheries involves, see Dill (1978); and for a discussion of the philosophies of optimum sustainable yield applied to freshwater recreational fisheries management, see Anderson (1975).

11.4.3 Mixed fisheries

It is often the case that a species of fish is the target of both commercial and recreational interest—lake trout are a good example in freshwater, and bluefin tuna are an example from the sea. Some anadromous species, such as salmon, are taken commercially and for sport (and for subsistence by native groups) in both salt and freshwater. For fisheries of this kind management is extremely complex, for it usually requires reconciliation of conflicting interests related to different kinds of uses.

Sportsmen clamor for a bigger share of the catch, usually referring nostalgically to their remembered catches of years gone by. Commercial fishermen are disinclined to spare ten fish so that anglers, in their hilarious ways, may catch five. Each group may have its own 'fisheries manager' to carry the acrimony to the highest levels of administration. In search of a common denominator to present a compromise that appears valid, each group may muster its own economic statistics, and the premises of the economic approaches may be dissimilar. The commercial fisheries manager usually speaks of landed values, subsequent value added (processing), and economic multipliers (what the fishermen and fishing companies buy). The fisheries manager for the sportsman's point of view never speaks of landed value, but more often of what it costs to catch the fish (usually a rather awesome amount), or of such things as what the angler might have earned if he had not gone fishing. Comparisons of the two sets of analyses, commercial and recreational, are appropriately unrewarding.

As in simpler situations, the proximate objective of management is more a matter of muddling through than of establishing once and for all a demonstrably optimum allocation of catch. This is especially the case when there is more than one level of government involved (federal and state or province), and perhaps more than one government at each level. The potential political trade-offs are many and various. As dead, no matter who catches them, the fish may be the biggest losers. The sad history of the fisheries of the Great Lakes is sufficient testimony to the difficulty of achieving objectives of management in complex jurisdictional and social settings.

11.5   SOME ENCOURAGING CONCLUSIONS

One might think from much of the foregoing that fisheries managers must be a discouraged lot. Some undoubtedly are, but the majority have a sufficient background in how their subject developed and what it involves, to be realistic in their expectations. Where circumstances are simplified, they may manage, with modest success, to accomplish *maximum sustainable yield*, and thereby gain the satisfaction that goes with the technical achievement of an ostensibly desirable objective. Attaining *maximum net economic return* may also be achieved, but usually only as a passing phase, for if there is no mechanism for effectively limiting fishing effort, it is usually expected that a fishery will go beyond the ideal economic state. The fisheries manager cannot responsibly indulge in single mindedness; he must aim for the mistier objectives of social optimization. It pays to look far into the future with respect to actions that, once taken, may not be reversible; for example, always preserving at least some remnants of each stock of each species. Except for such long-range considerations, however, the fisheries manager is generally best advised to *pursue whatever option is available in engendering greater public satisfaction* with his efforts. He should not, of course, be over-responsive to public whim and, given experience, should be able to explain his actions and show himself accountable in the face of transient hostilities. Where circumstances allow, he should experiment with new schemes of management, presenting them honestly as mutual ventures that may sacrifice some benefits to gain experience and new understanding.

In the pursuit of an image of such Solomon-like wisdom, almost all knowledge is germane; but for those who wish to know where to start, the most important things to appreciate are fish and their environments. It is often assumed in management discussions that we know all we need to know to make intelligent decisions. This is far from the truth. Much is known about fish, lakes and streams, and oceans, but in the past three decades alone there have been many major discoveries that have had profound effects on management. To name only a few: (1) in the 1950s it was revealed where the salmon went when they roamed the North Pacific, thereby creating awareness of international conflicts of interest; (2) in the 1960s there came the realization that environmental pollutants such as mercury and DDT became concentrated in fishes; (3) in the 1970s, it became widely realized that fisheries could have major impact on the genetic characteristics of fish populations; and (4) since the 1950s there has been world wide exploratory fishing, disclosing the existence of many major potential resources.

Despite the great fund of knowledge now available, there is much to be learned. The wise fisheries manager will invest in research to provide a better base of operation for the manager who will be in charge in the future. The decision concerning how much to invest in research is never easy, for the returns are uncertain; there is no end to what can be learned, and there are always other demands for funds. Since there is always research being done elsewhere, it is tempting to just keep watch on what others discover. Taking the broader view, scientific understanding proceeds from the pooling of effort through publication. When all contribute, the total effort brings results. The free use of the findings of others is matched by the obligation to produce findings for the use of others. This simple truth has built the present knowledge of fisheries science, and is its best assurance of greater competence in the future. The wise manager always has the objective of doing his job better than it has been done before.

11.6   REFERENCES

Anderson Richard O. (1975) Optimum sustainable yield in inland recreational fisheries management. *Amer. Fish. Soc. Spec. Publ.* 9:29–38.

Anonymous (1976) *Policy for Canada's Commercial Fisheries.* Ottawa: Fisheries and Marine Service. May 1976.

Baranov F.I. (1918) [On the question of the biological basis of fisheries.] *Nauch. Issled. Ikhtiol. Inst. Izu.* **1(1)**, 81–128.

Beverton R.J.H. and S.J. Holt (1957) *On the dynamics of exploited fish populations.* U.K Min. Agr. and Fish. Fish. Invest. (Ser. 2) 19.

Christy Francis T.Jr. and Anthony Scott (1965) *The Common Wealth in Ocean Fisheries.* Baltimore, Md.: John Hopkins.

Clark Colin W. (1976) *Mathematical Bioeconomics.* Toronto: Wiley.

Dill William A. (1978) Patterns of change in recreational fisheries: their determinants. *In:* J.S.Alabaster, (ed.) *Recreational Freshwater Fisheries: Their Conservation, Management and Development*, Water Research Centre, Stevenage Laboratory, Herts, U.K.

Dollinger Philippe (1970) *The German Hansa.* [Translated and edited by D.S.Ault and S.H.Steinberg from Dollinger, 1964.] London: MacMillan.

Durant Will (1939) *The Story of Civilization: The Life of Greece.* New York: Simon and Schuster.

Gordon H. Scott (1954) The economic theory of a common property resource: the fishery. *J. Political Economy,* **62**, 124–142.

Graham M. (1935) Modern theory of exploiting a fishery, and application to North Sea trawling. *J. Cons. Int. Expl. Mer.* **10**, 264–274.

Graham M. (1939) The sigmoid curve and the overfishing problem. *Cons. Int. Explor. Mer, Rapp. et Proc.-Verb.* **110**, 15–20.

Graham M. (1943) *The Fish Gate.* London: Faber and Faber.

Graham M. (1956) Concepts of conservation. *In Papers presented at the International Technical Conference on the Conservation of the Living Resources of the Sea.* p.1–13. United Nations, New York.

Gulland J.A. (1969) Manual of methods for fish stock assessment. Part 1. Fish population analysis. *FAO Man. Fish. Sci.* **4**.

Hjort J., G.Jahn, and P.Ottestad (1933) The Optimum Catch. Essays on Population. *Hvalråd Skr.* **7**, 92–127.

Larkin P.A. (1963) Interspecific competition and exploitation. *J. Fish. Res. Bd. Canada*, **20(3)**, 647–678.

Larkin P.A. (1966) Exploitation in a type of predator-prey relationship. *J. Fish. Res. Bd. Canada*, **23(3)**, 349–356.

Larkin P.A. (1977) An epitaph for the concept of maximum sustained yield. *Trans. Amer. Fish. Soc.* **106**, 1–11.

Larkin Peter A. and Norman J.Wilimovsky (1973) Contemporary methods and future trends in fishery management and development. *J. Fish. Res. Bd. Canada*, **30(12)**, Part 2: 1948–1957.

McFadden James T. (1969) Trends in freshwater sport fisheries of North America. *Trans. Amer. Fish. Soc.* **98(1)**, 136–150.

Nielsen Larry A. (1976) The evolution of fisheries management philosophy. *Marine Fisheries Review*, **38(12)**, 15–23.

Pledge H.T. (1939) *Science Since 1500: A Short History of Mathematics, Physics, Chemistry, Biology.* U.K. Ministry of Education, Science Museum. H.M.S.O. London.

Ricker W.E. (1954) Stock and recruitment. *J. Fish. Res. Bd. Canada*, **11**, 559–623.

Ricker W.E. (1975) *Computation and Interpretation of Biological Statistics of Fish Populations.* Fish. Res. Bd. Canada, Bull. **191**. Ottawa.

Roedel P.M. (ed.) (1975) *Optimum Sustainable Yield as a Concept in Fisheries Management.* Washington: American Fisheries Society.

Russell F.S. (1931) Some theoretical considerations on the 'overfishing' problem. *J. Cons. Int. Expl. Mer.* **6**, 3–27.

Russell F.S. (1942) *The Overfishing Problem.* Cambridge: Cambridge Univ. Press.

Swingle H.S. (1950) Relationships and dynamics of balanced and unbalanced fish populations. *Agric. Exper. Stn. Alabama Polytechnic Inst. Bull.* **274**.

Taylor H.F. (1951) *Survey of marine fisheries of North Carolina.* Chapel Hill, N.C.: North Carolina Univ. Press.

Thompson W.F. and F.H.Bell (1934) Biological statistics of the Pacific halibut fishery. (2) Effect of changes in intensity upon total yield and yield per unit of gear. *Rept. Int. Fish. Comm.* **8**.

United States Department of Agriculture, Forest Service (1977) *Proceedings: River Recreation Management and Research Symposium.* U.S.D.A. Forest Service General Technical Report NC-28.

Volterra Vito (1931) Variations and fluctuations of the numbers of individuals in animal species living together. *In Animal Ecology* (Royal N.Chapman), p.409–448. McGraw-Hill.

Yudovich Yu.B. and A.A.Baral (1968) [*Exploratory Fishing and Scouting*] *Izdatel'stvo 'Pischevaya Promyshlennost'', Moskva.* Translated by Israel Program for Scientific Translations, Jerusalem, 1970.

MANAGEMENT OF FISHERIES

Chapter 12
Management of Lakes, Reservoirs, and Ponds

RICHARD L. NOBLE

12.1 LENTIC RESOURCES

The lentic habitats, i.e. inland standing waters, consist of three major types—lakes, reservoirs, and ponds. An inconsistency of terminology has resulted from the attempt to categorize a habitat as belonging to one or the other of these groups. The term 'lake' has often implied 'natural lake' in contrast to large man-made impoundments termed 'reservoirs.' As a matter of further confusion, many natural lakes have been equipped with water level control structures, creating artificial lakes. A 'pond' is usually anything smaller, and the term has been applied to either natural or man-made. Further confusion is added by different lower bounds for 'lakes' and 'reservoirs.' A natural body of water as small as 4 hectares is frequently considered a 'lake,' whereas some consider 200 hectares as the lower bound on 'reservoirs.'

Terminology is certainly more important, in our context, as it categorizes bodies of water for fisheries management. As will become more evident, approaches to management of natural and artificial habitats differ markedly. Therefore, for purposes of clarification, lake will be used to refer to all natural bodies of water; reservoir to artificial impoundments greater than 2 hectares, but generally in excess of 200 hectares since relatively few fall in the 2–200 hectare range; and pond to artificial impoundments less than 2 hectares. Small lakes and reservoirs may in some respects be managed in the same way as are ponds.

12.1.1 Natural lakes
In North America, most natural lakes occur in Canada and the northern tier of states of the United States. A similar band of lakes occurs in northern Europe. These lakes are primarily of glacial origin and vary from deep and oligotrophic (nutrient poor) to shallow and eutrophic (nutrient rich), depending on glaciation processes, nature of soils, and length of growing season. Such lakes are frequently accompanied by shallow, relatively productive marshes. Fish communities in these lakes are dominated by coldwater or coolwater species.

Natural lakes occur to a limited extent in the southeastern U.S., particularly in Florida and Louisiana. These lakes are typically shallow and eutrophic. A variety of processes, including such diverse actions as limestone solution and tectonic movements (movements of the earth's crust) are responsible for these lakes. Large scale tectonic movements have also formed such lakes as the great lakes of the African Rift, the world's deepest lake (Lake Baikal in Siberia), and the Great Salt Lake in Utah.

Another major category of lake is the alpine lake, formed by glacial or volcanic processes. Due to the high altitudes, coldwater species are favored. A widespread type of naturally-occurring lake is the oxbow or cutoff lake, left when river channels change within a floodplain. With recent extensive channelization of rivers for flood control and navigation, oxbow lakes are frequently now man-made rather than natural. Many such lakes, because of occasional inundation by river waters, are not conducive to fisheries management. Frequently, however, flood control has been attained so effectively that oxbow lakes can be managed with reasonable assurance that recontamination with riverine species will not occur.

12.1.2   Artificial reservoirs

Most large reservoirs are instream, multipurpose impoundments. Although some have been constructed and are utilized primarily for a single purpose such as city water supply or power plant cooling, most serve a variety of functions additionally including navigation, flood control, irrigation, hydroelectric power generation, and recreation. Construction of reservoirs has contributed substantially to the lentic habitats available for freshwater fishes. The number of reservoirs in the United States has increased from about 100 in the year 1900 to about 1500 in 1980, resulting in approximately 4 million hectares of reservoirs. Development in Canada, with its abundant natural lakes, has been markedly less, resulting in somewhat over 200,000 hectares.

Creation of instream reservoirs has provided a lentic resource in many areas which previously had few natural lakes. Substantial reservoir development has occurred in California, Texas, Oklahoma, Tennessee, and the Dakotas. Type of fish communities in these reservoirs varies markedly, principally in relation to nutrient levels, length of growing season, and species endemic or introduced.

A small type of reservoir, the floodwater retarding structure, has been widely constructed in the U.S. for flood control. Over 13,000 structures, each resulting in impoundments of about 4 to 20 hectares, have been developed for selected watersheds throughout the United States. Because of the substantial fluctuations in water level which these structures experience, they physically and ecologically resemble multipurpose reservoirs.

12.1.3   Ponds

Ponds have been utilized for centuries to provide food fish through aquaculture (see Chapter 16). Small bodies of water have also provided substantial traditional capture fisheries. Frequently such ponds are constructed for purposes other than fishing, but increasingly they are being built specifically to provide fishing.

The term 'farm pond' has been widely applied to small man-made impoundments which are typically constructed by damming an agricultural waterway, gulley, or intermittent stream. Surface runoff thereby provides the necessary water supply to maintain water level. Farm ponds have most frequently been constructed for soil and water conservation and livestock water supply. Assistance from public agencies has provided a major impetus to farm pond construction in the United States. Most of these ponds are less than 2 hectares in size.

An increased affluence of American society and the proliferation of suburban housing developments has resulted in the construction of private ponds and reservoirs principally for recreation. In addition, many states without extensive lakes and reservoirs have developed public recreation areas around small impoundments. Intensive fishing, as well as other recreational demands, makes management of these ponds a particular challenge.

A widespread type of small artificial impoundment which may vary in size from a pond to small reservoir is the borrow pit. Borrow pits are formed in excavated basins as the by-product of borrow or quarry activities. Such impoundments include gravel and limestone quarry pits, stripmine basins, and highway and railway borrow pits. These basins fill either as a result of extending into the water table or from runoff. Borrow pits, because of their unique characteristics, offer unique opportunities to provide fishing, but also present very special challenges for management.

12.1.4 Characteristics of lentic habitats

Although large lakes and instream reservoirs may be similar in size, other physical, chemical, and biological features are markedly different. Such features include shape of basin, water exchange rates, extent of water fluctuation, thermocline depth, and species composition (Table 12.1). Attempts to explain variations in fish production have indicated that mean depth and total dissolved solids, combined into a *morpho-edaphic index* (Fig. 12.1), are principal factors determining fish production in both lakes and reservoirs, but that other factors have substantial effect. In particular, shore development, water exchange rates, outlet and thermocline depth, water level fluctuation, growing season and reservoir age have been shown to influence reservoir fisheries. Analyses have indicated that factors affecting standing crop are not always synonymous with those affecting harvest.

Fish communities in lentic environments are typically dominated by carnivorous, usually piscivorous, species which are target species for recreational

Table 12.1.   Comparison of general characteristics of glacial lakes and instream reservoirs.

	Glacial Lake	Instream Reservoir
Age	10,000 years	Less than 100 years
Distribution	Circumpolar	Ubiquitous
Outlet Barrier	Glacial Moraine	Dam
Shoreline	Uniform	Irregular
Basin	Shallow Near Outlet	Deepest Near Dam
Outlet Depth	Surface	Variable
Water Exchange	Low	High
Initial Eutrophication Rate	Low	High
Long-Term Productivity	Eutrophy	Oligotrophy
Water Level Fluctuations	Low	High
Establishment of Fish Community	Naturally Evolved	Artificial
Nature of Fish Community	Complex, Stable	Simple, Unstable
Opportunity for Fish Management	Low	High

and commercial fishermen. In the northern and alpine lakes, coldwater species, particularly salmonids, predominate. In more temperate waters percids and esocids are usually important. In North America, these fishes are complemented by centrarchids and ictalurids, whereas other catfishes and cyprinids are more common in northern Europe and Asia. In the warmer environments of the southern U.S., centrarchids, percichthyids, and ictalurids support most fisheries. Food resources for the piscivorous fishes vary from principally small percids, clupeids, and cyprinids in the north to predominantly clupeids and centrarchids in more southern areas of the U.S. Tropical communities of Asia, South America, and Africa differ markedly from those of more temperate areas. A great diversity of cichlids, characins (in the New World), and various catfishes dominate these tropical communities.

12.1.5   Inland fisheries

In North America, fisheries on lakes, reservoirs, and ponds, exclusive of the Laurentian Great Lakes, are almost entirely recreational. Estimates of U.S. recreational fishermen in 1975 indicate that about 30 million warmwater fishermen and nearly 10 million coldwater fishermen utilized lakes, reservoirs and ponds for fishing. At the same time, over 6 million people participated in sport

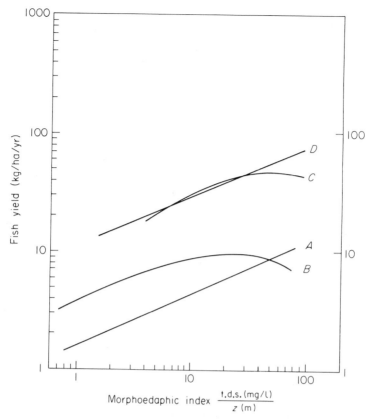

Fig. 12.1. Relation of morphoedaphic index (total dissolved solids/mean depth) and fish yield for four regional lake groups: *A*, north-temperate lakes of North America; *B*, Finnish north-temperate lakes; *C*, temperate and south-temperate reservoirs of the USA; *D*, tropical African lakes. Modified from R.A.Ryder, *et al.* 1974, *J. Fish. Res. Bd. Canada*, **31** (5).

fishing in Canada, primarily in inland fishing. Based upon recreational fisher-man expenditures and estimated values of recreational fishing experiences, economic value of the fisheries resources for recreation far exceeds that for commercial purposes (Chapter 9). This is in marked contrast with Asia, Africa, and South America where management of lakes and reservoirs is directed toward commercial fisheries and that of ponds toward aquaculture.

12.2   THE NEED FOR MANAGEMENT

Most natural lakes, except for oxbows, have existed for over 10,000 years. Consequently, the aquatic community has evolved so that available resources are apportioned among species, and an efficient, yet delicate, ecosystem has resulted. This complex system in lakes has evolved a high level of stability, which precludes the need for intensive management until the system is seriously disrupted. Disruptions of natural lake systems arise from environmental changes, often the unintentional result of human activities, and from fishing pressure itself which may upset predator–prey relationships of the fish community. Consequently, most management of natural lakes is carried out in relationship to environmental control, including adjustments for environmental change, or in relation to fishing.

Artificial impoundments, in contrast, have not existed until relatively recently. Consequently, opportunity for the evolution of efficient ecosystems has been time-limited. In the case of reservoirs, most have been constructed on streams and rivers, thereby beginning their fish community development with riverine species. Effective predator-prey systems and stability comparable to that in natural lakes seldom exist. Reservoirs represent an integration of both lotic (flowing water) and lentic environments to the extent that whether an ecosystem similar to that in natural lakes will ever evolve in large impoundments is questionable.

Artificial impoundments, particularly reservoirs, place an unusual demand on fisheries managers. Because they are constructed principally for purposes other than recreation, opportunities for fisheries management are frequently constrained by other uses. In fact, some operational procedures predicated by primary use of reservoir waters may be counter-productive for populations of important fishes and for fishermen.

Artificial impoundments also provide great potential flexibility in management. Constructional features, including location, can be planned to favor certain types of fish and to allow types of management not possible for natural lakes. In the case of the farm ponds and watershed reservoirs, it can even be possible to limit fish species in the impoundment. Such flexibility does, however, require knowledge of appropriate species combinations and their response to management. The simple systems which are so developed lack stability and are consequently very susceptible to being upset by fishing or environmental variation.

12.3   MANAGEMENT OBJECTIVES FOR RECREATIONAL FISHERIES

Because of the importance of recreational fishing, lake, reservoir, and pond management is usually oriented toward optimizing recreational fishing values for the participant. Values of recreational fishermen are abstract, and studies of the sociological aspects of fishermen indicate some disagreement on what is more important—fishing or catching fish. Therefore, management must be concerned with managing for the fishing experience, of which the fish population is only a part. Rarely will maximum sustained yield, in the biomass sense, be an objective.

Lakes, reservoirs, and ponds, because of their individuality, offer the potential for providing a variety of recreational fishing. An isolated mountain lake can provide the natural aesthetic setting where fishing can provide great satisfaction, even if catches are low. In contrast, a fishing derby at a municipal gravel pit stocked with hungry trout or sunfish can provide similar satisfaction for elbow-to-elbow youngsters. A major responsibility for fisheries management agencies is to be adequately aware of who their constituencies really are and what those constituencies need, so that those needs may be addressed through management.

The needs of fishermen which affect management objectives, beyond that of aesthetics such as natural beauty and solitude, pertain to whether size of fish or number of fish is being sought. Managing for harvest of large numbers and large fish simultaneously in a population is essentially impossible. If a resource is to produce trophy fish, fishing pressure on sub-trophy fish must be limited so that the probability of fish getting old and large is increased. Catch-and-release programs provide a reasonable attempt to overcome the conflict. The more common approach to resolving the dilemma, however, is to manage certain populations for large numbers, others for large sizes. Although public agencies have sometimes employed such 'zoning' of waters for management, private landowners with numerous ponds have sometimes been more effective. Certain ponds are managed for trophy fish through carefully regulated harvest of complex systems, others employ intensive fishing on simple single-species ponds which approach aquaculture in their management.

Decision-making based upon objectives of the user must be recognized by public fisheries management agencies. Limited funds for conducting management programs dictate a continual assignment of values to management results. The remainder of the chapter discusses four general approaches to management of fish populations—population manipulation, habitat manipulation, harvest regulation and impoundment design—applicable to lakes, reservoirs, and ponds.

12.4   POPULATION MANIPULATION

Population manipulation is the direct alteration of the abundance or species composition. This may be accomplished by reducing existing populations through selective removal, and by stocking individuals of new or already established species. Population manipulation includes total eradication of existing fish populations and stocking of new waters. It also includes stocking of catchable-sized fish in put-and-take fisheries. In general, however, this specialized form of management is exercised much less in standing waters than in coldwater streams.

12.4.1   Direct population reduction

Direct population reduction has usually been applied to undesirable fish species, a technique commonly referred to as rough fish control. Species of negligible value such as carp, suckers, gar, and gizzard shad frequently reach high population levels in lakes and reservoirs. Rough fish control crews were at one time an integral part of many fisheries agencies, utilizing netting, trapping, and electrofishing techniques to reduce rough fish populations. Other agencies have subsidized commercial fishermen to selectively exploit nongame species. Most such attempts have provided, at most, temporary control. Unless rough fish populations are totally eradicated, subsequent population explosions cause them to quickly return to, and sometimes even exceed, standing crops which existed prior to implementation of control measures.

One major exception is the sea lamprey control program which is conducted on the Great Lakes of North America. Following the development of a selective toxicant, TFM, tributary streams which serve as nursery areas for larval lampreys have been effectively treated and lake populations of lampreys have been significantly reduced. The reduction of lampreys has been instrumental in the reestablishment of salmonid fisheries in the Great Lakes.

In small ponds, thinning of overabundant, stunted fish, particularly centrarchids and clupeids, can be a feasible management technique. Although such thinning may be attainable by trapping or netting, it is usually accomplished by application of a toxicant at a low concentration. Rotenone is the most universal fish toxicant, but a variety of other chemicals have been used. Rotenone at low concentrations is toxic to small centrarchids and shad, but ictalurids and larger centrarchids are minimally affected. The extent of population reduction necessary depends on the abundance of predator fishes in the pond and whether the management program includes predator stocking in conjunction with the partial poisoning.

In many cases, undesirable species which are not conducive to thinning reach such abundance that total eradication of the fish community becomes the only reasonable recourse. Eradication, variously referred to as reclamation or rehabilitation, can be accomplished by use of toxicants or by draining. Ponds and small lakes and reservoirs are more conducive to reclamation than are larger bodies of water, but reclamation measures have been applied to large reservoirs through a combination of drawdown and application of toxicants. Because predator-prey ratios in ponds commonly become undesirable after about five years, some managers recommend periodic reclamation and restocking as a principal management approach for small ponds. Frequently reinvasion of reclaimed ponds by undesirable species occurs quickly resulting in only temporary benefits from reclamation. Because of the high costs of toxicants, alteration of spillway design and reclamation of other ponds in the watershed should be considered with the chemical reclamation.

When toxicants are used, either for reclamation or thinning, they should be applied during warm months when fish are more susceptible and when detoxification occurs more quickly (Fig. 12.2). If waters are thermally stratified, measures must be taken to insure mixing of the toxicant into the hypolimnion (cold bottom water). Special consideration should be given to possible down-

(a)   (b)

(c)   (d)

Fig. 12.2. Toxicants applied from boats may be sprayed into shallow waters (*a*) and mixed with outboard propwash (*b*, *c*), thereby assuring total mixing and eradication of all fishes (*d*).

stream effects in case overflow should occur before detoxification. Care must also be taken to comply with government regulations concerning acceptable toxicant concentrations and edibility of fish recovered from reclamation. Any chemical developed for use as a fish toxicant must undergo stringent testing to receive U.S. government approval, and limitations on its use will be fully described on the instructions for application.

12.4.2   Stocking

Stocking programs may be applied to either new or established aquatic communities. The addition of a species into a habitat where it does not at the time occur is termed an *introduction*. Introductions may utilize species which naturally occur in the area (endemic species) or species whose geographical range does not include the locality being stocked (exotics). Sometimes introduced species are unable to reproduce, and maintenance stocking is necessary. Supplemental stocking of individuals into an established population is sometimes done to compensate for insufficient reproduction. Often such stocking utilizes fingerling fish, but supplemental stocking of catchable-size fish may be employed where fishing pressure is high. Management of these 'put-and-take' fisheries relies upon immediate, high exploitation of the stocked fish.

Stocking programs are much more widespread on ponds and reservoirs, where initial stocking may be done in an attempt to establish populations of desirable fishes. Stocking programs on natural lakes have been used successfully when relatively permanent environmental changes have occurred. Introductions of new species on top of stable, existing populations in lakes have seldom been successful, probably because of the natural diversity and efficiency of those ecosystems. Introductions sometimes fail because there is no available niche for the introduced species to fill.

Some of the most dramatic successes of fisheries management have come through introduction of exotics but results have not been very predictable. Impacts of the introduction of the common carp, a species highly esteemed in parts of the world but now unpopularly ubiquitous in much of North America, have led to a conservative approach to the use of exotics. Unforeseen interactions between exotics and native species may lead to demise of desirable native species. Careful evaluation must precede experimental introduction before application to management programs is made.

Fish are usually stocked as fry or fingerlings, but successful programs have been conducted also using fertilized eggs and adults. Preference for size at which fish are stocked depends upon a combination of hatchery economics and post-stocking growth and survival. Use of hatchery-reared fish has met with many failures. By maintaining stocks in hatcheries over many generations, hatchery strains are developed which have been selected for culturability, but which may be poorly equipped to compete in the wild. Some of the more successful stocking operations have been those which take the eggs from wild stock each year, fertilize and hatch them in hatcheries adjacent to receiving waters, and stock them directly back into the waters from which the brood fish were collected. Stocked fish also may have low survival due to stress of hauling and abrupt water quality differences between hauling and receiving waters.

(*a*) *Stocking game fishes.* Stocking of fish into natural lakes has usually involved game species. Deterioration of spawning grounds due to eutrophication and watershed disturbances sometimes prevents or limits reproduction. Hatchery production of native fishes such as lake trout and walleye has effectively sustained such populations through supplemental stocking programs.

The Great Lakes salmonid stocking program has probably been the most dramatic fisheries management success in recent years. Following the invasion of the Great Lakes by the sea lamprey, native stocks of predatory fishes, principally lake trout, and their prey, a group of coregonids, declined sharply. Major changes in fish community structure followed, resulting in an assemblage of species much less attractive to fishermen than the native stock had been. The alewife, a small clupeid, became so abundant that obnoxious annual die-offs occurred. Following control of the sea lamprey by selective chemical treatment, desirable fish populations were established through the stocking programs. In the lower lakes where communities had changed most, Pacific salmon and steelhead were introduced to utilize the abundant alewives. Maintenance stocking of salmonids is conducted annually using locally available brood stock collected during spawning runs. In the upper lakes, lake trout stocking programs have been undertaken to reestablish spawning populations which eventually may sustain the species without supplemental stocking.

Fisheries deterioration in the Great Lakes also led to one of the earliest attempts to develop a hybrid fish for utilization in fisheries management. Because lake trout were so susceptible to lamprey attack, a similar fish was sought which would occupy an area of the lake where lampreys were less abundant. By crossing the lake trout with brook trout (also known as speckled trout), a fertile hybrid called 'splake' was produced which not only preferred a different depth, but also grew rapidly and matured at an early age. The subsequent control of lampreys and establishment of other salmonid programs has resulted in only limited use of splake in fisheries management.

The role of game fish introductions in reservoirs has been considerably greater than in natural lakes. Very early in the management of reservoir fisheries, it was recognized that new reservoirs produced exceptional fishing during the first several years, but by the time a reservoir was ten years old, yields to fisheries dropped off precipitously. The decline in recreational yield was accompanied by increases in rough fish populations. Factors such as nutrient cycling, differing population dynamics of game and rough fish, and forage availability have been considered to be involved in the evolution of reservoir fish communities into undesirable fisheries. With increased knowledge of the fish communities which develop in old reservoirs, stocking has become an effective management tool. New reservoirs can be initially stocked with species which will attain a longer term balance between predators and prey species, depending upon the conditions which develop.

Surveys of standing crops of fishes in U.S. reservoirs have indicated that clupeid fishes comprise nearly half of the total fish biomass in reservoirs (Fig. 12.3). Gizzard shad is usually the principal species, but shad contribute little directly to catches, and because of their large size and pelagic distribution they are not well utilized as prey species by native game fish. Consequently, a wide

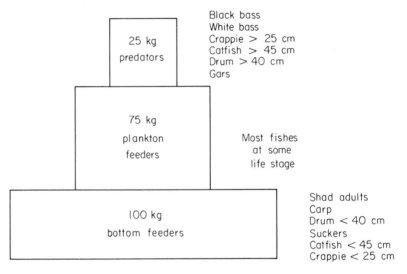

Fig. 12.3. Average reservoir fish biomass in kilograms per hectare, wet weight. Modified from Jenkins R.M. (1975), *In* R.H.Stroud and H.Clepper. *Black Bass Biology and Management.*

variety of predator species have been introduced, particularly in the southern United States. Walleyes, white bass, striped bass, and hybrids between white bass and striped bass have been widely introduced with reasonable success. Food habits studies on these species indicate a heavy utilization of clupeid fishes, with minimal impacts on native game fishes such as largemouth bass, crappies, and catfishes. Other predators which have been introduced successfully include northern pike, the Florida subspecies of largemouth bass, and some marine species including red drum and flounders. Conditions which favor these fishes are much more limited than for the previous ones. Where hypolimnetic conditions permit, 'two-story' fisheries have sometimes been developed. Cool, oxygenated waters of the hypolimnion support salmonids, principally rainbow trout, while warm surface waters support warmwater species.

The worldwide interest in game fishes is demonstrated by numerous examples of widespread introduction of game species. The brown trout, native to Europe, has been established on five additional continents. The largemouth bass, native to North America, has become equally ubiquitous through intentional introductions.

(*b*) *Stocking prey species.* Addition of prey species to natural lakes has been quite limited. Stocking of small species of fishes has been utilized successfully in some situations, but competition with endemic species has generally led to discouraging results. Some invertebrates, particularly crayfish and mysid shrimp (*Mysis*), have been widely introduced. Stocking of *Mysis* sp. has been particularly successful in coldwater lakes where trout often are the only fish species which have been established.

In reservoirs, stocking of prey species has been widely used. Initial stocking of new reservoirs frequently includes prey species such as minnows, silversides, and sunfishes to serve as food for initially-stocked game fishes. Throughout the southern United States, threadfin shad have been extensively introduced as a prey species. Because of their small adult size, they remain vulnerable as prey throughout their lives. Threadfins successfully compete with gizzard shad for food resources, and consequently, overabundant gizzard shad populations can be successfully reduced. Threadfin shad distribution is limited by low temperatures, so their use in management is limited to the southern half of the U.S.

(*c*) *Stocking farm ponds.* Ponds offer the greatest potential control over fish species composition of any aquatic habitats. Ponds can frequently be developed with assurance that no fish will be present, and, if constructed properly, with assurance that no unwanted fish will invade during periods of excessive runoff and overflow. Consequently fishes introduced into the pond will be the only ones present, and, following initial stocking, management can emphasize maintaining proper balance between those species. A good rule-of-thumb for ponds is to keep the fish community simple. Often stocking is limited to a combination of a single predatory species and a single prey species. Factors affecting choice of stocking combinations include pond size, climate, productivity, water clarity, and angler preferences.

The most widely-used pond stocking combination is largemouth bass-bluegill. Although stocking may utilize any size fish, both species are usually stocked as fingerlings. Both species grow rapidly to maturity and at age 1 or 2 spawn prolifically. Young bluegills serve as food for the bass, but compete only minimally with them when favorable ratios of the two species are maintained. A pond with a favorable ratio of the two species is said to be in 'balance,' and is additionally characterized by rapid growth rates and large individuals of each species. One major attribute of the bass-bluegill combination is that under balanced conditions, both species can provide fishing. Another attribute is that channel catfish can be stocked into this combination as an additional harvestable species. Because channel catfish occupy somewhat different habitat and utilize somewhat different food resources (principally bottom organisms), they have negligible impact on the bass and bluegill populations. Because of specialized spawning requirements of channel catfish and the high susceptibility of catfish fry to bass predation, most successful results are obtained by stocking fingerling catfish periodically into such ponds.

Unfortunately, the bass-bluegill combination easily goes out of balance. Under conditions of low fertility, short growing season, high turbidity, or excessive bass fishing, bluegills tend to overpopulate, become stunted, and limit bass reproduction. Under conditions of moderate fertility, very clear water, and underfishing for bass, the bass deplete their prey supply, become stunted, and limit their own recruitment through cannibalism. Although carefully controlled fishing may minimize the chances for these imbalances to occur, population manipulation techniques are frequently used. These techniques include, in addition to population reduction or total reclamation, modification of stocking recommendations.

In areas where overpopulation of bluegills is anticipated, stocking recommendations should be modified to provide a competitor for the bluegills. The redear sunfish, which is a less prolific spawner, has been the most successful for this purpose. In its presence, bluegill populations do not tend to overpopulate so readily and balance is more easily sustained. The redear also grows to a catchable size, but is somewhat more difficult for fishermen to catch than are bluegills.

Some bass-bluegill ponds which have become unbalanced can be restored to balance by supplemental stocking with advanced fingerling bass. This 'corrective restocking' utilizes approximately 15-cm bass stocked at about 100 fish per hectare. Size of bass and stocking rate are determined by the severity of imbalance, size at which bluegills are stunted, and whether population reduction techniques are used in conjunction with the restocking programs. Addition of these sub-adult bass in late summer leads to increased predation upon the sunfish thereby leading to successful bass reproduction the following year and additional control of sunfish reproduction by young bass. The stocked bass rapidly grow to catchable size and can be fished as soon as balance is restored.

In some areas where bass-bluegill combinations are unlikely to succeed because of overpopulation of bluegills, other prey species have been substituted. Bass-golden shiner, bass-yellow perch, and bass-lake chubsucker combinations are some such examples. Stocking of hybrid sunfish instead of bluegills has received wide application. Because hybrids frequently have reduced reproductive potentials, either due to sterility, fewer eggs, or uneven sex ratios, they are unlikely to overpopulate. They also exhibit hybrid vigor, characterized by rapid growth rates, large sizes, and aggressiveness, all of which make them desirable sport fishes. Performance of hybrids varies greatly, depending upon parent species and local conditions; consequently recommendations regarding appropriate hybrids to stock also vary regionally.

In clear, productive waters bluegills must be supplemented with an additional non-competing species. Gizzard shad and golden shiners have been used successfully. Because both species grow to sizes large enough to avoid predation by bass, their populations can be sustained under intense predation. Threadfin shad and smaller cyprinids have also been used successfully, but under intense predation their populations may be completely eliminated

Some alternatives to a combination of largemouth bass plus a prey species have been evaluated. In some areas bass are stocked alone. In the absence of another species of prey fish, the bass utilize invertebrates and smaller members of their own species for food. It is often recommended that warmwater ponds less than 0.4 hectares be managed exclusively for channel catfish rather than for largemouth bass. If so, increased catfish production can be achieved through stocking of a highly vulnerable prey species such as the fathead minnow. In water too cold for satisfactory largemouth bass populations, species of trout may be stocked. Production from both catfish and trout ponds can be substantially increased by supplemental feeding of commercial rations, which increase growth rates and allow greater densities of fish than can be maintained on natural foods alone.

12.5   HABITAT MANIPULATION

Alteration of fish habitats may serve a number of management objectives. Habitat manipulation may be effectively used to enhance fishing, without actually increasing populations or altering community structure. In some situations, habitat manipulation may be required to sustain fish life, or at least certain species of interest. This has been particularly necessary due to specific spawning requirements of many species which are not met in artificial impoundments or may no longer be met in lakes due to environmental degradation. Certain alterations are undertaken to provide competitive advantage of desirable species, others to increase total carrying capacity. Carrying capacity, the long-term maximum biomass (of fishes) which a habitat can support, is dependent in large part upon the food resources available. The amount of these food resources is, in turn, dependent upon the available nutrients and the efficiency at which they are utilized and incorporated into the fish food chain. Successful use of habitat manipulation is dependent upon correct diagnosis of the environmental factors which are limiting production of desirable fish species.

12.5.1   Fertilization

Nutrient limitations of fish production in impoundments occur because of natural infertility of waters, usually due to soil characteristics of the watershed, and incorporation of nutrients into ecological pathways which do not substantially benefit fish. Competition for nutrients between aquatic macrophytes (usually vascular plants) and phytoplankton continually occurs. Desirable fish populations seem to result where a balance between phytoplankton and macrophyte production is reached, resulting in moderate production of both phytoplankton and macrophytes.

In small bodies of water with nutrient limitations, addition of fertilizers has been utilized to increase phytoplankton production and total standing crops of fishes. Although organic fertilizers such as hay and manure have been applied, inorganic agricultural fertilizers are now most commonly used. Fertilizer formulations most widely used have a 4:4:1 ratio of nitrate, phosphate, and potassium, e.g., 8:8:2 and 20:20:5. Local variations in soil type may dictate marked deviations from these formulations, and in old ponds fertilization with only phosphates may be sufficient.

Fertilizers should be added to ponds starting early in the growing season when phytoplankton production normally begins its spring pulse. Maximal impacts, sometimes up to a tripling of fish biomass, can be attained by periodic fertilization at two to six week intervals throughout the growing season. However, increases of up to 50 percent have been obtained with a single application made early in the growing season.

Use of fertilizers to increase standing crops of fishes has met with variable success. In areas where natural fertility or use of agricultural fertilizers in watersheds is high, addition of fertilizers contributes little. Addition of lime (calcium carbonate) to soft (calcium deficient) waters may be necessary before effects of fertilization can be expected. Where waters are turbid, light limitations may preclude uptake of nutrients. Addition of nutrients too late in the spring, following the phytoplankton pulse and coincident with initiation of macrophyte growth, likely will lead to increased production of macrophytes, rather than phytoplankton. Effective timing of fertilization will accentuate and prolong phytoplankton production, resulting in a shading of submergent macrophytes and minimization of competition by macrophytes for nutrients. Excessive fertilization can, however, lead to excessive phytoplankton 'blooms,' which during periods of cloudy weather can cause such high respiratory demand that oxygen depletions occur and 'summerkill' of fishes occurs.

12.5.2   Aquatic macrophyte control

Aquatic macrophytes are an essential part of most fish communities. This vegetation provides shelter for young fish, substrate for food organisms, and concentrates fish for harvest. Nevertheless, excessive vegetation interferes with fishing activities, upsets predator-prey relationships by providing too much cover, causes water quality problems during growth and decomposition, and becomes aesthetically unpleasing (Fig. 12.4). Vegetation control may be undertaken using biological, chemical, and mechanical methods.

Fig. 12.4. Excessive vegetation is undesirable and may be controlled by biological, chemical, and mechanical methods. Photo courtesy of Texas Agricultural Extension Service.

Biological control methods have usually involved use of herbivorous organisms such as crayfish, white amur (grass carp) and some cichlids of the genus *Tilapia*. Both white amur and tilapia have been successful in controlling certain submergent macrophytes, at least in small bodies of water. Nevertheless, ecological interrelationships with endemic species and potential impacts on non-target habitats are poorly understood. Consequently, these exotic fishes are banned in some parts of the U.S. Fishes such as the common carp and black bullhead provide some biological control by disturbing littoral (shallow, shoreline) areas where vegetation is rooted, and simultaneously increasing turbidity, thereby causing shading of submergent macrophytes. Control of macrophytes with shading provided from fertilization programs also should be considered as biological control. Some research has explored use of organisms pathogenic to certain macrophytes, but with little success.

Chemical control methods—the use of herbicides—provide economical control of vegetation with greater flexibility than biological control methods, but effects are only temporary and continual control must be used. Treatment can be limited to selected areas of shoreline where vegetation is interfering with fishing or boat access. Sectional treatments can be conducted so that water quality problems related to excessive decomposition can be avoided. Because of increasingly stringent governmental control of chemicals and their use, combined with the high costs of conducting research necessary to prove their safety, few effective chemicals are available for aquatic macrophyte control.

Mechanical control methods for aquatic vegetation have been varied. Cutting techniques, sometimes combined with harvest or removal, and mechanical shading techniques have largely proven ineffective. Water level manipulation has been used to control submergent macrophytes, either by raising water levels so that existing vegetation becomes light-limited, or by lowering water levels so that the plants become exposed to drying. Like chemicals, mechanical methods provide only temporary control of aquatic macrophytes.

12.5.3   Water level manipulation

Besides its use in vegetation control, water level manipulation can be used to favor desirable species, both through control of spawning and effects on predation. Because reservoirs are typically constructed with features for water level control, they are particularly conducive to this method of fisheries management (Fig. 12.5). Water level manipulation is considered by some to be potentially the best management technique for reservoir fisheries. Unfortunately, water uses of higher priority limit use of water level manipulation to its full potential in fisheries management.

Differing spawning times and depths of fish species makes it possible to drop water levels, leaving eggs of undesirable species either exposed or isolated. This technique is particularly applicable to shallow water spawners like the common carp. Conversely, if spawning habitat for a desirable species is insufficient, raising of water levels may provide additional areas favorable to spawning. For example, the highly desired northern pike enters marshy areas for its spawning. Such habitat may be increased by raising water levels. Marginal marshes have sometimes even been equipped with water level control structures to maintain the marshes for an extended period after spring water levels in the lake decline.

Fall drawdown forces prey species from cover provided by aquatic vegetation and makes them more susceptible to predators. By concentrating the prey fishes, fall growth of predatory game fishes is enhanced, while simultaneously reducing prey species which are overabundant. For fall drawdown to be effective, water

(a)

(b)

(c)

(d)

Fig. 12.5. Drawdown of reservoirs concentrates prey fishes for utilization by predators and allows vegetation to establish on mud flats, which upon reflooding helps control turbidity. (a) Full reservoir before drawdown; (b) Drawdown, with original shoreline indicated by dashed line and drawdown by solid line; (c) Vegetation established; (d) Refilled reservoir.

levels must be lowered to the point that surface area approaches only about half of normal size.

12.5.4   Turbidity control

Excess turbidity, due to suspended clay particles, has numerous adverse effects on most recreational fishes. Direct adverse effects include interference with feeding of sight-feeding fishes and smothering of developing embryos by sedimentation. White crappies, most ictalurid catfish, and sauger, however, maintain productive populations under turbid conditions. High turbidity has the indirect effect of limiting light penetration such that photosynthesis by phytoplankton occurs only in the top several centimeters of the water column. Consequently, much of the available nutrients is not incorporated into the food chain. Low light penetration also limits the amount of aquatic submergent vegetation, resulting in little favorable habitat for small fishes which inhabit the littoral zone.

Turbidity can be reduced by providing abundant organic matter which, upon decomposition, causes the clay particles to precipitate. Green or dry organic matter may be added directly to a turbid pond, or plant growths can be established on exposed bottoms either before initial filling of reservoirs or in conjunction with drawdown.

Chemical methods, particularly addition of gypsum or alum, will give more immediate results, but water clarity is not likely to be maintained unless fertilization follows immediately. Such fertilization should produce a phyto-plankton bloom which will help produce further turbidity control through increased organic decomposition. Effects of these control methods are most apt to be long-lasting if increased water clarity allows aquatic macrophytes to establish. These macrophytes reduce shoreline production of turbidity by providing protection from wave action. Decomposition of annual macrophyte production additionally contributes to alleviation of turbidity problems.

Often turbidity control can be facilitated by good watershed management (Fig. 12.6). Shorelines should be well grassed and livestock disturbances minimized by limiting their access to only a portion of the water's edge. If agricultural practices are carried on in the watershed, grass waterways and a grass buffer zone between cultivated fields and the pond or reservoir and its feeder creeks should be maintained.

12.5.5   Artificial spawning structures

Artificial spawning structures may be necessary because of deterioration of natural spawning grounds in natural lakes or their tributary streams, or because artificial impoundments lack the conditions for spawning of introduced species. Frequently, amount of suitable substrate for spawning is insufficient. Addition of gravel along the shoreline at proper depths can provide spawning require-

Fig. 12.6. Watershed practices. Fencing pond banks to limit livestock access (a) will prevent trampling of shoreline vegetation and minimize turbidity and erosion (b). An uncultivated strip around the shoreline of a pond (c) will minimize erosion and siltation characteristic of ponds which have tilled land to the pond bank (d).

ments for many game species. Rip-rap areas of dams have inadvertently provided spawning areas for fish such as walleye in some southern reservoirs where rocky shoals are lacking. Artificial spawning channels have been constructed to meet stringent spawning requirements for trout where limited suitable spawning areas exist between the fall line and the lake they inhabit.

In small ponds, it may be necessary to provide structures which meet the specific spawning needs of game or prey fishes. Fathead minnows, commonly used as a prey species in small ponds, require a flat underside of a rock, shingle, or other material. Channel catfish usually need to be provided a barrel or similar container in which to spawn. Numbers of spawning structures provided can be used to exercise control over the amount of reproduction, thereby preventing overpopulation from occurring.

12.5.6   Provision of cover

This broad category of habitat manipulation includes techniques applicable to all types of lentic habitats. These techniques can be applied at the time of construction or in established lakes, ponds, or reservoirs. In contrast to the successes of increasing wildlife populations through provision of cover, fisheries management utilizes artificial cover principally to concentrate fish for harvest. Direct effects on fish populations are limited to protection of small fish from predation in systems fairly devoid of vegetative cover.

The principal methods of providing cover at the time of construction of artificial impoundments are to leave standing timber and brush in coves and to leave deep gulleys, creek channels, old roads, and bridges whenever possible. The technique applied to existing waters is the addition of artificial reefs. These structures include tire reefs sometimes assembled according to sophisticated designs, and brush or discarded Christmas tree reefs which are submerged and anchored. Some limitations upon the types of cover which can be used are imposed by the extent to which they interfere with other uses of the water. Standing timber must not interfere with navigation, and debris which interferes with turbines and spillways must be minimized.

Although provision of reefs in moderate amounts may effectively concentrate fish for harvest, reefs in excess have much the same effect as submerged vegetation in excess. That is, prey fishes may become overabundant and game fish will be so widely scattered that they are difficult to locate. Few data currently exist on optimal size of artificial reefs. For some species, small artificial reefs which provide cover for only several game fish are adequate since new individuals inhabit the reef almost immediately after any fish are removed by harvest.

12.5.7   Aeration

Adequate oxygen levels—above 5 mg/l for coldwater fishes, above 3 mg/l for warmwater fishes—are usually maintained in lentic waters by photosynthetic activity and exhange at the air-water interface. When opportunity for oxygenation by these processes is limited and the oxygen demands for respiration and decomposition are high enough, oxygen levels fall below tolerable limits for fish. Artificial aeration devices have been developed to overcome these conditions.

When respiration in water below ice exceeds photosynthetic production of oxygen, winterkill results. It typically occurs in shallow ponds with large amounts of organic matter when snow and ice conditions prevent adequate penetration of sunlight. Continued circulation by aeration devices such as pumps or compressed air lines prevents ice from forming over the entirety of the lakes, thereby facilitating aeration at the air-water interface and simultaneously eliminating barriers to the incidence of light.

Artificial destratification of waters during warm weather is similarly achieved. In waters of the southern U.S., thermal stratification may form as early as mid-April and extend into late October. In most waters, the hypolimnion will rapidly become anoxic. In small ponds, the thermocline may be as shallow as 1.5 meters, thereby excluding much of the pond's volume and bottom area from utilization by fish. Nutrients are trapped within the hypolimnion, additionally detracting from pond productivity. However, artificial destratification may cause increased clay turbidity which may offset the effect of increased nutrient availability.

A highly-specialized form of aeration has been applied to ponds in which the hypolimnion remains cold enough for trout, but oxygen levels become marginal. By pumping water directly from the hypolimnion, aerating it with minimal change in temperature, and pumping it directly back into the hypolimnion, aeration can be accomplished without destroying the stratification. If destratification were to occur, temperatures would become too high throughout the pond to allow trout to survive.

12.6   REGULATIONS

In inland recreational fisheries, regulations serve many functions. Regulations seldom are effective in protecting species from overharvest in public lakes and reservoirs, but they do help attain other management objectives. Closed areas, closed seasons, and creel limits serve principally to divide the resource among fishermen rather than protect the resource. Size limits may be more necessary to insure adequate number of large fish for the creel than to allow adequate numbers of fish to reach spawning size. Spring closed seasons in ice-fishing areas are an effective means of encouraging fishermen to stay off dangerous melting ice. Creel and size limits create psychological goals for the recreational fisherman. Regulations of commercial fisheries may be aimed at minimizing conflicts with recreational fishermen.

Use of regulations in management of public waters for recreational fishing has followed some definite trends. In the first half of the twentieth century, regulations became increasingly stringent. Size limits, creel limits, and closed seasons were in widespread use. At about mid-century, it became evident that these regulations, which had been promulgated with the intent of protecting the resource, were not necessary. On the contrary, they were judged to be imposing unnecessary restrictions on fishermen, and consequently widespread liberalization of regulations occurred. Recent years have produced a growing concern that regulations are perhaps overly liberal. This concern has grown as a result of increased understanding of exploited freshwater systems, increased fishing pressure, and greater emphasis on qualitative aspects of catch, particularly size of fish, in recreational fisheries.

Currently size limits and creel limits are broadly applied to public waters in an attempt to minimize the chance of overharvest. These techniques are most effective on bodies of water with low productivity, heavy fishing pressure, or widely fluctuating population levels. Although size limits have traditionally been set as minima, there has been some recent emphasis on use of a split size limit (protected size range) to promote production of large fish. This has proven effective for largemouth bass, which will produce populations with a higher proportion of large fish if harvest is limited to both very large and very small fish. Mid-size bass are released so they can grow to the large size.

A highly specialized form of creel limit is the 'no-kill' or 'catch-and-release' regulation. When fishing pressure is very high, sustained catch can be maintained only by returning fish to the water to be caught again (or by put-and-take stocking). Complementary regulations requiring the use of barbless, single-point hooks minimize injury to fish when hooked. Catch-and-release programs, when applied to appropriate situations, have resulted in annual catches far in excess of standing crop of the target species.

Small ponds are quite susceptible to overharvest of game species, consequently control of harvest can be a major management technique. Self-imposed regulation of harvest in small privately owned ponds can be instrumental in sustaining a balanced fish community and enjoyable fishing experiences. Total annual harvest (quota) can be set so that overfishing or underfishing does not occur for any species. Size limits can easily be varied annually depending upon the condition of the fish stocks and the fishing priorities of the pond owner.

12.7   DESIGN CRITERIA FOR ARTIFICIAL IMPOUNDMENTS

Major impacts on management can be achieved through proper design and construction. Some of these have already been discussed, particularly in relation to habitat manipulation. In large impoundments, structural modifications to enhance fisheries management opportunities are limited to outlet design, although methods of basin modification (or lack thereof) may be important. In small reservoirs and ponds, it is likely that impoundment design can be modified to enhance both fish populations and fishing—particularly access.

In large impoundments, outlet features can be designed so that hypolimnetic waters rather than productive surface waters are released. An additional downstream fisheries benefit often accrues below warmwater reservoirs due to the creation of conditions favorable for trout survival as a result of release of cool hypolimnetic water. These tailrace fisheries for trout are often created in areas far from the nearest available trout fishery. Hypolimnetic discharge may limit the opportunity for two-story fisheries, however, by reducing or even eliminating the hypolimnion by the end of the growing season.

In construction of ponds and small reservoirs, greater flexibility for design exists. Ponds should be built large enough—or small enough—to encourage the type of fish populations desired. Generally small ponds cannot be effectively managed for a predator-prey combination such as bass-bluegill. Pond design should also include provision for drawdown and draining, so that opportunities for reclamation and water level manipulation remain open (Fig. 12.7). Particular attention should be given to overflow provisions so that invasion of undesirable species is prevented from below. A vertical standpipe (principal spillway) to allow release of moderate inflows and a wide overflow area (emergency spillway) beside the dam are major assets. Preferably the emergency spillway should lead to a concrete drop-structure to prevent any upstream migration. Modifications to prevent downstream migration during overflows are generally unnecessary, since downstream emigration is not likely to be significant except for over-populated ponds.

Design criteria should also include provision of access (Fig. 12.8). For large, multipurpose reservoirs, adequate numbers of boat launching areas are needed. Planning of public parks for areas with moderately-sloping shoreline will substantially increase participation by shore fishermen. Access needs of handicapped persons can be incorporated into design of fishing piers and retaining walls along tailraces. In ponds and small reservoirs with gently sloping shorelines, earthen or rock jetties may be built at irregular intervals to provide shoreline access to deep waters.

Top view

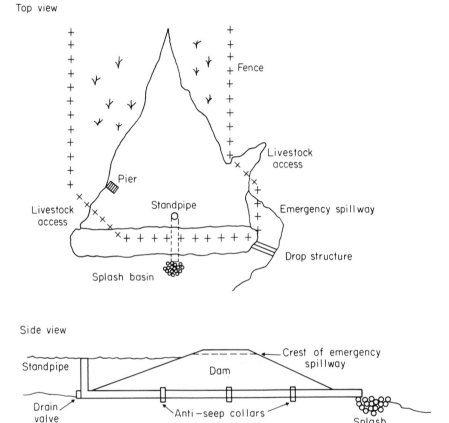

Side view

Fig. 12.7. Top and side views of farm pond demonstrating principal features conducive to fisheries management.

Fig. 12.8. Additional fisheries benefits can be attained through provision of adequate access. (a) Pond-bank retaining walls; (b) Fishing piers; (c) Boat launch site; (d) Tailrace guardrails. Photo credits: a and c: Texas Agricultural Extension Service; d: Texas Parks & Wildlife Dept.

12.8   GLOBAL PERSPECTIVES

Nowhere in the world is recreational use of fisheries resources of the high priority that it is in North America. Although recreational fishing, particularly in relation to tourism, is increasing in many parts of the world, the inland standing waters are generally expected to provide fish protein for human consumption through commercial fishing. Consequently, management is often directed toward fishes of lower trophic status than the predatory game fishes which support most recreational fisheries. In reality, management of the commercial stocks frequently discriminates against predatory fishes which detract from production of maximum fish biomass. At its greatest intensity, commercial fisheries management approaches aquaculture, where predatory fish are usually excluded entirely.

Although commercial, rather than recreational, utilization of fisheries resources prevails on a world wide basis, management needs and management techniques differ little from those appropriate to recreational fisheries management. Population manipulation, habitat manipulation, and regulations are typically employed in combination to provide effective management.

Commercial fishing is often intensive enough that stock depletion and shifts in fish community structure occur. Similar effects have occurred as a result of water quality changes, sometimes natural but often related to human activities. Such changes, for example, have been well documented for many European and Asian lakes and reservoirs where cyprinids have replaced percids as the predominant species.

Population manipulation has employed conventional supplementary stocking where natural reproduction is limited, introduction of exotics, particularly cyprinids and cichlids, and reclamation procedures. Reclamation of small lakes in the Soviet Union, accompanied by fertilization and restocking with simple combinations, has produced markedly improved fish yields.

Habitat manipulation frequently is limited to water quality control, particularly for large bodies of water. Moderate eutrophication is frequently beneficial to total fish production, particularly in oligotrophic waters, and consequently may be desirable to enhance commercial yields. However, excessive eutrophication often associated with domestic and industrial water discharge and agricultural runoff diminishes the available habitat for various life processes of fish and consequently reduces commercial fish production. In ponds and small lakes and reservoirs, fertilization may be employed to economically increase production. Various species have been widely introduced to provide control of aquatic macrophytes and algae, but these are valued additionally for their yields to commercial fisheries. Other habitat manipulation techniques are equally applicable to commercial fisheries management in these waters, but are not broadly employed.

Use of regulations in commercial fisheries management is widespread, but their applicability to specific situations is largely untested. Regulations are generally aimed at preventing depletion and protecting spawning stocks and juvenile fishes, rather than toward providing maximum yields. A combination of gear (mesh size) restrictions and catch limits are needed to achieve the objective of maximum yield, but such are seldom employed.

The opportunities for increased production from commercial fisheries are vast. It has been estimated that the 46 major lakes of the world are capable of producing twice the current yield simply through more efficient utilization of their existing stocks. Small bodies of water which are conducive to population and habitat manipulation have even greater potential for increased yields. Pond fish culture, utilizing intensive management, offers even greater promise for meeting the protein needs of the world's population of humans (Chapter 16).

12.9   CHALLENGE FOR THE FUTURE

Although fisheries management has been conducted on inland waters for nearly a century, the need for effective management is rapidly increasing. Some of the premises for effective management are the same as those facing all fisheries: increased fishing pressures, organized fisherman groups, demands of other users of the aquatic resources, and emphasis on accountability and cost-effectiveness. Nevertheless, some of the needs for management are unique or more specifically evident for inland fisheries.

Recreational fishing pressure is increasing rapidly. In the United States, numbers of recreational fishermen have been increasing at a rate of about three percent annually. Construction of new reservoirs has helped meet this growing demand, but similar increases in reservoirs are unlikely in the future. Consequently, more intensive management of existing waters will be necessitated.

Organizations of fishermen have proliferated in recent years. National and international fisherman clubs are reinforced by active chapters at the local level. Lake associations commonly have been formed out of concern for fisheries and water quality for recreation. These organizations greatly assist in the dissemination of management information to the fishing public, but also insist on intensive and well-justified management. In many cases they have directly supported research and management efforts; in others they have become involved as political activists to stimulate implementation of management programs. The growing influence of fisherman organizations dictates that fisheries managers not only carry out programs effectively, but carry on adequate public relations programs so that management objectives and techniques are well understood.

Composition of the fishing public is constantly changing. Until recently, most fishermen were from rural areas. Now over two-thirds of U.S. fishermen are from metropolitan areas. Urban fisheries—high intensity fisheries, often emphasizing non-predatory fishes—have become increasingly developed to meet the needs of these fishermen. Fuel shortages and escalation of travel costs are likely to result in further pressures for intensive management of urban fisheries and those in surrounding areas. Competitive fishing, usually expressed through fishing tournaments, has increased dramatically in the last decade. Studies have indicated that some reservoirs of the southern U.S. support several hundred tournaments per year. Currently, tournament fishing usually accounts for a small percentage of the total annual harvest, and many tournaments require careful handling and release of most fish. As tournament fishing increases, fisheries managers will be increasingly faced with integrating the needs of tournament fishermen into overall management programs.

The role of commercial fishing in inland waters may change markedly. Although it is unlikely that recreational fisheries will be replaced by commercial fisheries in the forseeable future, greater demands and higher prices for non-game fishes may stimulate interest in development of commercial exploitation. The fisheries manager must be prepared to evaluate the impacts, both direct and indirect, on game fishes, and to develop management strategies conducive to both recreational and commercial fisheries.

Relationships with other agencies comprise a fundamental part of management, particularly relative to reservoir management. Pre-construction considerations such as impacts on agricultural lands, historical sites and stream habitat, and flow regimes must be carefully coordinated. Possible impacts upon endangered species and archeological artifacts must be determined in accordance with public law. Downstream use, both present and future, will have significant effect on decisions for reservoir development. Reservoir development and operation even has effects on fisheries as far downstream as the oceans, where freshwater inflows into estuaries limit their productivity as nursery grounds.

Lakes, reservoirs, and ponds will continue to make a major contribution to fisheries, both recreational and commercial, throughout the world. A broad spectrum of management techniques is available, and the fisheries manager will be challenged to utilize these techniques to obtain maximum benefits from these limited aquatic resources. As more knowledge of aquatic environments, fish communities, and fisherman needs is obtained and integrated, substantially greater benefits can be expected than are currently being realized.

12.10 REFERENCES

Bennett G.W. (1971) *Management of lakes and ponds.* 2nd edn. New York: Van Nostrand Reinhold.

Benson N.S. (ed.) (1970) *A century of fisheries in North America.* Washington: American Fisheries Society.

Calhoun A. (1966) *Inland fisheries management.* Sacramento: Calif. Dept. of Fish and Game.

Colby P.J. (spec. ed.) (1977) Proceedings of the 1976 Percid International Symposium (PERCIS). *J. Fish. Res. Bd. Canada,* **34**(10).

Gerking S.D. (ed.) (1967) *The biological basis of freshwater fish production.* New York: John Wiley and Sons.

Hall G.E. (ed.) (1971) *Reservoir fisheries and limnology.* Washington: American Fisheries Society.

Lagler K.F. (1956) *Freshwater fishery biology.* Dubuque, Iowa: W.C.Brown.

Reid G.K. and R.D.Wood (1976) *Ecology of inland waters and estuaries.* New York: Van Nostrand Co.

Reservoir Committee, Southern Division, American Fisheries Society. (1967) *Reservoir fishery resources symposium.*

Stroud R.H. and H.Clepper (1975) *Black bass biology and management.* Washington: Sport Fishing Institute.

Chapter 13
Fisheries Management in Streams

EDWIN L. COOPER

13.1   SCOPE OF THIS CHAPTER

Before describing the fisheries resources of streams, it may be useful to outline the limits of this chapter, compared with other chapters in the text. For our purposes the major questions are: What is a stream?; and, What fisheries are to be included?

Ecologists and hydrologists prefer to consider a stream and its entire terrestrial watershed as a single ecosystem because of the many interactions between the land and the water draining from it. In this context, a river is considered only as a large stream. Natural swamps, ponds, river lakes, and man-made impoundments become intimate parts of this whole system and, in many ways, influence the biota and productivity of the stream. But, for pragmatic purposes it is convenient to consider separately *ponds, lakes, and reservoirs* in Chapter 12, *estuaries and coastal zones* in Chapter 14, and to reserve *stream* fisheries as the subject of this chapter.

We will concentrate on resident populations of fishes, or those populations for which the stream offers necessary habitat and support for a major portion of their life cycle. Chapter 14 will consider populations of anadromous forms for which the major part of the production occurs in large lakes or the ocean, but for which the stream offers only spawning or temporary rearing habitat.

13.2   USES OF STREAMS

Historically, man has used a river for many purposes. Indeed, the exploration, colonization, and industrial development of large land masses have been influenced by rivers which have often been used additionally as easily-defined political boundaries. Rivers have always been important for navigation, and when modified by dredging, locks and dams, and canals, remain as an efficient and inexpensive mass transport system. As the human population increases, the river provides other uses such as irrigation water and as water supply for domestic and industrial purposes. Waste disposal in rivers, utilizing their natural ability to dilute and oxidize unwanted materials, was a natural consequence of colonization of river valleys by man. In turn, engineers were tempted to transform free-flowing rivers to a series of impoundments with hydroelectric and multipurpose dams. Newer consumptive uses of a river such as cooling water for thermal-electric power plants or for nuclear-power generating stations extend the river's uses almost to the ultimate stage of development.

Against this background of man's increasing uses of a river for many purposes, fisheries have usually taken on a secondary role. Even though it is fashionable now to think that rivers should be managed so as to support a diverse natural biota, this Utopia does not exist, nor is it ever likely to when and where industrial man colonizes major river systems.

While this historical perspective of the value of river fisheries to man is somewhat pessimistic, it argues strongly for the idea that rivers should be managed in the future as ecosystems, and that all of man's needs and wants should be considered, even though many of these wants are aesthetic, personal, and difficult to evaluate in monetary terms.

13.3   STREAM HABITATS OF THE WORLD

Streams and rivers over the world differ as to the habitats which they offer to their biota. Patterns of rainfall, topography of the land, seasonal changes in temperature, and morphometry of the drainage basin all affect the character of the runoff. Vegetational responses to climate and soils of the basin may influence the quality of the water and help to determine the biotic productivity of the ecosystem.

Small stony streams in the headwaters of most of the major river systems are similar enough to warrant a generalization of their ecosystem. In contrast to large rivers, where the basis of production is detritus, the production of small stony streams is more often based on the diatoms of clean-water riffles. The benthic faunas of small streams are remarkably similar, with crustaceans and aquatic insects being the dominant forms. The fishes of stony streams have many secondary adaptations to living in fast water, but these specializations have evolved within many different major groupings of fishes. As a result, fish faunas in small stony streams are selected from the major groups available on zoogeographic grounds, but differ from one part of the world to another much more so than do the invertebrate faunas.

It is possible to recognize major differences in large river systems from different parts of the world, despite man's perturbations of these natural ecosystems. Temperate rivers like the Mississippi or Columbia draining North America have similarities to the Danube or the Volga draining Europe, but they are quite different from the great seasonal rivers in the tropics such as

Table 13.1. Seasonal changes in environment and fisheries of a tropical stream compared with a temperate stream.

Seasonal river in Tropical Zone (After Lowe-McConnell (1975) Fig. 4.2)	Item	Mississippi River in North Temperate Zone
High seasonal variation in amount; rainy season alternating with dry season	Precipitation	Moderate rainfall; nearly equal amounts spread over four seasons
Extreme annual flood, peaking early in dry season	Run-off	Steady run-off; may be minor flood condition at snow melt, or hurricane event; run-off moderated by dams on many tributaries
High temperature, low seasonal variation	Water temperature	Extreme variations in north, moderate variations in south
Amount depressed by high natural BOD coming from flood plain, concentrated in pools at low water levels	Dissolved oxygen	Amount depressed by heavy cultural BOD
Diel variation extreme at low water levels due to high temperature and high biotic activity		Diel variation greatest during summer due to high temperature and active photosynthesis
Most species move upstream to spawn in flooded areas; move downstream as flood plain dries up	Fish migrations	Minor in extent except for a few anadromous species. Timing based on water temperature
Uniformly high during rainy season, controlled by flooding and expansion of habitat	Fish reproduction	High in spring and early summer, controlled by temperature
High during peak flows, utilizing food resources produced on flood plain	Fish growth	High only for short periods in spring and summer at favorable temperatures
Intensive during migrations; fish very difficult to catch in flooded habitat	Fishing activity	Angling and commercial netting most effective during periods of good growth and favorable temperature; ice fishing good for some species

the Niger or Zambezi in Africa, the Mekong in Asia, or headwater streams of the Amazon in South America (Table 13.1). Temperate rivers are likely to have relatively stable hydrologic cycles compared with the annual rainy season-dry season cycles of large tropical floodplain rivers. Water temperatures in temperate rivers encourage high biological productivity for only short periods of the year, whereas in the tropics, regeneration of nutrients is rapid and biological productivity is high over much of the year. Against this background of seasonal changes, fishes have evolved spawning behavior, feeding and growing cycles, and migrations which are adaptive to these conditions. Man's exploitation of these fisheries resources also have evolved to take advantage of local conditions.

13.4   BIOLOGICAL ZONATION OF STREAMS

The classification of streams on a basis of dominant groups of fishes is a natural consequence of man's interest in harvesting fish populations. These schemes have been developed most extensively in Europe, and probably have the greatest use in characterizing temperate streams and rivers, where there are similarities in topography, soils, climate, and functionally-similar groups of fishes. They have not been very useful for describing the seasonal, flood-plain rivers of the tropics, but neither has man shown much interest in managing the fish populations of these tropical rivers.

Stream gradient along with water temperature are the dominant features delimiting the zones recognized as favoring certain groups of fishes. The increase in volume of stream flow downstream, and its hydrologic consequences of broadening the stream profile and decreasing the current, also changes the habitat and thereby changes the fish species best adapted to the habitat. Thus, from the eroding headwater areas to the sedimentary downstream areas of a typical stream several different zones are recognized (Figs. 13.1, 13.2).

The *trout zone* is usually located in small headwater streams maintaining cool temperatures and rapid current provided by the steep gradient. The water is well oxygenated at all times. The bottom is clean, with sand, gravel, and cobbles most often present. There is a lack of quiet water pools or backwaters which would support rooted vegetation.

The *grayling zone* generally is found in streams with rapid flow over riffles and moderating through deeper runs and pools. The water is usually well oxygenated throughout day and night, due to a lack of weed beds and fine organic sediments.

The *barbel zone* is found in larger rivers of moderate gradient and moderate current, riffles and rapids alternating with rather extensive pools. This permits greater warming of the water, and the establishment of weed beds along

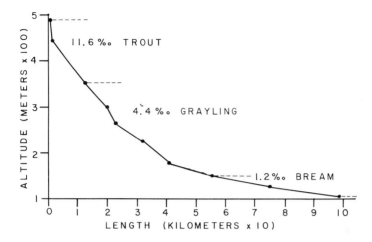

Fig. 13.1. Stream gradient and fish zones of the Lesse River in Belgium, which illustrate habitat selection by different fish species on a basis of water velocity. After Huet (1959).

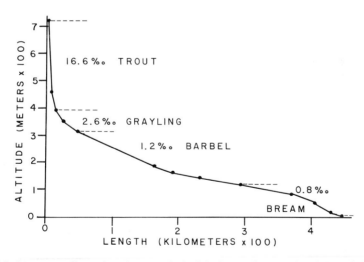

Fig. 13.2. Stream gradient of a sub-watershed of the Susquehanna River in North America with dominant European fish species expected in different zones. After Huet (1959). See Table 13.2 for dominant North American fish species found in these zones.

depositional meanders of the river. It provides some running water habitat in the riffles, but also some quiet water habitat in backwaters or eddies.

The *bream zone* consists of quiet waters of lower stretches of large rivers with only slight current. Very long, deep pools, open to the sun, alternate with shallow riffles, resulting in warmer temperatures and deposition of fine silts and organic matter. The amount of dissolved oxygen is usually less and may show high diel (24-hour) variation due to high photosynthetic activity of extensive weed beds.

These zones are characterized by different communities of fishes with a dominant species sometimes apparent. A comparison of the fauna of the River Lesse in Belgium with that in the Susquehanna River in North America (Table 13.2) shows some similarities of fish groupings. Perhaps because the number of fish species available for colonization in the Susquehanna River is considerably greater than that of the River Lesse, it is more difficult to single out dominant species in the Susquehanna.

13.5   FISHES UTILIZED BY MAN

Stream-inhabiting fishes have been reported from more than 90 families of fishes over the world and this number is probably not complete. Faunal lists from major river drainages are also extensive and diverse; more than 1300 species of fish have been described from the Amazon, about 560 from the Congo. In tropical Asia there are about 550 species compared with 250 in the Mississippi drainage of North America, and only 192 species in the whole of Europe. Not all of these constitute important fisheries, but selection and management of individual fish species seems to bear little relation to the number of species available.

A model of an ideal fish for exploitation and management of a fishery would have several characteristics. It would be good to eat with few nuisance bones, have high commercial value for an industrial use, or provide unusual opportunity for sport fishing. It should be intimately adapted to the habitat in question, and should be able to compete favorably with less utilized species. It should be an herbivore, and quickly grow to large sizes to maximize production from the primary productivity of the habitat. It should have a good public image in the sense of historical value, should not violate religious or cultural taboos, and should not be involved in the transmission of human diseases, parasites or toxic substances. Under modern management practices, it is advantageous, but not necessary, that this model fish be indigenous to the area.

From these criteria, it should be obvious that there are few ideal fishes for management, but several groups rate rather high. As more favored species become rare or extinct, the availability of another fish species increases its

Table 13.2. Fishes of recognized zones of the river Lesse in Belgium, compared with fishes in comparable sections of the Susquehanna River in North America.

	Fish Zones			
	Trout	Grayling	Barbel	Bream
		River Lesse		
Dominant fish	Brown trout	Grayling	Barbel	Bream
				Carp
				Tench
Other species	Sculpin	Barbel	Roach	Roach
	Minnow	Chub	Pike	Pike
		Hotu	Perch	Perch
			Eel	Eel
		Susquehanna River		
Dominant fish	Brook trout	White sucker	Fallfish	Spotfin shiner
		Stoneroller	Bluntnose minnow	Brown bullhead
		Tesselated darter	Hog sucker	
Other species	Sculpin	Creek chub	Rock bass	Walleye
	Blacknose dace	Cutlips minnow	Smallmouth bass	Northern redhorse
		Margined madtom	Common shiner	White perch
		River chub		

acceptability. As an example, it is reported that in the modern Thames River 'coarse fish' cyprinids are much prized by anglers. It is doubtful that they would be if it were possible to restore runs of Atlantic salmon in this river system to the levels of more than 100 years ago. Management thus becomes a dynamic art of fulfilling the wants of man within the capability of the environment to supply these wants.

In general, it can be surmised that many of the most productive species of fishes available in an area are, or can be, utilized in food or sport fisheries. On such an assumption, the following comparisons are made between broad geographic areas. The presence of different groups of fishes over the world is largely determined by zoogeographic features of their evolution and dispersal, combined to a small extent with man's introduction of exotics.

13.5.1   Eurasia and North America (Fig. 13.3)

Because these land masses were interconnected during early periods of differentiation of the major groups of fishes in the Tertiary Era (50 million years ago), the freshwater faunas of northern Europe and Asia show many similarities to that of North America. Cyprinids, esocids, and percids are abundant on both continents and are extensively used by man. Centrarchids, catostomids,

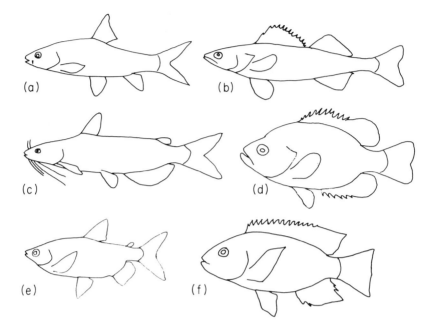

Fig. 13.3. Important fish groups in Eurasia (a) cyprinid, (b) percid; North America (c) ictalurid, (d) centrarchid; and South America (e) characin, (f) cichlid; which illustrate diversity in body form and general morphology.

and ictalurids are largely restricted to North America. These include several species important in management.

13.5.2   Central and South America (Fig. 13.3)
The freshwater fish fauna of Central and South America differs a great deal from that of North America, but has close affinities to that of Africa. Again, these similarities infer a land connection between the two continents for some time after the major differentiation of fish groups in the Tertiary Era took place. The characins and the cichlids (ecologically and morphologically similar to the cyprinids and the centrarchids, respectively) are dominant groups in South America with the greatest number of species found anywhere in the world. There are also numerous families of catfishes which include more than 1000 species in South America. Fisheries management in rivers of this area is negligible, but the potential for production is great.

13.5.3   Africa (Fig. 13.4)
Africa has many families of fishes in common with South America such as characins and cichlids. It also has representatives of families from southeast

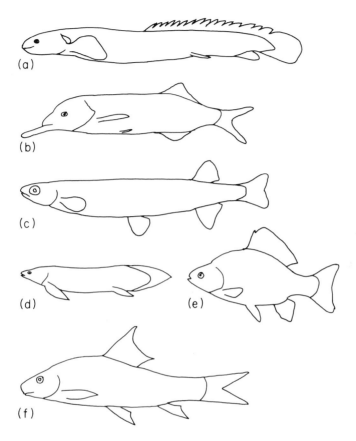

Fig. 13.4. Important fish groups in Africa (a) bichir, (b) mormyrid; Australia and New Zealand (c) galaxiid, (d) lungfish; and East Asia (e) goldfish, (f) cyprinid; which illustrate diversity in body form and general morphology.

Asia such as cyprinids, notopterids, plotosids, and anabantids. Families unique to Africa include bichirs, snoutfishes, and gymnarchids, among others. Salmonids and centrarchids have been introduced into some parts of Africa and are managed intensively for sport.

13.5.4 Australia and New Zealand (Fig. 13.4)
Stream dwelling fishes are almost absent from Australia except for the Australian lungfish and an osteoglossid with relatives in Africa and South America. The remaining species in rivers of Australia or New Zealand are either natural derivatives from marine families like the galaxiids, or are exotics acclimatized for sport fishing purposes like several salmonids. A high degree of management is afforded to these introduced species.

13.5.5 Eastern and Southern Asia (Fig. 13.4)

Cyprinids dominate the fish faunas of these two regions, with a few additional families closely related to African fishes. The culture and management of many cyprinids in Asia have been practiced for centuries, often in conjunction with other agricultural practices.

13.6 USES OF STREAM FISHES

Over the years, man has used the fishes of streams for human food or recreation, and for a variety of industrial uses such as animal food, fertilizer, and as a source of various raw materials such as oils. Man's uses of stream fishes have often changed as his local culture has progressed from primitive hunter through independent settlers to industrialized societies in which one individual is highly dependent on another for various needs and wants. Alternate choices of food and recreation compete for man's attention, frequently resulting in changes in his valuation of fisheries resources.

From such an historical perspective, we may sketch some of the changes which have occurred in the use of fisheries resources of the Mississippi River in North America. This is perhaps a model of the pattern of events that is taking place in different stream systems throughout the world, but is one in which large changes have occurred over a short time span.

The fisheries of the Mississippi River were important sources of food, ornaments, and trade goods to the prehistoric inhabitants of the area, judging by artifacts found in excavating ancient mounds in the valley. Mussel shells, freshwater pearls, otoliths of freshwater drum, and the bones of catfishes, turtles, geese, buffalo, and deer found in these mounds attest to their widespread use.

When western man colonized the area, the Indians were already adept at catching fish with nets of twisted willow bark or with hooks made from animal bones, and freezing or drying fish for future use. Catfish and the paddlefish were the abundant and choice species used by the Indians. The following quotation from Harriet Bell Carlander (1954, p.9) is illustrative of the importance of fish to these people. 'Many Indian legends and ceremonies developed around fish and fishing. Among the Sioux there was a feast called the raw fish feast. In this ceremony the medicine men performed a dance and then ate raw fish, entrails, scales, everything but the head and large bones. Tribes of the Great Lakes region thought that the knowledge of how to make nets for fishing was given to them by the gods.'

The early settlers quickly seized upon the fisheries resources as subsistence food, and a commercial fishery soon developed. This was usually restricted to local markets as fresh catfish, paddlefish, and buffalofishes because of the perishable nature of the catch. Sport fishing in the main river was fashionable there as early as the 1850s.

Deforestation, clearing and draining bottom lands, leveeing of river banks, dredging, channelization and lock-and-dam building on the river in the early 1800s started an unending controversy between conservation groups and land developers about their effects on the river fish populations. By 1870, yields had apparently declined over much of the nation to the extent that in 1871 a United States Commissioner of Fish and Fisheries was authorized, empowered by statute to investigate the fisheries, and to engage in fish cultural practices. Legislation by many states soon established state agencies for the same purposes, and propagation and stocking of many fishes, both indigenous and exotic, flourished.

As a natural corollary of the perceived necessity of restocking the depleted waters, another major management activity began on the river, that of fish rescue work. As the usual spring floods receded, pools and lakes along the stream were isolated from the main channel, stranding many fishes. Believing that these fish were needed in the river, or would be useful in other depleted inland waters, this rescue work was expanded and continued for the next 50 years. Crews seined many backwater pools and cut-off ox-bows and transferred millions of fish to more suitable habitats, either by small boat or truck, or later by specially equipped railroad cars which distributed fish far and wide. The work was described as very disagreeable and unhealthy, requiring men to work in a hot sun in mud holes, and sleeping in the river bottoms at night. In retrospect, it is probable that very little increase in production from the river fisheries resulted from these efforts, but it is difficult to evaluate what did happen.

The pearl button industry, centered in the upper Mississippi River around Muscatine, Iowa, had an interesting but brief history, beginning in 1889. This was a good example of a boom-and-bust exploitation of a natural resource, with depletion occurring many years before the plastics manufacturers made pearl buttons economically impractical. Several species of thick-shelled mussels were abundant over large shallow riffles of the river. These were harvested for their shells only, by ingenious crow-foot dredges with wire hooks. Mussels literally grabbed onto these wire hooks as they were dragged along the bottom. The mussels were cooked to extract the meat from the shells and the meat was discarded. From the shells, buttons were cut and polished in small local factories.

Shelling was so intense under no restrictive regulations that within 20 years depletion of the mussel beds provided part of the impetus for establishing the national Fairport Biological Station on the river, where methods for culturing mussels were soon developed. For a short time, restocking of depleted areas of the river, combined with closure of shelling, brought about an increase in the populations of mussels over wide areas. However, by 1930, pollution of various kinds combined with dredging channels for barge navigation spelled the eventual end of the pearl button industry in the area.

By the early 1900s increases in human population and the rapid industrialization of the area soon transformed major tributaries like the Ohio River almost into sewers which were used also as barge canals. The low point for aquatic environmental health in this large watershed was probably reached in the 1940s when the necessity of producing war materials for World War II took precedence over efforts of conservation groups to clean up the river.

Throughout this population boom and industrial growth of the area some sport and commercial fisheries survived. Following World War II various groups, such as the Upper Mississippi River Conservation Committee and the Ohio River Valley Water Sanitation Commission were organized to advise governmental agencies on multiple uses of the watershed. With continued public support, some progress has been made in restoring river fisheries to historic high levels.

From this short history of the Mississippi River fisheries there are lessons to be learned about the potential for managing stream fisheries. Firstly, developing societies probably consider non-fisheries resources of a stream to be more valuable than fisheries resources. Secondly, individual stream fisheries are at different stages in their historical development, depending upon the stage of society's demand on other resources of the watershed. And, thirdly, if and when the major needs and wants of a society are met, values of the fisheries to society increase, and support for management is gradually made available.

13.7   MANAGEMENT OF FISHERIES IN LARGE STREAMS

13.7.1   Water quality

The management of a river to control its water quality may embrace one or more practices such as soil and water management on the watershed, treatment of domestic and industrial wastes, temperature and water level control by planned storage and release from reservoirs, and many others. In most cases managers use a mixture of these practices, sometimes handled by different regulatory bodies with different goals. It is no wonder that systems management of rivers is often advocated, as now administered by river basin commissions or similar public agencies.

In a world-wide sense, attempts to maintain water quality in streams in their pristine state, or even of condition capable of supporting a diverse community of aquatic organisms, has been a losing battle. And, this trend is destined to continue so long as the consumptive or degrading uses of a stream, such as irrigation or sewage disposal, are considered to be more important than commercial or recreational fishing which have very little detrimental effect on stream quality. In only a few places in the world have we turned the corner and reversed this natural process of degradation.

The many avenues available for managing streams may be better illustrated by selecting a few streams of different characteristics, and relating some of the important features of their management history. In some cases, fish populations have been partially sacrificed in favor of other uses; in others fisheries and water-based recreation have been recognized as values well worth saving.

13.7.1(a) The Nile River of Africa has been managed for more than 5000 years for man's benefit. It is a còmplex watershed of natural lakes, tropical swamps, and connecting streams, flowing north to empty into the Mediterranean Sea. Its annual flood-and-drought hydrology has been studied for centuries, and man's greatest aim in this area has been to control the devastation of the floods and to use flood water for irrigation. During the present century, the river has been brought under control by the construction of many dams. With the final completion of the Aswan High Dam in 1971, this river has the distinction of becoming the first to be completely controlled and utilized by man. For several years no water has entered the Mediterranean Sea from the river, and, only in the event of an exceptional flood, is it likely that Nile water will ever enter the sea again.

The Nile fisheries resources are very productive for two reasons. High temperatures favor rapid turnover rates of metabolic processes, and many of the harvested fish species are herbivores whose food chains are therefore more efficient than predatory life cycles would be. Most of the production occurs in the natural lakes and dam basins, but many of the fish species must necessarily return to the rivers to spawn, and are frequently harvested during this migration. Such a life cycle offers one example of the impractical idea of distinguishing between stream fisheries and reservoir or lake fisheries.

Although fishing is important in the Nile basin, the value of the water for irrigation is much higher. Also, over most of the Nile watershed, where industrialization has scarcely begun, chemical pollution is not a problem. And, because domestic sewage is considered an important source of fertilizer for irrigated agriculture, there are only a few local problems of eutrophication in the Nile basin.

The loss of nutrient-rich, and sediment-rich water of the Nile River to the Mediterranean Sea has resulted in many ecological changes there, some of which are detrimental to fish populations. A productive sardine fishery at the mouth of the Nile has completely collapsed, and salt water is invading many delta lands formerly suitable for irrigated agriculture.

Thus the complete control and utilization of Nile River flood water will stand as a monument to man's ability to change the face of the earth. It is too early to say whether or not this control will be of ultimate net benefit to man.

13.7.1(b)   The Monongahela River joins the Allegheny River at Pittsburgh, Pennsylvania, to make up the Ohio River which then flows for 1579 km to join the Mississippi River at Cairo, Illinois. The 'Mighty Mon' as it is known locally, begins in wilderness areas in the Appalachian mountains of West Virginia and flows north gradually passing through a large coal-mining and steel-producing complex to reach Pittsburgh.

In the early 1900s the stream was transformed from an unpolluted, soft-water mountain tributary into a highly-polluted 'sulfur water' resulting from coal-mine drainage, with almost a complete loss of fishes and other acid-sensitive biota from the stream. This polluted condition was accepted by most people as an inevitable consequence of industrial progress and it was difficult to generate public support for treatment of sewage and industrial wastes. At the height of industrial progress coincident with World War II, the aquatic habitat was scarcely fit for man or beast. Domestic water supplies for the people in these industrial communities came from closely-guarded small tributaries out of reach of coal mining activities. And, due to the mountainous nature and sterile soils of the area, water for irrigation was never considered to be important.

After World War II public sentiment changed dramatically in favor of cleaning up the environment, and massive programs of neutralization of acid-drainage, and treatment of industrial wastes were started. Possibly as a result of these programs, and perhaps also due to a natural lessening of acid flow from abandoned coal mines, the river has made a remarkable recovery within the past 10 years. The water quality has improved to the point of supporting at least 45 species of fishes in the river near Pittsburgh, and fishing and recreational boating have greatly increased. Sensitive fishes like the smallmouth buffalo and the sauger, last seen in the river in the 1870s, have recently invaded the area, presumably from downstream refuges. This example points out that, contrary to pessimistic predictions of the unwillingness of the American public to finance environmental improvements at the expense of industrial progress, it is possible in some instances to make some headway in this direction.

13.7.1(c)   The River Thames in southern England is a good case history of learning to cope with cultural pollution in a stream (see Chapter 10). A low gradient stream flowing through London to the Thames Estuary, this river has been modified and managed by man with canals, locks and dams, and fish weirs for more than 1000 years. Attempts to reduce the sewage pollution, naturally accompanying human settlement, were recorded here as early as the year 1664. By the late 1800s an official governmental agency known as The Conservators of the River Thames was empowered to control pollution in the whole watershed, and in 1965 their authority was extended to prevent discharge of polluting substances into any underground strata within the catchment area. With increasingly strict effluent limitations, the water quality of the River Thames is improving.

The River Thames was once an important spawning and nursery stream for Atlantic salmon and brown trout, and supported an important salmonid fishery. It is now a highly-eutrophic water supporting a very productive fish community of roach, bleak, gudgeon, perch, and dace. Apparently, the modern angler on the Thames does not consider these 'coarse fish' as undesirable, and closed seasons and size limits which regulate their catch are designed to maximize the numbers and sizes prized by anglers. Fish competitions are frequently held on many parts of the river. Apparently, the predatory pike is regarded by anglers as an unwanted species in the community, on the belief that the pike is unnecessarily decreasing the potential supply of roach and bleak.

As described by Mann (1972, p.217), 'The general picture is of a river basin with a population of 3.5 million people, receiving all the effluent which this implies, yet supporting more angling and pleasure boating than any other in the United Kingdom. Many people swim in the river, although there are those who hesitate to do so, and it is more than 100 years since a salmon was caught in the Thames, although they were once plentiful.'

13.7.2   Culture and exploitation

A large stream is difficult to manage, both because of its size, current, and variation in discharge, and also due to the large number of species in the fish community. Examples of management of large streams therefore come mostly from case histories of intensive water control or pollution abatement such as described earlier. A few instances of management can also be cited for large rivers in the tropics where methods of fish culture or harvest are adapted to the prevailing conditions. The following three examples are considered as management in a very primitive sense.

Many large rivers in India and Burma undergo a seasonal flooding, at which time a diverse community of fishes become active, move upstream and spawn. Many of these species have respiratory adaptations for a semi-amphibious existence, or like lungfishes, aestivate in mud-cocoons during dry periods. As the flood recedes, water is diverted into rice fields for irrigation. Fish fry, spawned in the river, are naturally diverted into the rice fields, or are captured and stocked into grow-out ponds. Almost all species are then grown, harvested, and utilized for food. Some of the most prized species are the 'air breathers' which can be transported alive for long distances to fresh-fish markets.

Rivers are sometimes used as natural production facilities in another way. The River Tjibunut near Bandung in Indonesia is heavily enriched by raw sewage. Fingerling carps of several species are placed in small bamboo cages in the river and feed naturally on the rich culture of algae and invertebrates supported by the sewage. Growth is rapid and production is high under these conditions of high temperature, rich food sources, and short, efficient food chains.

Another ingenious way of increasing fish yields from large rivers was noted in Dahomey of central Africa. Fish parks (brush shelters, artificial reefs, or *acadjas*) are established along the shores which attract fishes in search of food or refuge. The brushy branches or growing vegetation increase the surface area for the production of periphyton, also increasing the potential fish yield from these areas.

13.8   MANAGEMENT OF FISHERIES IN SMALL STREAMS

13.8.1   Small coldwater trout streams

One of the best examples of managing fisheries in small streams is that of sport fishing for trout. There is an abundance of information available on life history of salmonids within their native range, or as exotics on all continents.

Because of its long historical development, the management of coldwater trout streams takes many forms, from maintaining or improving the habitat, through manipulation of the fish stocks, to regulating the yield by anglers by appropriate regulations or customs.

13.8.1(a)   Habitat improvement

Interest in streams and their improvement for trout and salmon was evident in early British literature, but was usually practiced in secret by game keepers on large private estates. With the settlement of eastern North America by English colonialists it was only natural that these management practices would be brought along as part of their culture. The same thing has happened in New Zealand, in South Africa, and elsewhere where the introduction of brown trout and stream management followed English colonization.

In North America, fisheries resources quickly became public resources managed by state and federal agencies to a greater extent than elsewhere in the world. Thus, although stream improvement devices were developed and installed in streams on private estates, like the low-head 'Hewitt Dams' in New York, stream improvement received a great impetus in the 1930s with the development of state programs on public waters in Michigan and Wisconsin.

The scope of trout stream improvement as now practiced covers all aspects of the environment that impinge on different stages of the life cycle of the fishes, and logically begins with a stream survey which analyzes factors which limit, or are detrimental to, the production of trout. As a consequence, the emphasis on management changes with differences between geographic locality. In Wisconsin, it is considered important to stabilize stream bank erosion by fencing

(a)

(b)

(c)

Fig. 13.5. Habitat improvement in trout streams in Wisconsin. (a) A trout stream damaged by ditching and cattle grazing. (b) Cattle-crossing for a trout stream graded and covered with crushed stone or gravel, providing access to water but controlling damage to stream banks and siltation of water. (c) Early stages of improvement of a trout stream, including bank stabilization and current deflectors.

cattle and installing proper in-stream devices (Fig. 13.5). In other areas, practices that minimize scouring by floods, and that encourage ground-water recharge might be the practices that produce best results (Fig. 13.6). Stream improvement programs often contain financial incentives for proper soil-and-water management on the watershed some distance above the permanent water courses as a practical means of reducing excessive erosion and sedimentation within the stream itself.

Several manuals are available on constructing in-stream devices that aim to maintain conditions described as natural. It is assumed that natural conditions provide for the proper mix of cover, nursery areas, spawning areas, and clean substrates for production of trout food. Where streams have been degraded by building and abandoning of beaver dams, or by ditching and straightening by man's bulldozers, the restoration of the stream to its original state is likely to be difficult. But evaluations of this improvement in Michigan and Wisconsin have demonstrated a fair degree of success (Table 13.3).

Table 13.3. Effects of habitat alteration on the brook trout population in Lawrence Creek, Wisconsin. Adapted from Hunt (1969) p.299.

	Percentage change	
	Improved section	Control section
Biomass	+ 40	+ 11
Food consumption	+ 28	− 1
Annual production	+ 17	− 9
Fishing pressure	+ 188	− 28
Yield	+ 196	+ 32

Natural conditions may have inherent limitations that sometimes can be remedied. One such example is the improvements made to increase spawning success and survival of young trout. Salmonids all utilize clean gravel in which eggs are deposited and incubated for long periods of time. In very cold climates, proper temperatures for incubation are assured only by an abundant source of ground water that percolates up through the redds. Perhaps in response to their evolution in these cold climates, brook trout carefully select stream sites where upwelling ground water is favorable to incubation (Benson 1953). Brown trout and other fall-spawning salmonids which have evolved in warmer climates do not show this predilection for upwelling spring water to the same degree as do brook trout.

Proper aeration is provided both by the original site selection in the stream and by the gravel-cleaning activity of the female in constructing the redd. Excessive siltation of the gravel during incubation is responsible for large losses

(a)

(b)

Fig. 13.6. (a) Abundant ground water inflow correlated with ice-free conditions at air temperature of $-35°F$; good brook trout reproduction here. (b) Insufficient ground water inflow to keep stream ice-free; poor brook trout reproduction here.

of young salmonids, especially where watersheds erode rapidly as a result of road building or logging. Another situation of low spawning success occurs as a result of a complex biological process. In limestone-rich streams which are excessively fertilized by run-off from agricultural lands, or enriched by sewage effluent, the proper spawning and aeration of the redds may be prevented by marl (calcium carbonate) formation. Rooted aquatic plants and periphyton, stimulated by abundant nutrients, deposit marl by active photosynthesis from bicarbonate-rich water. This marl encrusts the rocks and gravel, especially in riffles where photosynthesis is high, completely cementing them together. Only the very large female trout is able to break up this concrete in digging a redd; experiments have shown that she quickly will find and utilize a small artificial gravel area installed for spawning purposes.

Trout managers have used this information on spawning behavior in developing artificial spawning areas. As early as 1908, English river keepers reported that 'Where there are no natural spawning grounds in the runners and tributaries they can easily be made. The best way to do this is to make a long cut from 4 to 6 feet wide, parallel to the stream and cover the bottom with a foot or 18 inches of gravel about the size of a pea. Water can be drawn from the stream in such a way that it is always under control, and the outlet from this artificial bed must be allowed to enter the stream lower down. At spawning time a good current should be turned over any new spawning ground and the fish will find their way onto it. There are many beds of this kind working successfully, up and down the country' (Armitage 1908).

13.8.1(b) Stocking

The introduction of salmonids outside their native ranges coincides well with the establishment of trout hatcheries, for it was found that eyed-eggs were hardy and could be iced and shipped long distances during their extended incubation period. Soon after 1850, trout hatcheries were in operation at least in Europe, North America, and New Zealand. Various individuals, governmental agencies, and the unique Acclimatization Societies of New Zealand soon changed their original purpose of introducing new species to that of maintaining trout populations. Techniques of culture and stocking changed with further evaluation of results, and followed a historical time sequence of stocking eyed-eggs, then newly-hatched fry, then fingerlings, to the present emphasis on put-and-take stocking of catchable-sized fish. It is safe to say that more emphasis has been placed on management of salmonid fishes than any other group. This has been due to their widespread acceptability as both commercial and recreational species, and their ready acclimation to temperate environments throughout the world.

Although there are continued efforts to introduce exotic trout into new habitats, most stocking is now directed toward two goals: (1) maintaining existing populations at levels consistent with the stream's carrying capacity to fully utilize natural production of fish food; and (2) increasing the short-term recreational opportunity of anglers by releasing catchable-sized fish. Often, both goals are considered to be part of the same stocking program.

The first goal assumes that managers have considerable knowledge of the dynamics of the existing populations and the limitations of the habitat in producing all life history stages of the fish from naturally-spawned fry to adults. Corrective stocking can be very successful in increasing populations when this information is carefully evaluated. Thus, contrary to accepted practices, in some southern Wisconsin streams, where natural spawning was deficient, but food production was high, a program of stocking fingerlings was successful in producing more adult fishes for anglers. The lesson to be learned here is that efficient and successful management is dependent upon intensive monitoring of the habitat and the fish population. Simplistic programs designed for large geographical areas may be easier to administer, but they are likely to sacrifice important local opportunities for improvement.

The second goal of stocking to increase short-term recreational opportunity is easily met by most managers. It is easy to generate public or political support for a new fish hatchery, often at the expense of other programs which produce less visible results. Allocation of fish to be stocked, where and when they should be planted, and other socio-political aspects of stocking are day-to-day problems that managers solve adequately by involving the user in the planning process. So long as the habitat is safe for the trout for short periods, the only constraints are those determined by the consumer in the way of financial support, or the allocation of species, numbers, and sizes of the trout to be stocked.

Planting catchable trout almost always can be counted on to generate increased fishing pressure, which in turn generates support for increased stocking, the *fishing pressure-stocking* syndrome (Fig. 13.7). Although anglers frequently voice desires for solitude in their fishing experience, one only occasionally hears a demand that some streams be set aside for a *no-stocking policy* which would help to maintain low fishing pressure on these streams.

13.8.1(c) Control of predators
It is not necessary to convince anyone who has operated a fish hatchery that kingfishers, herons, mergansers, or otter can decimate a raceway full of trout in a short time. What about predators in natural habitats? Do they compete with anglers for the trout produced? Only a few studies have addressed this question adequately to offer even a tentative conclusion.

In New Zealand two species of native predatory eels are present in most streams in which brown or rainbow trout have been established. In several sections of a test stream, almost complete removal of the eel population by electrofishing was followed by increases in the yearling brown trout population

Fig. 13.7. Typical angler response to stocking large numbers of catchable-sized trout. Lack of sufficient access to trout streams makes it difficult to provide elbow room. Photo courtesy of the Pennsylvania Fish Commission.

by 3 to 10 times. However, this large increase in yearling trout was accompanied by a decrease in their growth rate, and also a severe reduction in survival of the two year classes of trout immediately following. Control of the predatory eels here was not considered to be entirely beneficial (Burnet 1968).

In severe climates of North America, the better trout streams remain open during the long winters, while lakes and streams with little ground-water inflow are completely ice covered. Mergansers and other fish predators concentrate on these open waters and annually take more trout than fishermen do, particularly the larger trout (Alexander 1977). For these unique situations, it is thought that predator control would make more trout available to the anglers, but at the expense of the duck population. As usual, fisheries managers are faced with a difficult compromise decision.

13.8.1(d) Control of competitors

In ponds and lakes, removing all native fishes with toxicants and restocking with trout is a well accepted practice in North America. For streams, there have been very few instances where this procedure has been shown to be effective in permanently changing the fish communities. Impassable falls, or man-made barriers may sometimes be used to control the reinvasion of the unwanted species.

Small stream sections in the southern Appalachians of North America have been reclaimed for the native brook trout, or for introduced rainbow trout, with some success. It has seldom met with success in large rivers. The eradication of all the native fishes in the much larger Green River, located in three western states, was done in 1962 in conjunction with the filling of the Flaming Gorge Reservoir in which an introduced rainbow trout population was to be established. While this large-scale management scheme was hailed as a success by federal and state fisheries management agencies, it was criticized by others for endangering small populations of at least three native species which are considered rare. It is not likely that large projects of this kind will be done in the future without an extensive evaluation of all the environmental impacts on the ecosystem.

13.8.1(e)   Fishing regulations

In dealing with fishing regulations on a world-wide basis, it must be remembered that man's use of tools and techniques reflect the development of his culture. For example, fly-fishing only for trout in North America would probably be considered an absurdity to Amazonian natives who are adept at using poisons extracted from native plants for fish-harvesting social occasions. The development of restrictive regulations must therefore consider not only the scientific merit of the regulation but also the interference with accepted social customs. Fishing codes in North America bristle with unnecessary taboos which serve no useful purpose except to alienate fishermen, or to make enforcement easier. They often appear to impose a parochial code of conduct upon society as a whole.

Fisheries managers must rely on two bases of data to act responsibly in setting regulations. First, they should be knowledgable about the aquatic ecosystem in question, and should be able to predict the response of the fish species to various changes in population parameters, or various schemes of exploitation by man. If fortunate, the manager will have access to experiment stations where ideas can be tested for confirmation of predictions. Second, managers should know something about the fishing clientele being served. In the broadest sense, this should include all potential users of the resource, not just the meat-hungry angler.

One of the most difficult tasks for the trout manager is to determine what anglers want in their fishing experience. Occasionally, attempts are made to sample public opinion on important issues, and in some places organized groups of anglers regularly inform the management agency of their desires through formal petitions. But more often than not, the rule makers rely on feed-back response from the public to correct or modify an unwanted regulation once it has been proposed. So long as angler's wishes on regulations do not seriously jeopardize the viability of the resource it would seem prudent to accede to these wishes and to accommodate as wide a variety of angling experiences as possible in the management program. Successful managers are likely to adopt the 'something unique for everyone' philosophy in dealing with an unknown, unpredictable, and changeable fishing clientele.

Management agencies sometimes involve the public in workshops, hearings, and opinion surveys to help them determine what measures are deemed socially acceptable. When these are coupled with scientific predictions of the effects of a proposed regulation, and perhaps testing in an experimental program, it is often possible to arrive at compromises that are mutually acceptable to both the public and the professional fisheries scientist.

Some examples of regulations can be cited as good or bad, but with little hope of widespread applicability. The Pisgah System (U.S. National Forest in southern Appalachia) of 'each day an opening day' attempted to scatter fishing pressure in time and space by restricting the number of anglers and angling days on each body of water within the constraints of natural productivity, stocking schedules and social pressures (Chamberlain 1943). Anglers generally do not agree with this much interference with their fishing opportunity, and such a program would be impractical for large geographic areas with many different streams. In other places, no closed seasons are in effect with considerable public support of such a regulation. Various intermediate open-season programs are in effect, often justified on a basis of local custom, interference with hunting seasons, and many other social factors. In general, closed-season restrictions are being relaxed to the limit presumably imposed by biological considerations of the fish species.

Various combinations of size-limits, creel-limits and lure-restrictions are debated long and vociferously by concerned anglers and fisheries managers. It is obvious that there are many options open to managers here, and great possibility for the manager to 'fine-tune' his program to the life history variability of different species, cultural differences in anglers, and different goals for different angling groups. These range all the way from fishing rodeos and lotteries in canvas tanks at fairs, to the fish-for-fun, no-kill programs desired by the contemplative angler. It is difficult to judge these different programs as scientifically right or wrong—they are simply not comparable. There is a lesson to be learned here; adaptability is a trait that is just as applicable to the successful fisheries manager as it is to the successful evolution of a wild trout.

13.8.2 Small warmwater bass streams

There are very few instances of intensive management of small warmwater streams for selected species. These attempts have been greatly complicated by the diverse fish communities usually present. Courtois Creek in the Ozark Mountains of Missouri is such an example; although the stream supports at least 64 species of fishes including many minnows, suckers, and catfishes, the dominant species in this stream are the smallmouth bass, rock bass, and longear sunfish (Funk and Pfleiger 1975). Such a fish community offers a complex challenge to the fisheries manager attempting to improve fishing.

Tests of experimental fishing regulations in these streams have indicated that exploitation may be too high to permit maximum yield of the preferred smallmouth bass; higher size limits, and a no-kill fish for fun regulation resulted in an increase in biomass of adult bass and a greater predatory pressure on competing species. But too little information is available on the population dynamics of these mixed populations to be able to devise the set of fishing regulations which would result in optimum yield from this community of fishes.

13.9   FUTURE OF MANAGEMENT IN STREAMS

The potential for managing an indigenous fish population obviously depends on the value attached to particular uses. Fish can be *consumed* by making them into fish flour for humans, meal for animal food, anchovy paste for the gourmet, fish fillets for fast-food chains, or mounted trophies for recreation rooms. They can also be used non-consumptively by culturing and enjoying them as art objects, or by catching and releasing them as a means of displaying athletic skill or socio-cultural values. In any case, the goals of management will largely determine the potential magnitude of the yield, be it meat, sport, or aesthetics.

There are inherent characteristics of fish species that also constrain potential yield, such as growth rate, position in the food chain, and acceptability by the public. Management will succeed to the extent that these constraints can be modified, and the desired biological productivity can be realized. It is sometimes more productive to select fish species which have favorable growth rates and efficient food chains, like many salmonids, hoping to convince the public of their value, than it is to try to manage a slowly-growing top predator just because it is popular. Here again the manager may be given an unenviable task of maximizing production of a very rare species, like the muskellunge, because of the high recreational value attached to its rarity. In general, it appears that most streams of the world are harboring large populations of fishes of the wrong species and the wrong sizes to suit public desires. This is especially true as we move from human cultures depending on fisheries resources for food to more highly developed societies where fish are prized more as art objects or means of recreation.

Given some goal such as producing the greatest amount of a particular species, what avenues are open to the manager to accomplish this? Within the past 100 years of fisheries science, only a few major techniques have been developed. Among these, culture and stocking, control of predators, and various techniques of reducing competition have shown the most promise. Selective breeding of superior strains of fishes is still an unfulfilled goal for most fisheries managers. There are also hopes expressed for increasing natural production of fish by matching the physiological characteristics of existing strains of fishes to individual habitats, but the results here are also not very promising.

13.10   REFERENCES

Alexander Gaylord R. (1977) Food of vertebrate predators on trout waters in north central lower Michigan. *Pap. Michigan Acad. Sci. Arts. Ltrs. Michigan Academician, n.s.* 10(2): 181–195.

Armitage, Wilson H. (1908) *Trout Waters, Management and Angling.* London: Adam and Charles Black.

Benson Norman G. (1953) The importance of ground water to trout populations in the Pigeon River, Michigan. *Trans. N. Amer. Wildl. Conf.* **18**, 269–281.

Burnet A.M.R. (1968) A study of the relationships between brown trout and eels in a New Zealand stream. *N.Z. Marine Fish. Techn. Rep.* **26**, 49p.

Carlander Harriet B. (1954) *A History of Fish and Fishing in the Upper Mississippi River.* Upper Mississippi River Conservation Committee.

Chamberlain Thomas K. (1943) Research in stream management in the Pisgah National Forest. *Trans. Amer. Fish. Soc.* **72**, 150–176.

Cooper Edwin L. (1974) Trout in streams. *In* A.R.Grove (ed.) *Trout* 15(1): 4–11.

Funk J.L. and W.L.Pfleiger (1975) Courtois Creek, a smallmouth bass stream in the Missouri Ozarks. *In* H.Clepper (ed.), *Black Bass Biology and Management,* pp. 224–230. Washington: Sport Fishing Institute.

Hubbs Carl L., John R. Greeley and Clarence M. Tarzwell (1932) Methods for the improvement of Michigan trout streams. *Bull. Inst. Fish. Res. Michigan,* **1**, 1–54.

Huet Marcel (1959) Profiles and biology of western European streams as related to fish management. *Trans. Amer. Fish. Soc.* **88**, 155–163.

Hunt Robert L. (1969) Effects of habitat alteration on production, standing crops and yield of brook trout in Lawrence Creek, Wisconsin, pp. 281–312. *In Symposium on salmon and trout in streams,* H.R.MacMillan Lectures in Fisheries, 1968. Univ. British Columbia, Vancouver, Canada.

Hynes H.B.N. (1972) *The ecology of running waters.* Ontario, Canada: Univ. Toronto Press.

Lowe-McConnell R.H. (1975) *Fish Communities in Tropical Freshwaters; Their Distribution, Ecology and Evolution.* New York: Longman.

Mann K.H. (1972) Case history: The River Thames. *In* R.T.Oglesby, C.A.Carlson, and J.A.McCann (eds.) *River Ecology and Man,* pp.215–232. New York: Academic.

White Ray J. and Oscar M. Brynildson (1967) Guidelines for management of trout stream habitat in Wisconsin. *Wisconsin Dept. Nat. Resources, Techn. Bull.* **39**, 64 p.

Whitton B.A. (ed.) (1975) *River ecology. Studies in Ecology,* vol. 2: Berkeley and Los Angeles: Univ. California Press.

Chapter 14
Coastal Fisheries

J. L. McHUGH

14.1 INTRODUCTION

The world catch of all fishes, shellfishes, and other products in 1976 was about 73.5 million metric tons. About 10.4 million metric tons were from fresh waters, about 63.1 million metric tons from the sea. It is not known exactly how much is taken at different distances from the coast, but in 1976 off the United States about 1.7 million metric tons were taken within 3 miles, 0.7 million between 3 and 12 miles, and 0.4 million between 12 and 200 miles. In addition, considerable quantities were taken by foreign fishermen in the 12 to 200 mile zone. A rough estimate of the amount can be gained from the catch in the four FAO statistical areas surrounding the U.S., about 5.9 million metric tons in the two northern areas, and 3.1 million in the two southern areas. The southern areas include Central America and part of South America also, so the total is somewhere between 7 and 9 million metric tons off North America including Canada.

Coastal fisheries exploit a large variety of species. Altogether, at least 300 species figure in total landings in the United States, but only about a dozen groups make up over 80 percent by weight of landings (Table 14.1).

It obviously would not be feasible to discuss this variety of coastal fisheries in detail in a single chapter. We have chosen to discuss a few U.S. fisheries which are as representative as possible and which illustrate the principal problems. Menhaden (*Brevoortia tyrannus* and *B. patronus*) stands out as a prime candidate; it accounts for roughly 35 to 40 percent of United States landings and is important on the Atlantic coast and the Gulf of Mexico. Sardines (*Sardinops sagax*) and anchovies (*Engraulis mordax*) are its counterparts on the Pacific coast. These species and other clupeioids make up about 20 to 24 percent of total worldwide marine landings. Second are shrimps (*Penaeus, Pandalus*, and some others), the most important shellfish, making up almost 10 percent of landings by weight in the United States but over 23 percent by value. Crabs (*Callinectes, Cancer*, and others) are third by weight in United States catches, tunas (*Thunnus, Euthynnus*) are fourth, and salmons (*Oncorhynchus* spp.) fifth. Together, these five kinds make up nearly 73 percent of the United States total catch by weight. Tunas are largely offshore resources and are treated in Chapter

323

Table 14.1. Species or group of species making up the first 12 groups in United States landings by weight, 1977. Weights in thousands of pounds.

Species	Weight landed in 1977
Clupeioids	2,236,956
Shrimps	476,654
Crabs	398,539
Tunas	368,474
Salmons	335,642
Flounders	187,291
Cods and cod-like fishes	168,071
Jack mackerel	110,246
Clams	96,160
Croakers and croaker-like fishes	57,557
Oysters	46,026
Ocean perch	40,723
Subtotal	4,522,339
Total all species	5,198,100

15. The oyster industry has special features worth treating here, even though it is eleventh in order of weight. Striped bass (*Morone saxatilis*) and bluefish (*Pomatomus saltatrix*) in the United States also are of special importance to recreational fishermen, and deserve special mention here for that reason.

14.2   THE FISHERY CONSERVATION AND MANAGEMENT ACT

In 1976 the United States made an important change in direction of fisheries policy by passing the Fishery Conservation and Management Act (FCMA) by which it took jurisdiction over nearly all resources out to 200 miles, with the exception of tunas, and extended its authority over salmon to wherever the resource migrates on the high seas. Canada and Mexico also at about the same time extended their fisheries jurisdiction to 200 miles. These actions have vastly changed the patterns of management of the resources around our coasts. The principal features of the United States law have been to allow foreign fishing only when it is clear that United States fishermen cannot take and sell the entire allowable catch, and to begin to bring all resources in the zone under effective management. This means for the first time that American fishermen must and will be managed.

The definition of management given by the Act is those measures that will produce the optimum yield, that is those that will provide the greatest overall benefit to the nation, which is prescribed on the basis of maximum sustainable yield from such fishery, as modified by any relevant economic, social, or ecological factor. Earlier chapters have described the multi-dimensional nature of management under this type of definition, and the great difficulty of making decisions which address the optimum yield from a fishery. The full impact of this policy has not yet been felt by the fishing industry, and the agencies which manage coastal fisheries are presently in a status of change and adjustment.

The U.S. FCMA is implemented by a series of Regional Management Councils which are such an integral part of the fisheries management scheme that they should be described in detail. Eight management councils exist (Fig. 14.1); the Mid-Atlantic Council will be used for illustration. It represents the six States from New York to Virginia inclusive, and is composed of 19 voting and 4 non-voting members. Voting members are the six heads of the agency

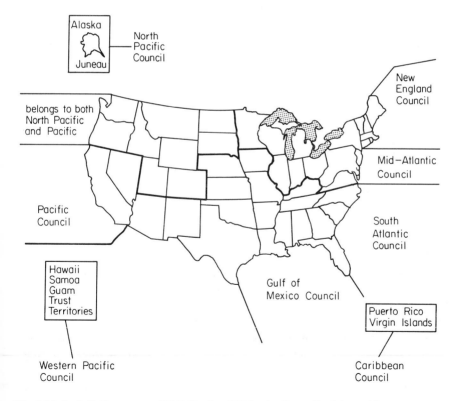

Fig. 14.1. Jurisdiction zones of U.S. Regional Fisheries Councils. Adapted from 'Establishing a 200-mile Fisheries Zone,' U.S. Government Printing Office (1977).

responsible for fisheries in each state, the Regional Director of the National Marine Fisheries Service for the area, six members chosen from lists submitted by the governor of each state, one from each state, and six members chosen at random for their particular expertise, also from lists submitted by the governors. The four non-voting members represent agencies with a particular interest, the Department of State, the Coast Guard, the Fish and Wildlife Service, and the Atlantic States Marine Fisheries Commission.

The primary task of the Councils is to prepare fisheries management plans for the fisheries in their area, and to coordinate these plans as necessary with neighboring Councils. These plans must conform with national standards laid out in the Act, must contain certain required provisions, and may contain other provisions such as license fees, limitations on fishing, and so on.

The plan must then be submitted to the Secretary of Commerce for approval. The Secretary must consult with certain other officials of the government, and must either return the plan for further work or approve it for implementation within 60 days of receipt. Public hearings must be held or opportunity must be granted to organizations or individuals to comment on the plan and regulations.

14.3   KINDS OF COASTAL RESOURCES

There is some merit in a broad classification of fisheries resources into four subcategories of food finfishes, industrial fisheries, shellfishes, and recreational fisheries. Each has a different history, probably because it goes to different users. Food finfishes are those used entirely or primarily as direct food for humans. Industrial fisheries are those which produce products for uses other than human food, such as oil and meal, bait, animal food, and so on. Shellfishes are primarily crustaceans and molluscs, but also turtles and others. Recreational fisheries are all fisheries that catch fishes for sport, although some are sold commercially.

14.3.1   Food finfisheries
This category has been declining for over thirty years in the United States. The peak was about 1945 with about 2.1 billion pounds landed (Fig. 14.2). Thereafter, landings dropped irregularly, but fairly steadily, to 1974 when they reached about 1.45 billion pounds, a drop of about 33 percent. Some stocks fell by several orders of magnitude during this period, others rose to partially compensate. At the same time the population of the country was rising steadily. There is no question that the supply per capita of food fishes was dropping substantially during this period. It is difficult to avoid the conclusion that, by and large, they were being overfished, taken in increasing quantities by recreational fishermen, or both.

Fig. 14.2. United States landings of industrial, food, and shell fish, 1929–1974.

This decline was mostly in the older fishing areas, the north Atlantic and north Pacific Oceans. Landings to the south, especially in the Gulf of Mexico, were rising. From 1951 to 1974, Gulf of Mexico landings as a whole rose from 695 million to almost 1.8 billion pounds, almost tripling the catch. Most of this increase was industrial fishes, but food finfishes rose over 50 percent. Thus, there has been a shift in the distribution of food fish production in the United States from the older to the newer fishing grounds where the stocks are still in good condition.

14.3.2   Industrial fisheries

Over the same period, from 1929 to 1974, industrial fisheries rose substantially, from roughly a billion pounds at the beginning of the period to about 2.5 billion at the end (Fig. 14.2). The location of the fishery has shifted also, first from the Pacific coast, where Pacific sardine was the principal species until about 1950; then to the Atlantic coast, where Atlantic menhaden was the major species until 1962; and finally to the Gulf of Mexico, where Gulf menhaden was the principal species. It is clear that the Pacific sardine was overharvested, from a combination of increasingly heavy fishing pressure, natural fluctuations in abundance, and failure to take drastic action until far too late. The symptoms were typical, a rising fishery for an increasingly important species, increased production as the declining stocks were stimulated to maximum biological productivity, a decline from natural causes, assisted or not by the declining stocks, and a rapid increase in effort to supply the reduction plants with fish. Overfishing could hardly have been avoided in such circumstances, especially since the operators successfully fought any serious attempts to manage until far too late. Indeed, it is not at all certain that management measures, e.g. quotas or other restrictions on total catch, would have worked. Although the fishery has been very severely restricted for more than 10 years, it has not recovered.

Essentially the same series of events has affected the Atlantic menhaden industry, although the resource has not suffered the same sharp decline in abundance yet. Catches rose to a peak in the decade 1953 to 1962, then dropped to about half peak production. The symptoms were similar, a decline in the range of the species from north to south, and a concentration of the fishery in the Chesapeake Bay area and to a lesser extent in North Carolina. The catch in Chesapeake Bay has held up remarkably well so far, in fact between 1973 and 1978 it was higher than in any other 5-year period. North of Chesapeake Bay the fishery is almost non-existent. Concentration of the fishery on young fish in the southern part of the range is not likely to stimulate production, but the picture is not all that clear. The likelihood is that further declines are inevitable, but that cannot be certain as yet. The menhaden industry as a whole has been equally resistant to controls.

The Gulf of Mexico menhaden industry is even newer, and the fishery is presently in its decade of highest production. The industry has been no more willing than it was on the west coast or in the Atlantic to be controlled by quota or other limits, and of course the situation is not as critical at present. It is interesting that all three fisheries have produced over a billion pounds of fish for about ten years, the sardine from 1934 to 1944, the Atlantic menhaden from 1953 to 1962, and the Gulf menhaden briefly in the early 1960s, then from 1969 to 1976. These fisheries are almost entirely conducted within 3 miles of the coast, hence are not managed by the Councils but by individual states. The migratory nature of the resource, especially north-south along the Pacific and Atlantic coasts, has not been conducive to easy action by the states.

An important species on the Pacific coast also is the Pacific northern anchovy, *Engraulis mordax*. It is as abundant or even more so than the sardine, but by the time its abundance attracted the attention of purse seiners a major controversy was underway over the rights of purse seiners to harvest resources indiscriminately. Sport fishermen asked, and got, strict limits on purse seining, and the catch has been strictly limited by quota to 300 million pounds or less, slightly exceeding that figure in 1975. It is too early to say what the outcome of this management program will be.

The danger of overfishing clupeioid stocks generally is emphasized by the fate of other clupeioid fisheries around the world. The history of the Peruvian anchovy (*Engraulis ringens*), Atlanto-Scandian herring (*Clupea harengus*), and Japanese herring (*Clupea pallasi*) have been equally disastrous. It is prudent, therefore, to be cautious in exploiting such populations. The only one which has been approached cautiously, of the several species discussed here, is the Pacific anchovy stock. This managed fishery will bear careful watching in future.

14.3.3 Shellfisheries

On the surface, it would appear that the shellfisheries are a bright spot among coastal fisheries. Landings have been increasing rather steadily since 1929 and now are about 3 times as large as they were (Fig. 14.2). Molluscs and crustaceans have also been increasing, on the whole, worldwide. Catches of oysters, in contrast, have been dropping steadily since the beginning; bay scallop reached a peak of 4.5 million pounds in 1931, soft clam peaked at 16.7 million pounds in 1937, and hard clam at 21.0 million pounds in 1950. All the other major species of invertebrates have reached peak landings since 1961, most in the 1970s. The invertebrate fisheries are different in that they are pursued for the most part with gears that catch little or nothing else, and thus are single-species fisheries. For this reason they should be easier to manage.

14.3.4 Recreational fisheries

Statistics on U.S. recreational fisheries are much less complete. Only three full surveys have been made, in 1960, 1965, and 1970, and they did not include estimates of any shellfish catches. A partial survey was made in 1974, which included some useful estimates of shellfish catches also, but that survey covered only the North and Mid-Atlantic regions. The recreational fisheries take mostly food fishes, and the catch of food fishes alone was somewhat greater in 1970 than the total commercial catch of food finfishes. Since some species are taken in very small amounts or not at all by the recreational fisheries, but in considerable quantities by commercial fishermen. it follows that some species are caught mainly by recreational fishermen. In 1970 over 24 times as many bluefish were taken by recreational fishermen as by commercial fishermen, 12.9 times as many cod (*Gadus morhua*), 11.6 times as many winter flounder (*Pseudopleuronectes americanus*), 6.8 times as many striped bass, 3.4 times as many summer flounder (*Paralichthys dentatus*), and 3.2 times as many weakfish (*Cynoscion regalis*).

If these figures are correct they mean that we cannot write meaningful fishery management plans for these fisheries because we do not know what relative or absolute proportion of the stocks are being taken. It is imperative, however, to know if such stocks are underfished or overfished in order to allocate the catch equitably between commercial and recreational fishermen and to work out a method of controlling recreational catches.

14.4   EXAMPLES OF MANAGEMENT

The best way to show how management has or has not worked is to give examples from coastal fisheries. These have been selected to illustrate differences between fisheries almost entirely within the three-mile limit and subject to state jurisdiction; fisheries mainly within the three mile to 200 mile limit, thus subject to Council jurisdiction; fisheries that extend from the coast to wherever they migrate on the open sea, thereby subject to state and Council jurisdiction; and fisheries primarily used for sport. Each has special problems that make management difficult.

14.4.1   Local jurisdiction
The clupeioids offer a good example of how local jurisdiction has failed to protect important fisheries. There is no question that clupeioid landings as a whole in the United States have declined substantially in the last 80 years or so, and this is true generally around the world. The principal reason is a lack of effective management measures until too late, and when management was finally tried the damage was already done. The story begins with the Pacific sardine on the west coast, continues with the Atlantic coast menhaden, meanwhile shifting also to the Gulf of Mexico, where another species of menhaden is fished, and ends with the recent exploitation of Pacific northern anchovy under somewhat more sophisticated methods (Fig. 14.3).

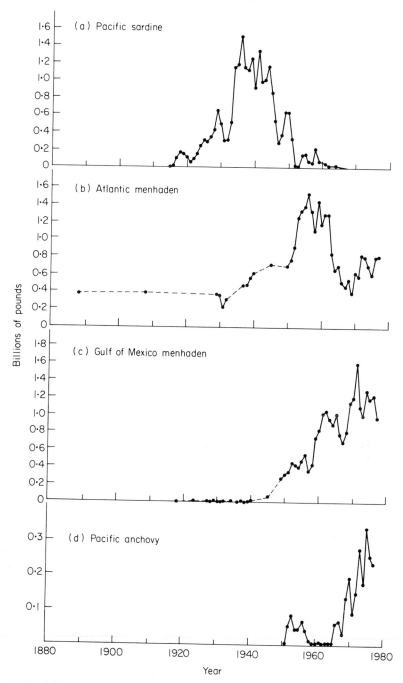

Fig. 14.3. Total U.S. landings of selected clupeioids, 1880–1977, based on Fishery Statistics of the United States.

The sardine fishery began in the late 1800s off California and became a thriving industry when World War I created large demands for canned sardines (Ahlstrom and Radovich 1970). About one-half of each sardine was wasted during canning, and reduction processes soon developed to convert the waste into fish meal and oil. As the reduction process became more profitable, increasing quantities of whole fish were used directly for reduction, a situation which caused legislative action to control the industry. Various California laws defined the quantities of fish which could be landed and what proportions could be reduced.

The fishing industry responded to these restrictions by creating floating reduction plants outside the 3-mile jurisdiction of the State of California. These plants operated during the 1930s, when catches of sardines peaked. In response to the depressed national economy of the 1930s, California laws were liberalized and reduction of sardines was permitted on shore. The fishery remained stable during the early 1940s, then declined to very low levels by the 1950s.

In the early 1950s the menhaden industry on the Atlantic coast began to increase, spurred on by the market for fish meal. The menhaden industry was much older than the sardine industry, having produced close to 400 million pounds as early as 1888, but being older, was less efficient, and could not compete with the developing sardine fishery until it began to collapse. In the 1950s the menhaden fishery began to rise, reached a peak in 1956, and also remained fairly high for about a decade. Then, like the sardine fishery, this fishery also began to drop off rapidly, declining to about 50 percent of the maximum level by the 1970s. The decline of the menhaden resource was not as complete, however, and from Chesapeake Bay south has remained remarkably productive. This may have been a matter of unusually favorable environmental conditions to the south, and it is difficult to believe that this high production can last. In Chesapeake Bay and to a lesser extent in North Carolina the fishery in recent years has been more productive than ever, but this has been achieved by catching smaller fish, and it is improbable that this can last.

In the early 1970s the menhaden industry in the Gulf of Mexico reached a peak, and also has remained high for almost a decade. This fishery shows no signs of declining yet, but it is prudent to wonder whether the yield can be maintained. In a recent paper on Atlantic and Gulf of Mexico menhaden fisheries Schaff (1975) presented convincing evidence that the Atlantic menhaden fishery is overfished and that Gulf menhaden might well be approaching that point. He showed that Atlantic menhaden probably would produce a maximum sustainable yield of 560,000 metric tons (on the average) with about 630 vessel weeks of effort. By the early 1960s this had clearly been exceeded, and the catch dropped to an average of 266,000 metric tons from 1963 to 1973. Effort was reduced in 1969, and by 1972 catches had risen to about 350,000 metric tons. In the Gulf of Mexico, the maximum sustainable yield was about 478,000 metric tons with about 460,000 vessel ton-weeks. Using these figures, he predicted that further effort would probably reduce the yield in both fisheries, and that the yield could be increased to at least 400,000 tons by reducing effort and by certain other measures in the Atlantic fishery.

Meanwhile, in California, as the sardine declined it has been well documented that a parallel increase has occurred in abundance of the Pacific northern anchovy. The sardine industry would have turned to this resource if it could. In fact, it did briefly in the 1950s but was soon limited by the sportfishing groups, which were determined that what had happened to the sardine should not be repeated for the anchovy. Canned anchovies were not in great demand and fishing for reduction was not allowed. In the late 1960s limited reduction was allowed, and in the 1970s the quotas were increased. Whether these increased quotas will be sufficient to maintain a viable industry, and whether the resource, regulated by quota, will be able to maintain itself, is too early to tell.

14.4.2 Extended jurisdiction

Stocks that extend beyond 3 miles from the coast, either in their entirety, or in part, are subject to control by the Councils in partnership with the Secretary of Commerce. Within the Mid-Atlantic region, the resource most urgently in need of management was the surf clam (McHugh 1978). The need for management was not brought about by foreign fishing, for the resource had been protected from the first by the provisions of the Continental Shelf Convention of 1958, which declared the surf clam a creature of the shelf, belonging exclusively to the coastal state, and foreign fishermen had respected this declaration. The damage was done entirely by the domestic fleet, which grew rather rapidly after the industry was established in the latter years of the second world war, and extended its operations from modest beginnings off Long Island in the mid-1940s as far south as the Virginia Capes by the early 1970s (Fig. 14.4).

When the Mid-Atlantic Council selected the surf clam as a candidate for its first fishery management plan, it took over a plan that had been developed by the Surf Clam Sub-Board of the State-Federal Fishery Management Board, modifying it somewhat before submitting it to the Secretary of Commerce. Essentially the plan placed a quota on the annual harvest of surf clams and ocean quahogs and a moratorium on the entry of new vessels into the fleet after a certain date.

Administration of the plan has been beset by problems since the beginning —an expected result as the many details of management have been worked out. For example, to keep a more or less even flow of raw materials to the factories ashore, it was decided to split the year into 4 quarters with separate quotas for each. Later, it was necessary to reduce the number of fishing days to two or even one day per week, to avoid reaching the quarterly quota too

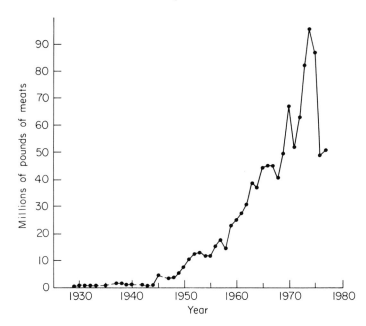

Fig. 14.4. U.S. landings of surf clam, 1929–1977, from Fishery Statistics of the United States.

soon. Recently, there have been requests for make-up days to compensate for days lost for bad weather. The intent has been to make it as equitable as possible for vessels of different sizes and capabilities, but that has been only partially successful. For example, some vessels can not operate as far away from home base as others, and some should stay closer to port in case of bad weather. The ocean quahog resource lies farther offshore than surf clams, so some vessels have a better opportunity to substitute ocean quahog when the surf clam quota is caught. These and other considerations have required close watch and almost constant revision of the plan, and all problems have not yet been resolved. Furthermore, despite all these restrictions, the surf clam has been harvested at a rate higher than the quotas, and a reduced quota may be required some time soon. As if these difficulties were not enough, substantial quantities of surf clams lie within 3 miles of the coast, hence do not come under the jurisdiction of the Council, and must be managed by the respective States. It cannot be said at the moment that the fishery is fully under regulation, or that the industry is satisfied so far. There are far more boats in the fishery than are necessary to take the allowable catch, and if economic return is a major objective of fisheries management then the vessels in the fleet must eventually be reduced in numbers substantially.

14.4.3   Salmon

Salmon are a special case under the Fishery Conservation and Management Act because those stocks that spawn in U.S. waters are protected wherever they happen to be on the high seas, even beyond 200 miles. The commercially valuable stocks are essentially Pacific salmon, and their value is considerable; the 336 million pounds landed in 1977 were worth $222 million to the fishermen. These fishes have not been notably protected by the domestic conservation regime, although there has been a multitude of regulations. The peak catch of all species was in 1936 at about 640 million pounds, by 1974 it was just over 200 million (Fig. 14.5). Average catches appear to have stabilized in the 1960s and 1970s, but annual catches are highly variable.

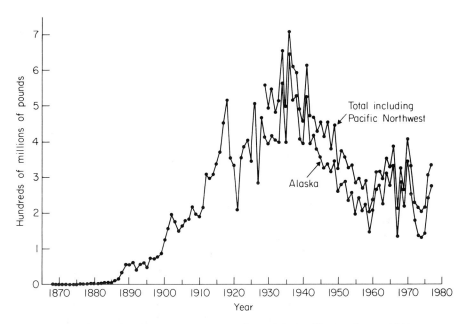

Fig. 14.5. Landings of Pacific salmon from Alaska and the Pacific Northwest, 1868 to 1977, based on Fishery Statistics of the United States.

The migratory habits of Pacific salmon, taking them from spawning grounds in coastal rivers across the entire North Pacific and back, complicate management (Larkin 1970). Early exploitation was concentrated in rivers during upstream spawning migrations, and traditional management has been oriented in that area. Limitations on harvest, including prescribed gear types and fishing times, and construction of fish passage devices around dams were designed to allow a desired number of adults to reach spawning grounds. Hatchery production augments natural production with the intention that increased numbers of young will cause increased spawning runs and harvest in the next generation. The only species which has clearly increased in abundance is silver or coho salmon in the Pacific northwest, as a result of a massive and enlightened program of hatchery production. There is some increase in Columbia River chinook salmon also. It appears that, despite large sums being spent on salmon rehabilitation, the best that can be done has been to prevent further declines in production.

Management has been complicated further by the development of high-seas fishing for salmon by Japanese in the North Pacific. Attempts to regulate this fishery resulted in a 3-way treaty between Japan, Canada, and the United States and the creation of the International North Pacific Fisheries Commission. That commission has been superseded by the FCMA, and it remains to be seen whether that Act will improve the situation by reducing Japanese catches of salmon on the high seas. In Alaska and the State of Washington programs of limited entry have been in effect for several years, to reduce effort in the fishery. Despite the controls, effort has not been significantly reduced yet, because the efficiency of the remaining fleet has increased in various ways. This program too has not been in effect long enough to produce detectable results in terms of increased catches.

14.4.4   Commercial-recreational conflicts

Conflicts between commercial and recreational fishermen are important features of management, whether the resources are inland or offshore. Two fisheries are particularly difficult in the United States, striped bass and bluefish. Although the statistics are far from satisfactory, they suggest that both species, but particularly bluefish, are taken in much greater quantity by sport fishermen.

Striped bass is a creature of coastal waters, and thus does not come under the FCMA. In the past striped bass management has been complicated by opinions, half-truths, and much misunderstanding of the issues, which have complicated management and made the issues more political than scientific. For example, one claim is that striped bass have been decreasing in abundance and thus need protection, whereas the facts show that the species has been increasing in abundance for nearly 50 years (Fig. 14.6). This increasing trend, however, has been marked by major fluctuations in abundance. At present the species is on a downswing in abundance which has lasted for some time, and there is no assurance that previous high levels of production can be regained. There is indirect assurance that striped bass has not been affected by fishing until recently, because it has always recovered from past declines, and in fact gone to higher levels of production. The critical issue is to find out what harvest striped bass can sustain and develop a program that will maintain that yield. The conservation issue is not a question whether the harvest is taken by

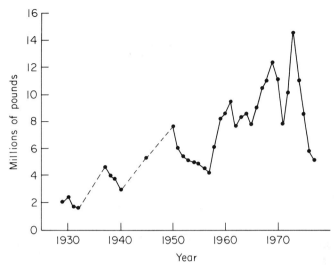

Fig. 14.6. Landings of striped bass by commercial fishermen, 1929–1977, based on Fishery Statistics of the United States.

commercial or recreational fishermen, but that the safe catch not be exceeded. The argument between commercial and recreational fishermen is a socio-political argument which must be solved by other means (Chapter 8).

The bluefish issue is in some ways more difficult and in some ways easier. The species is abundant, perhaps more abundant than it has ever been, so there is no problem at present with the supply. There is no major commercial fishery because the price is relatively low, and the market limited. How much the market is limited by the large sport catch is not at all clear, but it probably is, because the sport fishery certainly sells large numbers of fish. The fish is oily and does not freeze well, so its sale at present is limited to fresh fish. Recreational fishermen are understandably opposed to an expansion of the commercial fishery, especially if the fish are to be exported, for that will cut into recreational catches, and the recreational fishery at present is by far the largest in the Middle-Atlantic region. Another issue that is only dimly perceived at present is the varying supply, which if commercial landings are any indication, can vary widely. In the early 1940s bluefish were very scarce, and a return to that low level would affect sport fishing substantially, unless some other species should replace it.

14.4.5 Town and county jurisdiction
Of all resources to come under the lowest possible level of jurisdiction, clams are the most important, especially soft clam (*Mya arenaria*) and hard clam (*Mercenaria mercenaria*). Soft clam has been harvested mostly in Maine and later in Maryland. Hard clam was once most important in Rhode Island, now

it is New York and New Jersey. Most of these resources come under counties or towns for management, especially now when Maine is the principal producer of soft clams, and New York is the principal producer of hard clams. Local jurisdiction probably is best for these inshore resources, but it is far from perfect. The advantage of town control is a closer interest in the resource in places where it is most abundant; the principal disadvantages are an unevenness in town funding and control, a lack of coordination between managing units, and the effect of local politics on programs. Yet these two species have held up remarkably well for 100 years or more, under alternate neglect and attention, growing pollution of the waters, and inadequate enforcement. Total production of soft clams has declined only slightly, and is now about two-thirds of its level in the 1930s. The total harvest of hard clam has declined to about three-quarters of its level in the 1950s, and is remaining remarkably steady. This has come about despite closure of beds because waters are becoming increasingly polluted. In fact, there may be some merit in pollution if the closures can be adequately supervised, because that maintains a reserve of clams which can be transplanted to clean waters under supervision, to be harvested later, after they have cleansed themselves.

These clam beds also may be favored by local laws which ban the more efficient yet possibly damaging mechanical gears. This is not certain, for it has not been adequately investigated. Yet it is suggestive that these resources have stood up well under what might be considered difficult conditions whereas the offshore surf clam resource has been seriously overfished in an area in which conditions are better, but the older prohibitions against more 'efficient' gear were not imposed. The surf clam resource has been seriously overharvested in a much broader area of the continental shelf in less than a third the time of the apparently much more successful inshore clam resources. The reasons are not at all clear, but they are intriguing and should be investigated.

14.5   SPECIAL ASPECTS OF MANAGEMENT

14.5.1   Environmental impacts

The resources most vulnerable to environmental impacts appear to be anadromous species, which must come into rivers to spawn and rear their young, or estuarine sessile species like oysters and clams, which cannot escape coastal waters. All fisheries resources are subject to hazard from pollutants to some degree. The issue is complicated by a variety of special considerations, each contaminant is a special case and each probably has a certain concentration below which it is harmless (Chapter 10). Above that level, it may or may not be possible to detect damage even if damage is done. Adding to that is the well-known variability of the natural environment in estuaries, which sometimes has large effects upon the survival and health of organisms and cannot usually be distinguished from man-made effects. In addition there are the effects of fishing, which further complicate the situation, so that a clear distinction of the cause of a change in a resource is often almost impossible to detect even after years of data gathering.

There is no question that salmon are affected by pollutants of various kinds in rivers, and that formerly productive areas have been denied them by pollution of various sorts. On the east coast shad have been affected by river pollution, even though shad are known to vary widely also from natural variables. It is clear that oysters have been adversely affected by pollution, which affects reproduction and other bodily processes, but is is equally clear that pollution has been blamed for oyster losses that were probably caused by natural environmental change. It is much easier to place the blame on human intervention in one form or another because then, at least theoretically, something can be done about it, but with a natural phenomenon there is no recourse.

14.5.2 Political expediency

The oyster industry is not one of the largest in the United States now, but it once was, and it very likely could have been helped to stay that way if the industry had listened in time. It may have been a disservice to the industry to have started so early, when human populations were relatively small and shellfish bottom could be preempted so easily. In 1880 the population of the United States was about 50 million. By 1977 it was almost 217 million. In that period, oyster production dropped from about 30 million to about 10 million bushels, an average decline in annual per capita consumption from about 0.6 bushels or 150 oysters, to about 0.04 bushels or 10 oysters per year.

The events that led to this decline were complicated, and varied somewhat from place to place. In general, however, they were quite similar. The early oyster growers, whether they owned or leased bottom and worked it or harvested oysters on the public grounds, did little or nothing to conserve the resource. Oysters and cultch (the shell on which they were growing) were taken in great quantities from the bottom, and little or nothing was returned. Oysters were transplanted widely from various parts of the coast to other places, with little if any thought as to their suitability in the new environment, and with little thought to the predators, diseases, or competitiors that also might be introduced. Advice of scientists was usually not heeded.

The most detailed study of the oyster industry has been made in Virginia, which was once by far the largest oyster producing state in the U.S. This study found that the oyster industry was riddled with archaic practices and attitudes, economic and political conflicts existed between segments of industry and between the fisheries and other users of the environment, the industry and the State lacked firm and consistent purpose and practice toward achievement of realistic and improved management, and legal restrictions and economic practices continued which actually mitigated against and prevented improvements. Recommendations for revitalizing the fishery included leasing of unproductive public grounds, methods of improving seed production, methods of improving the public repletion program, evaluating the resource and improving utilization, and research which would benefit public and private participants. Whether they will work or not depends largely on a much closer degree of cooperation between all elements. A quotation from a report by the Virginia Marine Resources Study Commission dated November 27 1967 aptly describes the situation as it exists today:

> 'The planting and harvesting of oysters is taken for granted by oystermen and natives in tidewater Virginia in the same manner as citizens of rural areas consider farming; it is a livelihood and a way of life ... few persons have a comprehensive knowledge of the mechanics or the complexity of this phase of Virginia's economy.'

14.5.3   Resources that depend primarily on environment

Some resources are short-lived, and depend almost entirely on environmental variables in the local area for their welfare. Of the major fisheries resources of the United States, shrimps are outstanding in this respect. Although not the major seafood in weight landed, they are by far the most valuable in dollars, worth to fishermen $355.2 million in 1977, or about 23 percent of the value of all fisheries products.

Shrimp fishing developed first in the south Atlantic States and in the Pacific Ocean (Fig. 14.7). The catch in the South Atlantic peaked in 1945 as did many of the food fish species, and has declined since. The present catches are about 25 million pounds a year, with no particular trend. The catch in the Pacific was relatively small, averaging 5 million pounds or less, until the late 1950s, when rapidly increasing catches began in Alaska. South Pacific catches also rose after 1956, but only moderately compared with Alaska. The Gulf of Mexico shrimp catch has been rising since before the turn of the century, reached a peak of over 235 million pounds in 1954, and has been fluctuating around about 200 million since that time. But in 1977 it reached the highest peak of all, over 265 million pounds.

Although a few shrimp were landed in New England before 1960, the fishery did not develop seriously until the mid-1960s and reached a peak in 1969 at over 28 million pounds. Since that time it has declined drastically. The 1977 catch was less than one million pounds, and in 1978 the fishery was closed completely.

Fig. 14.7. Landings of shrimp from five geographical regions, 1880–1977, based on Fishery Statistics of the United States.

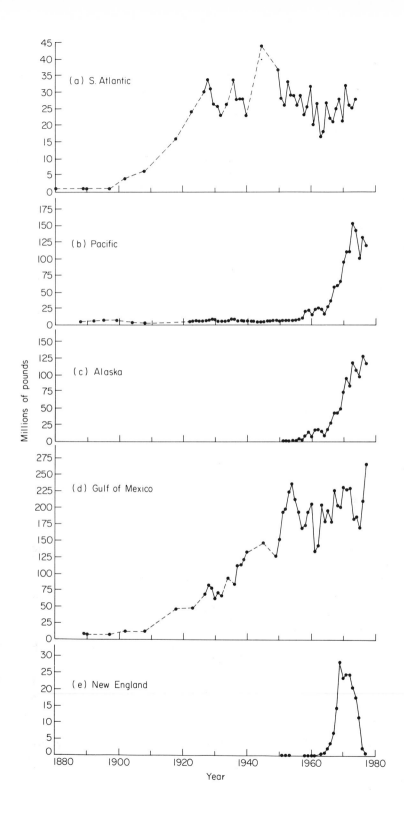

Environmental influences on shrimp may be illustrated for the Gulf of Mexico fishery. Shrimp there have a brief life span, probably not more than $1\frac{1}{2}$ to 2 years. Thus there are few cohorts to provide a stable population if reproduction and recruitment are poor in a particular year. Each cohort is cropped essentially in one year, and populations could fluctuate suddenly and widely from variations in factors such as salinity and temperature. Relatively stable catches in the Gulf probably occur because the Gulf supports many separate sub-populations and some nursery grounds are likely to have favorable conditions each year.

Although little can be done to avoid the annual variation in catch, management in these cases is somewhat simplified. Shrimp fishing is controlled primarily by size and zone restrictions, with two goals in mind. The first goal is to protect young shrimp until they reach spawning size. Since most adults will spawn only once, they can all be harvested after spawning without damage to the population when conditions are relatively favorable. The second goal is to protect stocks which are low in abundance or to prevent excessive harvest during vulnerable periods. So far these controls have worked fairly well, although they have not been applied equally in all parts of the range, and in some parts have not been started soon enough. If these laws were enforced effectively, it is possible that no other management measures would be necessary, at least in the southern regions.

14.6   FUTURE OF FISHERIES MANAGEMENT

Most coastal fisheries have had no effective management at all, either because they are not important enough locally to command attention by the States, because there is not enough biological information to point clearly to what must be done, or because the subject is so controversial that legislators or State administrators who are supposed to make decisions cannot decide what needs to be done. With some species the situation is even more unsatisfactory, because laws have been passed or regulations made on the basis of uninformed public pressures that hamper rather than assist conservation.

The past is not very reassuring that fisheries maanagement will improve. In the United States, especially, the form of government makes it difficult. Jurisdiction is divided among many subdivisions of government, without adequate methods of coordination. The industry is fragmented, with many local problems that often work against the common good. Commercial and sport fishermen have different objectives which often are impossible to solve without compromise. Scientific information often is inadequate, difficult to get, and often not heeded even when present. Environmentalists often have imposed constraints that are unnecessarily restrictive, difficult or impossible to achieve. What are the chances for improvement?

14.6.1 Problems of jurisdiction

Jurisdiction between the Councils and the States is perhaps the most pressing problem. At present the Councils have jurisdiction from three to 200 miles for most species, farther with salmon, but not at all for tunas. The States have jurisdiction from three miles to and including inland waters. The FCMA provides that the Federal government may take over if the States do not satisfy certain requirements of cooperation and performance, but the criteria are rather stringent and the likelihood of arbitrary action rather unlikely.

How well are the States doing? Fairly well insofar as their ability to act is clear. For example, in the surf clam managment plan, the States are trying to coordinate their laws with the management plans of the Councils, so that the part of the resource that lies within three miles is equally well protected. All States do not have the necessary laws yet, but there appear to be no serious obstacles to passing such laws.

There are jurisdictional problems between Councils that are difficult to solve. Where a resource is abundant in the waters of more than one, the problem is partially solved by having one Council take the lead in preparing a fishery management plan. Coordinating this plan between Councils is problematic, and sometimes there are special conditions that are different between Councils. The groundfish plan is a good example. This plan covers cod, haddock, and yellowtail flounder, and was prepared by the New England Council. Two of the three species trouble the Mid-Atlantic Council because they migrate seasonally into the south in winter, and can be caught by New England fishermen before they are available to Mid-Atlantic fishermen. Presently, there are proposals by fishermen from New York and New Jersey that they receive a special quota that will allow them a fair share of the resource. This, however, would not assure them of that quota, because the Act specifically states that no fisherman shall be given special consideration.

A third kind of jurisdictional problem concerns recreational fisheries and the consideration that they must receive in management plans. This will be especially important when plans are implemented on species like bluefish, summer and winter flounder, scup and others. These species are much more heavily fished by sportsmen than by commercial fishermen, and recreational interests will have to be recognized. At present, e.g. in the Atlantic mackerel plan, the sport catch is estimated and set aside before commercial quotas are established. This appears to be working satisfactorily for mackerel, because the commercial fishery is not important. But when commercial catches are large, and there is high demand for fish, as with flounders and scup, an equitable balance between sport and commercial fishing will have to be decided, and recreational fisheries as well as commercial will have to be regulated so that quotas are not exceeded. Without better information on recreational fisheries, this will be very difficult if not impossible.

In Canada and Mexico these problems appear smaller. The central government has greater power and can exercise control by executive order or similar means. Fishermen are just as vocal in those countries, however, and government cannot act capriciously if it wishes to remain in power. Generally speaking, problems very much like those already cited exist in one form or another in most parts of the world, and their solution will depend on the willingness of government to make them work.

14.6.2   Improved biological information
Biological information will be essential to the success of fisheries management if present management plans are to succeed. As a minimum, better information will be needed on standing crops, recruitment and growth, and natural and fishing mortality. The most urgent problem will be to develop adequate information on catches and fishing effort in the recreational as well as the commercial fisheries. This information requires cooperation from all people involved and it is therefore the most difficult to get accurately and completely. If it is not obtained with reasonable accuracy from all segments of the fisheries, the other information will be relatively useless, and management plans will be little more than guesswork.

14.6.3   Switch to multispecies management
A growing trend in fisheries management is to think in terms of total biomass rather than individual species. There are several good reasons for this, among them the fact that total biomass varies less than that of individual species. Fisheries that concentrate on individual species may aggravate the situation by generating greater fluctuations in individual species that are not harvested. Thus, the optimum strategy would seem to be for fishermen to be as flexible as possible, taking advantage of natural shifts in productivity.

While desirable in theory, multi-species fisheries demand more complex management. In general, there must be a total effort to spread the fisheries over as wide a resource base as possible. It means having adequate information on all species rather than a few and having that information sufficiently in advance that fishermen can receive and use it to their advantage. When less important or less desirable species are abundant and more desirable species scarce, markets must be prepared that allow fishermen to sell all types of fishes at a reasonable profit. This requires a strategy of test-marketing and development of new products that will be acceptable to the consumer. Governmental encouragement may be needed, for example, disincentive taxes on capture of scarce species and incentive payments for less desirable varieties.

14.6.4 Resistance to management

One of the greatest obstacles to progress in fisheries management is the lack of understanding of even the basic principles of management by the people to whom it means the most. Generally speaking, the fisherman has no desire to be managed. He wants to be left alone to catch what he can where he can. If he thinks about management at all, it is to think that it should be applied to someone else, not him, that in the absence of all the others he would have no problems. In the narrow sense this is true. The resource is common property, and virtually all who wish to can take it. If there is only one fisherman, his catch would be insignificant and conservation problems would never arise. When there are many fishermen, the individual sees things the same way, but in fact he is no longer free of constraint. He is competing with other fishermen for particular species, and that is a constraint, and he is not entirely free in turning to other species because other fishermen have equal opportunity. Yet he does not perceive this at all clearly. Moreover, commercial fishermen as a group, even though they are individualists and reluctant to come together, tend to unite to some degree when they come in conflict with recreational fishermen. Their objectives are different, to some degree incompatible, and they tend to prefer confrontation to compromise.

The only possible remedy for this resistance to management is some form of education designed to convince fishermen that they will do better if they work together rather than apart. It is not at all certain that education will work, that it will get to enough people to have an impact, or that there will be enough teachers with the ability and dedication to really get the message through. The development of extension personnel, whose primary task is the translation of technical and scientific information into the language and lives of fishermen, is essential to this purpose. In the U.S., the creation of the Sea-Grant program is developing educational programs for these purposes as one of its chief objectives.

14.7 REFERENCES

Ahlstrom E.H. and J.Radovich (1970) Management of the Pacific sardine. *In* N.G.Benson (ed.) *A Century of Fisheries in North America*, pp.183–194. Washington: American Fisheries Society.

Burd A.C. (1974) The north-east Atlantic herring and the failure of an industry. *In* F.R.Harden Jones (ed.) *Sea Fisheries Research*, pp. 167–191. New York: Wiley.

Larkin P.A. (1970) Management of Pacific salmon of North America. *In* N.G.Benson (ed.) *A Century of Fisheries in North America*, pp.223–236. Washington: American Fisheries Society.

McHugh John Laurence (1978) Atlantic sea clam fishery: A case history. *In Extended Fishery Jurisdiction: Problems and Progress, 1977.* Proc. N. C. Governor's Conf. Fish. Mgmt. under Ext. Jurisdic., Off. Sea Grant, NOAA, U.S. Dept. Commerce and N. C. Dept. Admin., UNC–SG–77–19: 69–89.

Schaff William E. (1975) Status of the Gulf and Atlantic menhaden fisheries and implications for resource management. *Marine Fish. Rev.* **37(9)** 1–9.

Chapter 15
Open Ocean Resources

J. A. GULLAND

15.1 THE NATURE OF THE RESOURCES

15.1.1 Distribution of major fisheries

While the dividing line between the resources dealt with in this chapter and the coastal resources discussed in Chapter 14 is not distinct, there are some important differences between typical coastal and open ocean resources that have significant impacts on the way they are managed. Coastal resources are affected, actually or potentially, by a whole range of human activities other than fishing—land reclamation, waste dumping, etc.—while fishing itself is often carried out by a variety of small-scale fishermen, usually restricted to fishing close to their home port or landing place. Management of these fisheries is likely to become a complicated business of balancing various social and economic pressures, including many considerations outside the fisheries themselves. Open ocean resources, on the other hand, are typically larger in absolute magnitude, and are exploited by large-scale, industrial fishing fleets. Analysis of the impact of these fisheries on the stocks is, on the whole, rather easier to carry out, and the management practices are, at least in principle, more likely to be based on relatively simple biological or economic analyses largely restricted to the fishery. The last three decades since the end of the World War II have seen an enormous expansion in ocean fisheries. It would be impossible to outline them all in this chapter. Instead we will look at four fisheries (or more strictly groups of fisheries) which can serve as typical examples. These are the cod fisheries of the North Atlantic; the Peruvian anchoveta fishery; the tuna fisheries; and whaling.

Because industrial-scale open ocean fisheries require good catch rates to be economically viable, their development has been concentrated in those parts of the world where natural conditions produce good catches. This requires some or all of the following conditions to be satisfied—a high level of primary production; an efficient transfer from primary production to fish; a favorable ratio of standing stock (biomass) to production; and the concentration of this production in convenient packages for efficient harvesting.

The variation in primary production in the ocean, and the contrast between

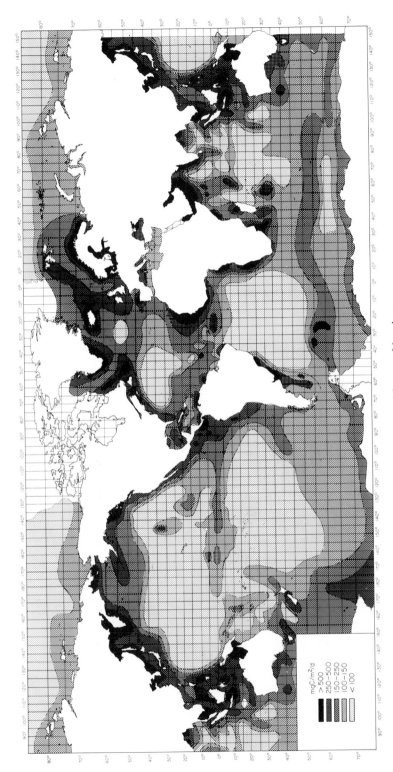

Fig. 15.1. Distribution of primary production (mg C per day) in the world ocean. Note low values in mid-ocean. (from FAO 1972)

oceanic 'deserts' such as the Sargasso Sea, and the richness of coastal upwelling areas such as the waters off Peru is well known (Fig. 15.1). The combined effects of differences in primary production and in the efficiency in transfer to fish (fewer steps and smaller losses at each step) on fish production, has been well demonstrated (Ryther 1969). He showed that the greatest potential for fishing occurs in the upwelling areas, followed by the coastal areas including most of the area over the continental shelves, with the open oceans being much less promising.

(*a*) *Upwelling areas.* The most important upwelling areas are those off the sub-tropical parts of the west coasts of the continents, and match the areas of major deserts on land (the two are generated by the same steady pattern of winds and currents). Upwelling occurs elsewhere, e.g., along the equatorial divergence, but the coastal ones are, in fishery terms, the most important. Of these the Peruvian system is probably the best known and, after the California current system, the best studied. At the height of the Peruvian fishery, catches of a single species, the anchoveta (*Engraulis ringens*), accounted for around 15 percent of the total world catch, and produced enough fish meal to contribute up to one third of the total exports of Peru (Fig. 15.2).

Off Peru and northern Chile, the cold Peru (or Humboldt) current flows northwards and then swings away from the coast. This brings about the upwelling of nutrient rich deep water, which supports a very rich bloom of phytoplankton. This high plant production in turn promotes the existence of the large stocks of shoaling pelagic fish (herrings, anchovies, and their relatives) which are typical of all coastal upwelling systems. In most of these systems these fish feed on zooplankton, at one step removed from the plants, but the Peruvian anchoveta (at least the adults) feed to a large extent directly on phytoplankton. The shorter food chain means that more of the original primary production appears as fish. In addition the efficiency of fishing is helped by the concentration of fish in schools.

(*b*) *Shelf areas.* The other major section of the ocean favorable to fisheries is the coastal areas; that is, the areas over the continental shelves, where the depth of water is small enough to allow fairly regular mixing and recycling of nutrients (Fig. 15.3). For fishermen these areas have another advantage in that the bottom forms a boundary on top of (and to some extent within) which there is a favorable concentration of fish (and of animals on which the commercial fish feed). Though in all but the shallowest water plant production is limited to a relatively shallow surface layer, the later stages in the food chain are not confined to this layer. At each stage in the food web of the surface layers—from phytoplankton to small zooplankton to large zooplankton to pelagic fish—there is a substantial proportion of waste material that falls to the bottom. This can support a rich community of bottom invertebrates feeding on the detritus, which in turn support commercially attractive fish. In the tropics the variety of these fish is vast

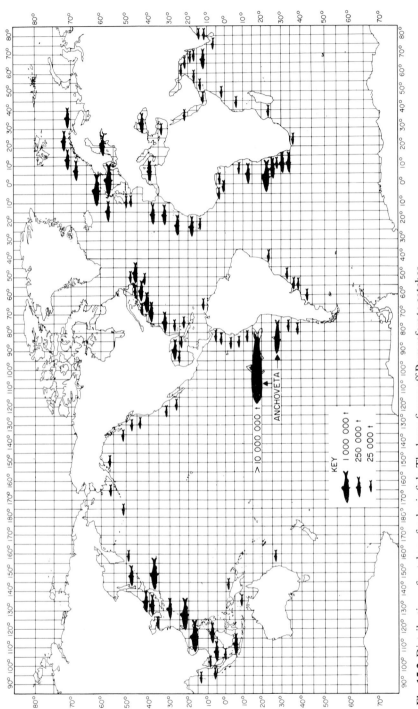

Fig. 15.2. Distribution of catches of pelagic fish. The large figure off Peru refers to catches of anchoveta, in a 'normal' year, before the collapse of the stock in 1972. (from FAO 1972)

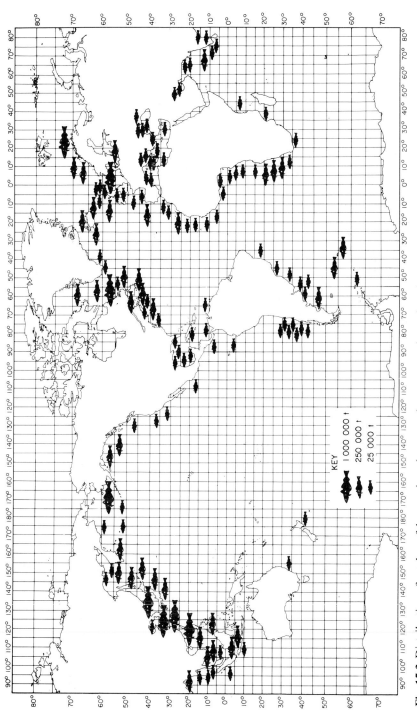

Fig. 15.3. Distribution of catches of demersal catches; note the concentrations of catches in certain areas, noticeably the North Pacific and North Atlantic. (from FAO 1972)

KEY

1 000 000 t

250 000 t

25 000 t

(perhaps 50 species may be caught in a single haul and one or two hundred in a single fishery), but this variety falls off in temperate and sub-arctic waters. Since it is easier to find markets and appropriate processing methods for one or two species rather than one or two hundred, this makes the colder waters more attractive. Another reason is that the fish in colder waters are longer lived. Thus the standing stock of cod in the Barents Sea may represent the accumulated production of up to ten years, whereas that of demersal fish in the Gulf of Thailand only of one or two years. The biomass in the former area—and hence the catch per unit effort—is therefore correspondingly higher for a given annual production. Another reason why the continental shelves are more attractive to fishermen, all other things being equal, is that because they do not have to disperse their efforts through the whole water column, their work is much easier. Most of the common methods of fishing—trawls, lines, or traps—are operated wholly or mainly on the bottom, not only for bottom-living fish (cods, snappers, etc.) but also to a considerable extent for pelagic fish (herring, mackerel, etc.) at times when they are concentrated on the bottom.

Among the bottom-living fish the cod (*Gadus morhua*) is, in terms of the historical importance of its fisheries and the volume of research into its biology and exploitation, pre-eminent. At the same time the problems of managing the cod fisheries are in many ways typical of those of most demersal fisheries. It is distributed widely in the North Atlantic (with a closely related species in the North Pacific) from Spitsbergen and the English Channel in the east to Labrador and New England in the west. In most parts of its range it has supported major fisheries (Innis 1954) and some of the earliest studies of fisheries management (e.g., Russel 1942; Graham 1948).

(*c*) *The open oceans.* While the primary production per unit area, or the density of fish stocks, is lower away from the continental shelves or the major upwelling areas, these parts of the open ocean are very large. The total production at each stage of the food chain is therefore not negligible. Fishermen can profitably work in these less rich areas if the productive system and the biology and behavior of the animals enable them to harvest the production from a wide area with little effort. This is the case for tunas and whales.

Being large predators, at the end of a lengthy food chain, tuna are not common. However, the smaller tuna in particular, including the smaller individuals of the larger species, form surface schools that can be spotted (or more usually the flocks of birds or porpoise schools associated with them can be spotted) at a considerable distance; most larger tunas seem to be less sociable, but are very active, covering a wide area in a few hours. Put the other way, a baited hook on a long-line has a good chance of catching any tuna occurring within a considerable area around the hook. In either case a fishing vessel can in a day hope to catch most of the tuna in an area of some hundreds of square kilometers, compared with the area of a fraction of a square kilometer covered by a trawler in the same time. Tuna occur and are fished throughout all the warmer parts of the ocean.

Sperm whales, like tuna, are high level predators, feeding mainly on squids—themselves important predators—and also occur mostly in the warmer seas, though the older solitary males do migrate into the cooler waters of the sub-Arctic and Antarctic. Though occurring in most of the oceans, there are clear concentrations in the areas of high primary production, especially the upwelling areas, both those off the continental coasts and in the equatorial current systems. Charts of the location of nineteenth century sperm whales reflect accurately the distribution of high primary production (Townsend 1935).

The baleen whales, which feed mostly on zooplankton, are part of a shorter food chain; this means that less of the original production is lost along the way, but also that it has less likelihood of being suitably concentrated. The main concentrations of baleen whales therefore are more limited. The outstanding example is the Antarctic, where the krill, an euphausid feeding principally on phytoplankton, forms dense swarms, very suitable for the great whales.

15.1.2 Migrations

In addition to feeding and behavior (which help determine how abundant they are and how readily they can be caught), the aspect of the general biology of ocean-living fishes that is particularly relevant to their management is their migration. All of them, to a greater or lesser degree, are active animals, and this mobility can cause a stock to be exploited by different groups of fishermen at different stages in their migrations. Clearly it is no use for one country to apply strict controls on the exploitation of a stock of fish in its waters if the same stock migrates into other areas where the fish are heavily exploited without controls. Successful management of a stock must involve coordinated action wherever it occurs.

The extent of the migrations varies. To the extent that migrations are undertaken annually to take advantage of seasonal differences in food supply, they are likely to be greatest when seasonal patterns are strongly marked. Thus the great whales move into the Antarctic (and also Arctic) during the summer to feed, but retreat considerable distances into milder, often sub-tropical waters during the winter; for example, humpback whales move between Alaska and Hawaii each year. The annual migrations of fish are, on the whole, less marked, but can still be significant from the management point of view—for example, some of the pelagic fish off north-west Africa move north and south over long enough distances to take them off the coasts of four or five different countries during the course of the year. Fish also undertake movements so that different life stages (newly-hatched larvae, juveniles, and adults) are at appropriate places at appropriate times (Harden-Jones 1968). The best known example is the migration of salmon between the spawning grounds in rivers and lakes, and the main oceanic feeding grounds, but the longest movements are probably made by some species of tuna. Albacore (*Thunnus alalunga*) are known to migrate across the North Pacific, and the southern bluefin tuna (*Thunnus maccoyii*) spawns in the tropical waters off north-west Australia, the young fish (two-to-three years old) are found along the south and east Australian coasts, and the older fish feed in cold waters of the southern Indian Ocean, moving as far west as off South Africa and into the Atlantic.

The demersal fish of the continental shelf undertake less spectacular movements (Fig. 15.4). Generally the adults move against the main currents just enough so that their eggs and larvae drift back again to the preferred feeding areas. These movements are not generally long. Thus although cod are distributed over much of the northern Atlantic, most fish (with a few odd exceptions, Gulland and Williamson 1962) stay within relatively restricted areas,

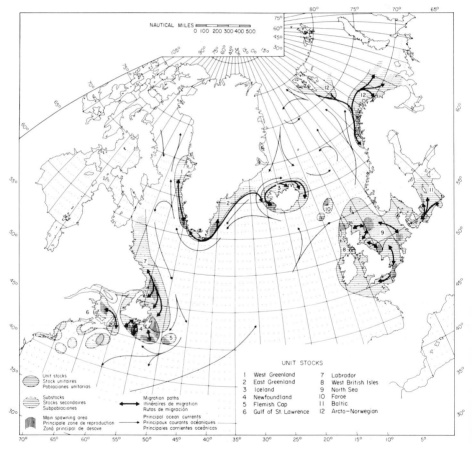

Fig. 15.4. The migration and stock structure of cod in the North Atlantic. (from FAO 1972)

so as to form distinct stocks, e.g., in the North Sea, around the Faroes, off Labrador, etc., with little mixing. Within these stocks the longest and most distinct migration is that of the Arcto-Norwegian stock between the spawning grounds in the Lofoten Islands, off north-western Norway, and its feeding grounds in the Barents Sea and off Spitsbergen.

Thus there is a gradation between stocks that are of interest only to one country, and spend all their lives within the 200-mile limit of that country (e.g., cod at Faroes or off Labrador); those that are of interest to a few adjacent countries (Arcto-Norwegian cod, sardinella off West Africa) and those of interest to several countries, as well as occurring outside the economic zones of any country (some whales and tunas). The distinction between the extremes has been acknowledged by the U.N. Conference on the Law of the Sea. The draft texts for a new Law of the Sea emerging from that Conference distinguish a special group of 'highly migratory species'—mostly tunas and whales, though it may be noted that some stocks or species in this group (e.g., some smaller tunas) may migrate over less distances than other species (e.g., the Norwegian herring which moves between Norway and Iceland) not included in the 'highly migratory' list.

15.2   ASPECTS OF MANAGEMENT

15.2.1   Markets

Besides good catch rates, the major open ocean fisheries depend on a large market to sell their catches; this means being able to transport the fish to big cities and other concentrations of people, which may be a long way from the fishing grounds and often also from the landing ports; this in turn requires some effective method of processing and preservation. The history of fisheries is therefore as much a matter of technological advances in methods of processing as a story of the actual catching. For a long time the only ways to keep fish edible for a period and to allow transport over significant distances were drying and salting. Dried cod and salted herring were the staples of important fish trade in Europe, and elsewhere, until near the end of the last century. Changes in the distribution patterns of herring in the Baltic and off Scandinavia gave rise to corresponding shifts in economic and political power between the coastal cities. The cod stocks of the Grand Banks of Newfoundland and other parts of the northwest Atlantic were almost as an effective magnet in attracting Europeans northward across the Atlantic as was the gold of Eldorado further south.

Up to the nineteenth century, however, the total demand for fish did not outstrip the potential supply from the local resources, and no management problem arose. This picture changed in the last quarter of the nineteenth century with the arrival of several new factors; the ability to manufacture quantities of ice cheaply allowed fish to be kept fresh for many days after capture; the railways permitted these fish to be distributed quickly to inland markets; and, rather later, the application of steam power to bottom trawling for the first time relieved the fisherman of the need to rely almost solely on his own muscles to take his catch. The result was the growth of trawling for high to medium value demersal fish as the typical methods of commercial, open ocean fishing.

In the last two or three decades the widespread use of freezing at sea and the use of fish meal as an important supplement to the diet of chicken and other farm animals have greatly changed the way fish catches are used. Freezing at sea, on board the catcher vessel, or an accompanying factory vessel, has virtually freed fisheries of any close ties to a particular base. With ice, three to four weeks was the maximum length of trip. Freezer vessels can stay out almost indefinitely, and in the extreme case of modern Japanese tuna vessels, supplying the high price sashimi (raw fish) market with hold temperatures of $-30°C$ or less, the fish can be put ashore virtually as fresh as at the moment it left the sea. Without freezing at sea many of the biggest trawl fisheries, as well as most of the tuna fisheries, could not exist in their present day form.

Many of the very large fisheries on shoaling pelagic species (anchoveta, etc.) are now based on the use of fish meal in animal feeding stuffs, especially broiler chickens. For good growth the chickens need some protein to supplement the cereals which supply the bulk of their feed, for which fish meal has proved extremely suitable. Indeed for a long time it was believed to contain some unknown growth factor which gave a better growth than could be explained on the basis of its protein content alone. In terms of weight the fisheries for abundant but low valued fish (menhaden, sandeels, etc.) are among the biggest in the world, and have accounted, in some years, for approaching one-third of the world catch.

In the early days of fishing, when each fishery supplied only a limited local market, the size of that fishery was determined by the size of the market. Typically, the resource could supply that demand without difficulty, and could, to that extent, be considered as being inexhaustible. This belief in an inexhaustible resource provided a logical basis for the doctrine of the freedom of the seas put forward by Grotius in the early seventeenth century. At that time the belief was in full accordance with experience. However, as fishermen became able to supply a wider market (effectively a world market in the case of frozen or canned fish or fish meal) the relation of supply and demand became reversed. From the point of view of a particular fishery for fish meal or other product with a world market (apart from the very big ones like the Peruvian anchoveta), the demand is virtually unlimited. The limit to the size of the fishery is set by the resource. Ocean fisheries, once established, therefore tend to grow until the abundance of the stock is reduced, and falling catch rates make the fishery economically unattractive (see Chapter 9). At this stage the more adventurous fishermen look further afield for alternative grounds. Up to recently these have usually been found, allowing the excess pressure on one stock to be relieved by movement (either of actual vessels, or of the capital and other resources) onto other stocks.

In most oceanic fisheries the shift of interest to other resources has prevented very great increases in fishing effort, and the yield curve (the relation of sustained catch to fishing effort) is sufficiently broad and flat-topped, that the effort that remains results in a reasonably productive stock and a reasonable total catch. That is, the fishery settles into a stagnation stage in which the fishing effort may be in excess of that determined by any rational objective (attainment of MSY, OY, etc.) and the net economic benefits negligible, but with a total catch not much less than that obtained toward the end of the expansion period. It is easy to demonstrate that these fisheries would perform better if there were better management, but until very recently the practical and political obstacles to implementing any such management were too formidable.

Very recently the changes in the Law of the Sea and the general introduction of a 200-mile limit have meant, first that many of these resources are now under a single national jurisdiction, and second (and equally important) that attention has been paid to problems of managing fisheries at high levels. The result has been that some fisheries (especially some around North America) are now beginning to enter a phase of being rationally managed (see Chapter 14).

15.2.2   Present fisheries

(*a*) *North Atlantic cod.* The basic pattern of development is most clearly shown by the North Atlantic trawl fisheries, especially those for cod (Gulland 1974, Chapter 3). Cod have been fished by hand-line and salted down aboard ship, or on shore close to the fishing grounds, for centuries. This method has continued into the modern age of fisheries and Portuguese fishermen were setting out in their dories from traditional schooners off West Greenland well into the 1970s. However, these fisheries have had little to do with the modern cod fisheries, which started with English steam trawling in the North Sea in the late nineteenth century. By the end of the century it was clear that catch rates in the North Sea were dropping, and fishermen began to look further afield. (It may be noted that the fall in catch rates of cod and also of plaice and other demersal species also caused people other than fishermen to consider whether the sea was in fact inexhaustible. This led to the first serious scientific examination of fish stocks, and to the establishment in 1902 of the International Council for the Exploration of the Sea, discussed more fully later).

Following the fall in North Sea catches the first move was to Iceland, where English trawlers were fishing before World War I. The inter-war period saw the development of the long-range fisheries around Iceland and also in the northeast Arctic (the Barents Sea and around Bear Island and Spitsbergen), where English trawlers were joined by those from Germany. The post-war period saw the full development of the northeastern Atlantic, and then the movement westward. It also saw the massive influx of trawlers from eastern Europe (especially the U.S.S.R. and Poland), as well as the replacement of the salt-cod fleet from southern Europe by trawlers (especially from Spain) and the growth of modern fisheries in Canada.

The activities of all these vessels had brought the cod stocks to a low level. However, since 1970 strong efforts have been made to rebuild some of the stocks to a more productive and economic level. This has involved a short-term drop in total catch since it takes a little time for the stocks to adjust to the lower level of fishing effort.

(*b*) *Whales.* The history of whaling is similar in being one of continual expansion onto new stocks, but different and less happy as regards the fate of the stocks after they became heavily exploited. Unlike cod (and most other fish) which have a very high fecundity (hundreds of thousands of eggs per female) so that a small adult stock has the potential to produce good recruitment, whales have only a limited potential to increase their reproductive rate in response to exploitation. A fall in the adult stock of whales, if accompanied by continued high catches, can lead to a disastrous fall in the number of young.

Almost at the same time as Grotius was basing his principles of freedom of the seas on their inexhaustible resources, Dutch and English whalers in the first act of ocean whaling were finding that the stocks of whales in the Spitsbergen area were by no means unlimited. These were the right whales, so named because the larger rorquals were too big and active to be caught from open boats with hand harpoons and were thus the wrong whales. As the stocks declined the whalers moved further afield, to both coasts of Greenland and later into the Pacific and sub-Antarctic. By the nineteenth century New England had become the center of world whaling, and interest had turned to the more abundant sperm whale (which provided good oil for lamps, etc.).

The mid-nineteenth century saw one of the few examples of an ocean fishery declining for reasons other than reduction in stock. The growth of the petroleum industry ruined the market for sperm oil for lamps, etc., and for nearly half a century whaling stagnated. (It is not clear whether the decline in the sperm whale fishery was not also helped by decreasing stocks and falling catch rates). Then the invention of the harpoon gun allowed the hunting of the larger baleen whales— an excellent source of oil for human consumption in margarine, etc.—and the last phase of whaling expansion began. This started in the North Atlantic but the richer grounds of the Antarctic were soon opened up. The history of Antarctic whaling has attracted much attention but has not always been well understood (but see, for example, Schevill 1974, and Gulland 1974, Chapter 2). To a large extent it has followed the classic pattern, with first attention being paid to the largest and most attractive species (the blue whale) and then shifting to successively less attractive species (fin, sei, and then minke whales). Thus 29,000 blue whales were killed in the 1930–1931 season (nearly three times the number of fin whales caught in that year), but never more than 18,000 in any later year, while by 1938 the catches of fin whales were nearly twice those of blue whales. By this time those concerned with whaling had become aware of the dangers of unrestricted whaling, and in 1946, among other expressions of better intentions characteristic of the immediate post-war era, the International Whaling Commission was set up. The subsequent history of whaling is very much the history of IWC, and is discussed later in this Chapter.

(*c*) *Peruvian anchoveta.* Unlike the other fisheries discussed here, those on small shoaling pelagic species do not lend themselves to the steady geographical expansion of a fleet looking even further afield as the local stocks become reduced. A load of anchovy or menhaden is not worth much, and to work profitably, a vessel has to make frequent trips. Large factory ships exist for processing fish meal a long way from their home ports (mostly converted whale factory ships), but they are very few. Most of these fisheries depend on working close to their home ports. There is less opportunity therefore of switching from one stock to another when the first declines, though there may be opportunities of changing from one species to another. Thus the fishery for pelagic fish round Japan has been maintained at a high level by switching from one species to another (Nagasaki 1973). Other fisheries are dependent on a single species, and, lacking the safety valve of switching to other species, need more careful management, a need that is increased by the technical difficulties of monitoring the changes in abundance of these stocks from the normal catch and effort statistics of the fishery.

The biggest pelagic fishery, and one that illustrates very well the problems of managing these fisheries, is that for Peruvian anchoveta. Once it was discovered, about 1955, that anchoveta could be caught easily in a purse seine, converted into fish meal, and sold abroad, the fishery expanded rapidly. Every year from 1955 to 1960 the catches approximately doubled and continued to expand rapidly to reach nearly 9 million tons in 1964. This was close to the productive capacity of the stock. Regulations were then introduced, starting with a closure during the spawning season and increasing in complexity and effectiveness. These limited the catches, but not the capacity of the fleet or of the fish meal plants. By 1970 there were enough fish meal plants to turn the entire world catch into fish meal and during the short open season the weekly catch approached the entire annual catch of the U.S. in the Pacific. Such over-capacity could not be efficient even though the boats and plants were individually highly efficient tools for catching and processing the fish. The fishery was then in economic difficulties even before the crisis of 1972.

The exact causes of this crisis are not well understood, though most of the basic facts are clear enough. In 1972 the ocean climate was unusual (an 'el Niño') with high surface temperatures. The older fish were concentrated and easily caught, while the recruitment from the spawning in 1971 was extremely poor. After high catches in March and April 1972, the stocks and the catches fell to a very low level. Since then the adult stock, the recruitment, and the catches have all remained at what, by Peruvian standards, is a low level. The total catches are carefully regulated, but with fish meal being a major export of a country with a serious shortage of foreign exchange, the pressure to take high catches is strong. It seems that recent catches (2 to 4 million tons annually) have been sufficiently low to prevent further collapse of the stock, but not sufficient to allow it to rebuild. The puzzle remains on the cause of the poor recruitment in 1972—the timing seems wrong to be simply a result of 'el Niño', since the warm water was not apparent until the strength of the recruitment was well established.

(*d*) *Tuna*. While the tunas are extremely widespread—all of the major species (except for the southern bluefin, *Thunnus maccoyii*) occur in all the warmer oceans—the tuna fisheries are largely carried out by a handful of countries and supply a restricted market—essentially for canned tuna in the U.S. and southern Europe, and for fresh consumption (particularly in the high valued sashimi market) in Japan.

Three major types of gear are used in tuna fishing—long-lines, pole-and-line, and purse-seines. Long-lining originated in Japan, and its history since the war provides a classical example of a fishery expanding geographically and changing from species to species in response to falling catch rates. By the 1960s Japanese long-liners had spread to all oceans, and onto all the large tuna species. Thereafter total catches have changed little, but Japanese fishermen are being steadily replaced by fishermen from Korea and Taiwan.

Japan has also been one of the major centers of pole-and-line fishing (or 'live-bait' fishing, since the method depends on attracting a school of skipjack or other small tuna into a feeding frenzy by supplying small live fish). The stocks of tuna concerned are larger, and seemingly less readily affected by fishing, than the big fish exploited by the long-lines. There has been less incentive to expand the fishery from Japanese home waters (and keeping the bait-fish alive during long trips, or catching them locally have in any case made this difficult). Expansion has only taken place comparatively recently, both in the form of a steady expansion of the Japan-based fleet southwards into the western Pacific and through the establishment of locally-based fleets, e.g., in Papua-New Guinea.

The home of tuna purse-seining is Southern California, where this method replaced pole-and-line fishing in the 1950s. Successful purse-seining is more dependent on favorable hydrographic conditions than the other two methods. It has been therefore mainly confined to the eastern Pacific and eastern Atlantic, where the thermocline is shallow (less than the depth of the net). The colder water thus acts as a floor, discouraging the fish from escaping downwards while the net is being closed. In these areas the fisheries are expanding further offshore so that now a significant proportion of the catches are taken beyond the 200-mile limit, especially in the Pacific.

Both the two surface fisheries (pole-and-line and purse seiners) seem to have some room yet for expansion, particularly in the Indian Ocean. Even where the tuna fisheries are fully developed (e.g., the long-line fisheries generally, and the surface fisheries along the Pacific coast of western America) there are few indications of a threatened drop in catch (except for bluefin tuna, including southern bluefin, which has such a potential to grow to a big size, that heavy fishing on the younger fish can appreciably reduce the yield per recruit). The main management problems at present are those of allocation (e.g., between those fishing on different sizes of bluefin) and participation. In particular, the developing countries around the tropical and sub-tropical seas wish to receive greater benefits from the tuna fisheries in these waters.

15.3 SCIENTIFIC STUDY OF THE STOCKS

15.3.1 Regional collaboration

Nearly all the stocks discussed here, like most other ocean fisheries, have been of interest to several countries. Extensions of national jurisdiction will reduce the number of countries interested in many of the stocks, but in many cases (e.g., whales, tuna, several cod stocks) more than one country will still be concerned because of the migrations of the animals concerned. Management of these fisheries must involve international collaboration. This collaboration must include, in addition to agreements on any management measures actually implemented, cooperation in the study of the stocks and the assessment of their state of exploitation, and in particular the compilation of certain basic data covering the whole fishery, e.g., statistics of the total catch, without which reliable assessments cannot be made. The history of the study of these stocks is therefore essentially a history of international collaboration through the substantial number of international commissions that have been established with responsibility for particular areas or particular groups of species (see Cushing 1972 for a discussion of these bodies, particularly from a North Atlantic viewpoint).

The precise functions of these bodies vary, depending on whether or not they include both scientific studies and recommending management measures, and on the degree to which the scientific work is carried out by scientists in national institutions as opposed to the staff of the regional body itself (Table 15.1).

The best example of the purely scientific body is the International Council for the Exploration of the Sea. Founded early in the century, ICES has been concerned with most aspects of marine research, but with an emphasis toward fisheries, and one of its earliest and longest activities has been the production of

Table 15.1. Major international fisheries commissions.

Body	Area	Species	Issue statistics?	Scientific committee?	Recommendations for management measures	Remarks
A. Independent Body						
International Commission for the Northwest Atlantic Fisheries (ICNAF)	Northwest Atlantic	All	Yes	Yes	Mesh sizes Catch quotas	Replaced by new organization in 1979
International Council for the Exploration of the Seas (ICES)	Mainly Northeast Atlantic	All	Yes	Purely scientific —has many working groups	—	Provides advice to NEAFC
Northeast Atlantic Fishery Commission (NEAFC)	Northeast Atlantic	All	No	No	Mesh sizes Catch quotas	
International Whaling Commission (IWC)	Global	Whales		Yes	Catch quotas Size limits	

				Own scientific staff	Catch quotas for yellowfin
Inter-American Tropical Tuna Commission (I-ATTC)	Eastern Pacific	Tuna	Yes	Yes	Catch quotas for yellowfin
B. FAO Bodies					
Fisheries Commission for the Eastern Central Atlantic (CECAF)	Gibraltar to Congo River	All	Yes	Yes	Mesh sizes for hake and sea-breams
General Fisheries Council for the Mediterranean (GFCM)	Mediterranean	All	Yes	Yes	Mesh size for trawls
Indian Ocean Fisheries Commission (IOFC)	Indian Ocean	All	No	No	—
Indo-Pacific Fisheries Commission (IPFC)	Indo-Pacific (not exactly defined)	All	No	Yes	—

detailed and comprehensive statistics of the fisheries in the northeast Atlantic. As the needs for management have increased, ICES has been increasingly involved in providing detailed advice to the two bodies in the region with direct responsibility for management (the Northeast Atlantic Fishery Commission and the International Baltic Sea Fishery Commission). This is mostly done through various working parties concerned with particular stocks (e.g., northeast Arctic cod, North Sea herring, etc.) composed of scientists from the countries interested in the stocks concerned. The responsibilities of ICES are mainly to provide meeting facilities and the necessary statistical and similar compilations, but with several dozen groups now meeting each year, this alone is quite a large task.

The opposite extreme, of the regional body carrying out all the work from the original collection of statistical data through its analysis, to the recommendation of catch quotas or other measures, is provided by the Inter-American Tropical Tuna Commission (I-ATTC), responsible for the tuna (essentially skipjack and yellowfin) in the eastern tropical Pacific. The staff of this Commission issue log-books to the tuna seiners, who report in them detailed statistics of daily operations; they analyze these log-books to assess the state of the stocks, carry out fundamental research on the biology of tunas, and make specific proposals on the regulations for each season. Most other bodies are intermediate. For example, both the International Whaling Commission and the International Commission for the Northwest Atlantic Fisheries (ICNAF) have scientific committees which have responsibilities for providing advice very similar to those of ICES, and in addition have direct management responsibilities (these are now disappearing in the case of ICNAF, which is being replaced as a result of changing jurisdiction).

On a global scale the Food and Agriculture Organization of the United Nations (FAO) has a special role—or rather a number of roles (Fig. 15.5). Within its framework FAO has established regional bodies in a number of areas, whose activities include the provision of scientific advice on the state of the stocks, as well as holding discussions on various other aspects of fisheries management. Their activities are therefore becoming increasingly similar to those of the independent bodies.

One of FAO's major responsibilities is to assist developing countries in all aspects of food and agriculture, including building up their capacity to manage their fisheries. This assistance is supported by the United Nations Development Program, as well as by a number of individual countries (notably Canada and the Scandinavian countries) who channel much of their assistance to developing countries through FAO. It is implemented by a large number of projects on specific topics (usually of limited duration) located in individual countries, as well as regional or inter-regional projects covering large sea areas (e.g., the Indian Ocean). The activities relevant to fisheries management can range from assistance in setting up a major research institution (e.g., the Instituto del Mar del

FAO

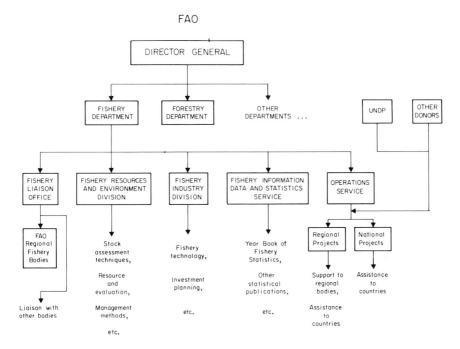

Fig. 15.5. Outline of the organization of the work of FAO and related bodies in relation to fisheries.

Peru), involving several expatriate scientists for periods of years, to short visits of individual experts to advise on some particular assessment of a stock.

FAO's other activities, as the major body concerned with fisheries, include the general dissemination of information, ideas, and techniques concerning fisheries management. For example, FAO has issued a number of manuals on stock assessment and related topics; each year it publishes a Yearbook of Fishery Statistics covering the world's fish catches; and in cooperation with many regional fisheries bodies it helps establish standards and definitions used in collecting and compiling statistics.

15.3.2   Methods of analysis
The basic methods of population analysis have been described in Chapter 5. All are potentially useful in advising on the management of open ocean resources, but the special conditions in providing this advice tends to narrow the choice in methods used in any particular situation. The main modifying condition is that the advice must be clear and readily understood by the administrator and, as far as possible, by the fishermen and the fishing industry. Also, where more than one country is interested, the scientists from each country must be able to agree on the conclusions reached. This puts a premium on simple methods, and also on the

continued use of the same methods once they have been accepted in the analysis of a given stock. Another reason for using well-established methods is that analyses often have to be carried out in a limited time, e.g., during the four or five days of a working party meeting. Therefore, although scientific groups actively engaged in providing management advice have been responsible for originating or developing many of the basic techniques—notably I-ATTC in respect of different forms of the surplus production model approach (Schaefer 1954; Pella and Tomlinson 1969)—the pressures on these groups in recent years have resulted in fewer new ideas being used and a greater emphasis within each group on a single method, familiar to the scientists and to those using their advice. Thus assessments of tuna stocks are increasingly made in terms of one or other form of the surplus production model, even though age-structured models have been shown to be usable, and in fact are particularly suitable to those tuna stocks where different fisheries take different ages of fish. In contrast, in the northeast Atlantic recent assessments of most of the individual stocks have been done almost solely by the so-called cohort or VPA analysis (Gulland 1977). This approach is based principally on the historical records of numbers caught at each age, and little use is made of other data, e.g., catch per unit effort data, or other models.

Another potential weakness in most current analyses is the range of species considered. Most scientific advice concerning management has been expressed in terms of a single species or a single stock of that species, e.g., the allowable catch for North Sea cod or for sei whales in the Antarctic. There are two reasons for this. Firstly, most current regulations are expressed in terms of a single species, e.g., as a total allowable catch for North Sea cod. Managers therefore demand as a matter of priority the single species analysis in order to update these regulations. Secondly, while it is possible to construct models that describe these multispecies interactions, and several have been constructed, these models are data-hungry. It is therefore rare to have enough data to estimate the parameters in the models with enough precision to give reliable and quantitative assessments of the effects of management actions concerning one species on the catches of another species. The multispecies analyses are thus at this stage not immediately helpful in setting quantitative regulations.

In general, therefore, and with some exceptions, the scientific analyses on which actual management decisions have been taken have been simple, and representative of the established and well-understood techniques existing a decade or more ago. This is not necessarily a bad thing. If, as in many fisheries, there is far too much fishing, it is much more valuable to have a simple analysis that gets the message across and acted on, than a much more sophisticated analysis that shows with great precision the degree of over-fishing, but is not understood by administrators and fishermen.

Also, when sophisticated analyses could be useful, the failure to use them should not necessarily be held against the scientists concerned, who would usually be only too glad to use them, given the time and facilities. The enormous expansion in the demands for advice, especially since the advent of extended jurisdiction and the introduction of exclusive economic zones out to 200 miles, has not been matched by any significant increase in the resources allocated to providing this advice, let alone a proportionate increase. As a result those scientists in government laboratories are fully engaged in routine resource evaluation and there is a real risk of a gap opening up between them and the scientists in academic institutions with more time to pursue new ideas.

15.4 IMPLEMENTATION OF MANAGEMENT

15.4.1 Targets of management

Earlier chapters have shown that fisheries management is a multi-disciplinary problem and that managers can and should have a variety of objectives, including the economic success of the fishery and the greatest possible net contribution to the national economy, as well as the health of the fish stock. The following sections therefore examine separately the degree to which management of ocean fisheries has been successful in respect of biological objectives (maintaining the stocks in a productive state) and of economic objectives. For the former it is convenient to distinguish 'growth over-fishing' from 'recruitment over-fishing'. Growth over-fishing occurs when fishing is so intense or begins at such a small size that the yield from a given group of fish (e.g., a year-class) falls because the fish have not had the opportunity to achieve their full growth potential. Given reasonable estimates of growth and mortality it is easy to detect, and to determine the appropriate measures (e.g., size limits, or reduction in the amount of fishing) to deal with it.

Recruitment over-fishing occurs when the adult stock is reduced to the level at which there is a significant fall in the average recruitment. It is much more serious than growth over-fishing and can lead to the complete collapse of the fishery, but is much harder to detect. Except for whales, all the animals discussed here have very high fecundities, so that a very small adult stock has the potential to produce a good recruitment, and in some cases clearly does. Further, there are usually very large natural fluctuations (up to two orders of magnitude) in recruitment, which make anything less than a very serious decline in average recruitment difficult to detect except over a long period.

15.4.2 Growth over-fishing

(*a*) *Protection of small fish.* Growth over-fishing has been dealt with in two ways—selective control of the catches of small fish, and a general reduction of the amount of fishing. The obvious way of approaching the former is to prohibit the capture of small fish; or more accurately, prohibit the landing or retention on board of small fish, since it is only these actions that can in general be controlled. This is effective if fishermen can avoid catching small animals while still catching the larger ones. This is only possible in a few cases, notably when each animal is taken individually, e.g., whales. Size limits are effective in controlling the catches of young whales, and have been used by IWC since its inception.

In most other fisheries it is less easy for fishermen to avoid catching a given size of fish. Setting a size limit means that they cannot land small fish, but the limit will not, in itself, mean that the small fish are not caught, so the benefit to the stock (and to future fishing) may be negligible. For example, in the Atlantic tuna fisheries it is clear that if small yellowfin tuna were not caught, the yield in weight would increase. ICCAT* has introduced a 3.2-kg size limit, but it seems that it is not producing much benefit because the small fish are mixed with legal-sized yellowfin and skipjack. The fishermen catch as many as before, but have either to discard them, dead, or land them illegally. On the other hand, for bluefin tuna there is a clearer separation between different sizes of fish, and a size limit has been effective and beneficial by forcing fishermen to avoid the schools with predominantly small fish.

The most productive approach to protecting small fish has been the use of gear that will let them escape while retaining the larger fish. Management of the cod stocks in the North Atlantic for the two decades up to the late 1960s was concerned with the use of a suitably large mesh size in the cod-end of the trawls. While the changes from the meshes used in 1950 (generally between 40- and 80- mm stretched mesh, depending on the fishery), to the sizes best suited to cod (up to 150 mm, again depending on the particular stock) were essentially simple, and involved no interference with the fisherman's other practices and his ability to fish where and when he liked, the actual implementation of these measures was a long process.

The difficulties that had to be overcome included the need to determine what sizes of fish should be protected for each stock, and the appropriate mesh size for that size of fish, taking account of the varying selectivities of different types of material; the fact that in many cod fisheries (especially the more southerly ones) other smaller species (haddock, whiting, etc.) are also caught, and large meshes suitable for cod would release too many of these species; and the short-term drop in catch that would occur following an increase in mesh size until the fish released would have time to grow. All this involved a great deal of work by the scientists of the international bodies concerned, as well as much political negotiation. As a result increases in mesh size only took place gradually, and is only now approaching the optimum size in the northern stocks, while the current mesh in use in the North Sea and other southern mixed fisheries is well below the optimum for cod. Nevertheless the increases that have been achieved have been effective in increasing catches; even if these increases are small and difficult to

* International Commission for Conservation of Atlantic Tuna.

demonstrate, relative to the magnitude of the research effort involved, an increase of 5 percent compared with what could have been taken with the earlier small meshes (a conservative estimate of the benefit), represents over 100,000 tons of cod annually.

(*b*) *Reduction in the amount of fishing.* In theory a reduction in the amount of fishing can be effective in increasing the yield per recruit (i.e., dealing with growth over-fishing) by increasing the average expectation of life of the individual fish, and thus increasing the average size of the fish caught. In practice the difficulties in implementing a limit on the amount of fishing have usually been too great to overcome in major multi-national fisheries when only an increase in yield per recruit has been in question. With the general extensions of national jurisdiction, and the greater attention being paid to management, it may be expected that restrictions on the amount of fishing in order to increase the yield per recruit will become more common.

15.4.3   Recruitment over-fishing

Control of recruitment over-fishing requires a reduction in the rate of exploitation (i.e., the percentage of the stock harvested each year) at a time when, due to falling recruitment, the size of the stock is decreasing. Successful control therefore very often involves a very substantial and rapid reduction in catches, which can be very difficult to impose on fishermen. These difficulties are increased by the uncertainties concerning the nature of the relation between the abundance of the adult and the average size of the resulting recruitment, so that the scientific advice often has to be expressed in terms of probabilities, rather than the certainty that a given measure will restore the stock. It is therefore not surprising that the history of fisheries management shows several examples of the failure to deal with recruitment over-fishing.

(*a*) *Whales.* Whales provide the obvious and best example of such failures. However, the current reputation of IWC should not obscure its real achievements and the difficulties it had to face. Depletion and near-destruction of whale stocks is not a new phenomenon of the twentieth century. That the gray whale does not exist in the Atlantic may possibly be due to the early Basque whalers. Certainly by the nineteenth century whalers had reduced the right whales in all oceans to a small fraction of their original numbers, so that the last Dundee whalers returned from Greenland in 1911 with only 7 whales among 8 ships.

It was because of the awareness of past whaling history and the concern that the same thing should not happen in the Antarctic, that industry and governments agreed to set up the IWC and to accept the severe restrictions imposed by the Commission from its beginning. The first years of the Commission from 1946 were among the earliest and most significant steps in the rational use and conservation of natural resources. For almost the first time a major world industry had its output kept well below what would have been possible (around half the possible output in the early 1950s) in order to preserve the resource.

All would have been well if there had been adequate machinery to adjust the catches taken to the current productivity of the stocks. The quotas in 1947 were close to the right figure, and it would have taken only a small adjustment downwards, and the proper balance between species, to have maintained stocks and industry in a healthy state. Unfortunately the handful of scientists advising the IWC had, in its early days, neither a theoretical framework to determine the desirable catch levels, nor enough data to build on that framework and produce good estimates of the allowable catch from each stock.

The lack of a framework has been the easier to rectify. Because recruitment falls off sharply as the adult stock declines, or the rate of exploitation increases, the curve of yield of whales as a function of stock or fishing approaches much more closely the simple theoretical curves, e.g., that of Schaefer (1954) than do those of most fish stocks, for which there may be no well-defined maximum or clear decline beyond the maximum (at higher fishing rates, or lower stock). Also the net production of whales in a particular year is more closely related to the current abundance than in the case of fish, with their great variation in year-class strength. Thus, despite its weaknesses, maximum sustainable yield (MSY) does provide a reasonable guide to whale management.

The big problem in applying MSY (or any other similar guidelines) to whales has been that of estimation. Except for those whales (gray whales, some humpback and right whales) that pass close inshore it is difficult to count whales directly, and the changes in regulations, the size and composition of the fleet, and species preferences make catch per unit effort data extremely difficult to interpret. Estimation of the abundance of the stocks in absolute terms and relative to the MSY level, and of the sustainable yield, has therefore to be somewhat subjective. The problems this can cause can obviously be seriously exacerbated by the growing polarity apparent within the IWC and its Scientific Committee between 'exploiters' and 'environmentalists.' It is therefore difficult to obtain a generally agreed assessment of what is happening to whale stocks. Nevertheless there has been a clear change in the knowledge of the stocks, and in the relation between catches and sustainable yield between 1965 (when the IWC received its first comprehensive and quantitative advice from the Committee of Three Scientists) and the present day.

Before 1965 there were no clear estimates of the abundance of the whale stocks, or of their sustainable yield, but it is now clear that the catches had significantly exceeded the sustainable yields, and both blue and fin whales had been depleted well below the MSY level. Now blue and fin whales are protected, and are almost certainly increasing, though by the nature of things there is not enough direct evidence to prove this. The stocks of sei and minke whales now being exploited in the southern hemisphere are certainly close to and probably at or above MSY, but the catches certainly do not exceed the sustainable yield by more than a small amount (not enough to seriously affect the stock unless continued for a long time). Other stocks of these species are believed to be probably below MSY; these are protected and are presumably increasing. More doubt surrounds the sperm whale stocks, where the complex social structure may make the stocks more sensitive to exploitation, and especially to the killing of males currently without harems, than was previously supposed.

The relatively minor uncertainties noted here—whether a stock is a little below or a little above MSY—do lead to major differences in the allowable catches under IWC rules. The big changes in recent quotas from year to year, resulting from changing interpretation of the status of the stocks and the arguments surrounding them give an impression of a much greater degree of uncertainty about the status of the stock than is the case. In fact for most whales (bowhead whales and possibly also sperm whales are an exception), the present situation is relatively satisfactory, and is steadily improving. The same unfortunately cannot be said about the whaling industry.

(*b*) *Fish.* Historically managers of fish stocks, i.e., excluding whales, have not done much explicitly about the problems of recruitment over-fishing. After an early period when it was felt that protection of spawning fish was a good thing regardless of actual scientific evidence—reflected, for example, in a common prohibition on the landing or sale of berried female lobsters—there was a generally accepted assumption that the fecundity of fish was so high that there would always be a sufficient spawning stock. So far as most ocean fisheries were concerned, i.e., excluding elasmobranchs and salmon, this feeling was for a long time consistent with the evidence. Even now there are few, if any, individual stocks of open ocean fish for which a difference in average recruitment at different stock sizes has been clearly demonstrated. (For a general discussion of the stock/recruitment problem see Cushing 1977 and Parrish 1973.)

In practice, management of several stocks has been based on an implicit assumption about the effects on recruitment. For example, I-ATTC has set catch quotas on yellowfin tuna in the more easterly part of the eastern tropical Pacific, based on the Schaefer-type model. Given the growth and mortality rates of the fish, the decline in total catch at high fishing rates predicted by the models used must imply a significant decrease in recruitment. Whether in fact this would occur, and hence whether the regulations have been strictly necessary, has not been clearly demonstrated. The main practical effect of the regulation is however clear—the fleet has been forced, after the closure of fishing on the traditional grounds, to move much further offshore onto new stocks and this has led to a considerable increase in the catch of yellowfin from the eastern Pacific as a whole.

The easy assumption that recruitment will not be affected has most clearly ceased to be tenable in relation to the clupeoid fisheries (see Murphy 1977). Too many stocks of these fish, including many of the herring stocks in the North Atlantic, as well as the Californian sardine and Peruvian anchoveta, have collapsed partially or wholly after periods of sustained heavy fishing, for much reasonable doubt to remain that fishing can reduce recruitment (see Chapter 14). Equally, there is clear evidence that many of these stocks can undergo big natural fluctuations independent of fishing. In the Californian current evidence from fish scales in bottom deposits show fluctuations over periods of centuries. Support for the importance of non-fishing causes also comes from the recovery of a few stocks (the Japanese sardine is an example) in the absence of definite management action.

Given the large year-to-year variation in recruitment—clearly due to natural causes—it has been extremely difficult to be sure in any given case that 'recruitment over-fishing' is taking place, at least until the decline in recruitment has been maintained for so long that the scientists are engaged in a post-mortem, rather than advising on the management of a sick, but still living, fishery. For example, arguments on the decline of several herring stocks in the North Atlantic continued while they declined to a level at which the fisheries more or less disappeared. These scientific doubts were disastrous for the management of these resources, especially in ICNAF and NEAFC, where the responsibility for taking decisions was diffused among the dozen or more member countries, with unanimous agreement being virtually essential for positive action to be taken. Inevitably, in the absence of clear advice from the scientists, decisions tended to be taken too late, if they were taken at all. This tendency for late, or inadequate decisions, has been helped by practice of expressing advice in terms of catch quotas. Without good estimates of current absolute abundance of the spawning stock and with no estimate of what the desirable abundance should be, it has been difficult to determine what the catch quota should be. Further, when the stock is rapidly decreasing, a moderate reduction in catch quota often appears to be a further strengthening of management, when in fact the result (taking account of a reduction in stock greater than the reduction in quotas) will be to allow the fishing mortality to increase. The result has been a collapse of many of the major herring stocks, a series of events that can be counted among the least successful episodes in fisheries management. In this respect, some of the stocks managed by a single authority have fared better; severe restrictions were applied to the British Columbian herring fishery when it appeared to be in decline around 1967, and this stock has recovered (Murphy 1977). On the other hand, the Peruvian authorities, while restricting catches since 1972 to levels that have probably been low enough to prevent further collapse, have been unable to accept the very severe restrictions (possibly no fishing at all) that would give the best chance of a rapid and full recovery of the stock.

15.4.4   Economic and social objectives

Commercial fishermen go to sea to make a living. Equally the common objective of countries in managing fisheries is the well-being of the national economy. The well-being of the fish stocks is only a means to an end, rather than an end in itself. Whales provide an exception, since the most vocal interests now concerned with whale management are the various environmental groups who at best tolerate the commercial harvesting of whales, and some of whom actively oppose it on principle. Despite this, and the fact that the importance of economic and similar factors have long been recognized by international bodies (e.g., ICNAF 1968), as well as by national governments, very few of the measures taken to manage oceanic resources have led to improved economic performance. If anything, the reverse has often been true, in that many measures that have been successful in maintaining a healthy fish stock have achieved this partial goal at the price of imposing economic inefficiency on the fishery (e.g., by short open seasons)—though this does not mean that in the absence of these measures the decline in the stocks would not have resulted in an even worse economic performance.

As shown as early as 1954 (Gordon 1954), unless access is restricted, it is very difficult to realize significant long-term economic benefits. For example, the overall catch quotas set by I-ATTC for yellowfin in the eastern tropical Pacific have done nothing to discourage the growth of what appears to be an excessively large fleet. The details of the regulations, which allow ships at sea at the end of the open season to land a full load, have also probably encouraged the development of super-seiners with uneconomically large carrying capacity to take advantage of this 'free' trip.

The biggest difficulty in obtaining much economic benefit—other than the prevention of the economic disasters that would occur if there were a failure to maintain the biological productivity of the stocks—is that management (and particularly management to achieve economic goals) has nearly always been actively considered only after the fishery concerned has expanded well beyond the economically most desirable level. Once too many boats have been built the best opportunity for saving on capital costs has gone. Once too many fishermen have entered a fishery, diverting them into other employment may be very difficult without great social disruption. In these cases the conflict between long-term and short-term considerations is very clear. Whatever the advantages of using fewer vessels, and hence less maintenance and other costs, as well as fewer crews, it can be extremely uncomfortable to attempt to arrive at this promised land at a faster pace than that provided by the natural wastage of ships and retirement of fishermen.

A clear example is provided by the Peruvian anchoveta fishery. Even before the collapse of the stock in 1972 it was clear that the level of investment and employment in the fishery did not make economic sense. The over-capacity was, of course, even greater after 1972, when the available catch was much reduced. In 1973 the Government nationalized the fishing industry, setting up a single state enterprise, PESCAPERU. Apart from other more general social and political objectives, this served to facilitate some degree of rationalization of the industry, and withdrawal of surplus boats and fish meal plants (particularly the latter). Nevertheless, the problems of finding alternative employment for the fishermen, especially in towns like Chimbote situated on the edge of a desert with little economic activity other than fishing, meant that this was a very slow process— much slower than the original attraction into the industry of people who, to a large extent, had never seen the sea before joining the anchoveta fleet. For several years the problem of keeping fishermen employed was tackled by dividing the fleet in half and allowing one half to fish in one week and the other half the next— and even then the fishery was closed to any fishing at all for much of the year.

15.5   THE FUTURE

15.5.1   Extended jurisdiction

The most significant event for management of ocean fisheries in recent years, and one which will dominate events for a few years to come, is the general extension of jurisdiction over fisheries to 200 miles. Although these extensions have mostly been taken in the context of the negotiations at the U.N. Conference on the Law of the Sea, they have been taken independently of the conclusions of that Conference, and largely pre-empt them.

In itself an extension to 200 miles does not solve any management problem, but does provide a framework within which solutions may be easier to find. This is made more likely because fisheries and ocean affairs generally have caught the attention of the public and of high level policy makers to an extent unheard of in the past, and unlikely to occur again. Those with day-to-day responsibilities have, for a fleeting moment, a chance to catch the eye of those who can make fundamental policy decisions, e.g., on whether everyone has the right to go fishing when and where he likes, subject to conservation controls applicable to everyone, or whether fisheries resources will become the 'property' of one or other group of interests.

The effect of extended jurisdiction on oceanic fisheries is to divide resources into three classes—those the exclusive interest of one country; those that are shared by more than one country (e.g., Arcto-Norwegian cod, sardinella off Mauritania/Senegal); and those that can be taken partly or wholly outside national jurisdiction (e.g., tuna).

The first group clearly now fall under a single authority that can take decisions on the basis of what appears best to that authority, without waiting for comprehensive scientific advice, or for the consensus of all participants that was necessary under the old arrangement (Chapter 14). The second group still comes under multiple authority, but the authorities responsible for a given stock are limited in number (usually not more than two or three adjacent nations) and are clearly defined. It should not be too difficult for these authorities to get together and agree on suitable measures. This may be contrasted with the typical situation before the general introduction of 200-mile limits when a dozen countries might be interested in a single stock (e.g., Grand Banks cod); further, these countries might at any time be joined by other countries, and management measures had to take account of the possibilities of these new entrants.

The third group is least affected by the changes, and to the extent that tuna or other species can be harvested beyond 200 miles the problems of the indeterminate authority and ownership that have made management difficult in the past will still remain. On the other hand, even the most oceanic and highly migratory species such as bluefin or albacore are taken to a large extent within 200 miles and—irrespective of the formal decisions taken at UNCLOS regarding highly migratory species—coastal states (e.g., Australia in respect of southern bluefin, the U.S. in respect of North Atlantic bluefin, or Mexico or Costa Rica in respect of Pacific yellowfin) will expect to have a much greater say in the management decisions concerning these stocks.

15.5.2   Scientific problems

The ability to make management decisions quicker and more easily arising from changes in jurisdiction places additional responsibility on the scientists to provide timely and appropriate advice. While, as already noted, much of this advice will continue to be based on well established methods, especially the single species models typified by the studies of Schaefer (1954) and Beverton and Holt (1957), these will not always be adequate. Even where only one species is being exploited these models are not adequate when there is a question of the size at which the adult stock should be maintained in order to ensure adequate recruitment for the future. Better resolution of the whole stock/recruitment problem is a matter that seems bound to attract increased scientific attention in the future.

One obstacle is that fish stocks and fisheries are affected by changes in many factors other than fishing. These factors may have effects lasting in duration from a sudden storm interrupting fishing to shifts in climate that can establish or eliminate a complete stock (e.g., the cod at West Greenland; Cushing and Dickson 1976). Despite these variations the commonly used models of the biology and economics of fishing deal almost exclusively with steady states, and average conditions. This is clearly not satisfactory; the variability of catches is an important factor in determining the economic value of a fishery. For example, steady catches of 5000 tons a year are likely to be worth considerably more than catches of 1000 tons one year and 9000 the next. It can be expected therefore that considerably more attention will be paid in the future to models, and scientific studies generally, that deal explicitly with variability and the effects of changes in the natural environment.

The other scientific problem that will undoubtedly require more attention is that of the interaction between fisheries on different species. Like the stock/recruitment it is easier to demonstrate that the problem exists than to put forward methods for tackling it (FAO 1978).

15.5.3   Allocation

The extension of national jurisdiction, and the growing realization that fisheries management should serve economic or social goals, as well as biological ends, is likely to focus attention on problems of allocation—that is who gets the benefits from fisheries resources. Extension of limits has effected a first stage of allocation to the coastal states in respect of all but the third category of resources (tunas, etc.). Where two or more states share the same resource, there will be a considerable demand on scientists to provide advice on the migrations and distribution of the stocks in order to provide a basis for allocation between the countries concerned.

Allocation in the fisheries outside 200 miles presents special difficulties. In practice, with the decline in whaling, these mean the tuna fisheries. Here there is a clear division between the rich developed countries of the temperate areas currently doing nearly all the fishing for tuna, and the poorer countries near (or fairly near) to whose coasts much of the tuna is caught. Almost certainly discussions on allocation of the benefits in tuna fishing (using the terms in a wide sense) will bring in wider considerations, such as the concept of a New International Economic Order, and a better balance between rich and poor. How far this will go and the degree to which the small island states of the southwest Pacific, or the coastal countries of the Indian Ocean will benefit from the tuna in the oceans off their shores remains to be seen. At this moment all that can be said is that present arrangements will come under very close examination.

15.5.4   New resources

Just as the falling catch rates of cod in the North Sea drove the English trawlers to Iceland three-quarters of a century ago, falling catch rates among currently fished stocks, and the restrictions necessary to manage these fisheries, coupled with the demand for fish from a growing world population, will lead to a search for new stocks to exploit. Despite the growth in world fisheries in the last couple of decades, there still remain a few stocks of familiar types of fish that can produce greater catches (e.g., pelagic fish in the north-western Arabian Sea, blue whiting off the British west coast). However, most of the stocks are of the less valuable fish, and are more scattered and expensive to harvest than stocks already being fished. Altogether the present catches of around 60 million tons of marine fish might be expanded to perhaps 90 million tons without large-scale exploitation of the so-called unconventional species. (Perfectly managed and fully exploited the conventional resources might yield around 100 million tons (Gulland 1971), but it is unreasonable to expect as much to be squeezed out, bearing in mind the difficulty of perfect management, and the existence of low-valued and scattered stocks which will not be worth exploiting fully). Attention is therefore shifting to the less familiar types of fish; prominent among these are the Antarctic krill, and the small meso-pelagic fish. Reviews of the current knowledge of these stocks (Everson 1977; Gjosaeter and Kawaguchi 1980) show that very considerable catches (tens of millions of tons) could be taken from these resources. However, this does not necessarily mean that such harvesting will occur. Catching krill in large quantities (several tons an hour) is possible, but requires large and expensive ships; processing to produce a cheap product that can command a large-scale market presents comparable problems. It is by no means certain that a large-scale fishery for either meso-pelagic fish or krill (apart from a small speciality market for krill in Japan) will exist in the near or medium-term future. What is surer is that such fisheries, when they begin, will not replace the conventional ocean fisheries, and will in no way reduce the need to manage these fisheries rather better than in the past.

15.6 REFERENCES

Beverton R.J.H. and S.J.Holt (1957) On the dynamics of exploited fish populations. *Fish. Invest. Minist. Agric. Fish. Food G.B.* (*2 Sea Fish.*), **19**.

Cushing D.H. (1972) A history of the international fisheries commissions. *Proc. R. Soc., Edin.* **73(36)**, 361–390

Cushing D.H. (1977) The problems of stock and recruitment. *In* J.A.Gulland (ed.) *Fish Population Dynamics*, pp.116–133. London: Wiley-Interscience.

Cushing D.H. and R.R.Dickson (1976) The biological response in the sea to climatic changes. *Adv. Mar. Biol.* **14**.

Everson I. (1977) *The Living Resources of the Southern Ocean.* Rome: UNDP/FAO, GLO/SO/77/1.

FAO (1978) Expert consultation on management of multispecies fisheries, Rome, Italy, 20–23 September 1977. *FAO Fish. Tech. Pap.* **181**.

FAO (1972) *Atlas of Living Resources of the Sea.* Rome: FAO.

Gjøsaeter J. and K.Kawaguchi (1979) A review of the world resources of meso-pelagic fish. FAO Fisheries Technical Paper (in preparation).

Gordon H.S. (1954) The economic theory of a common property resource: the fishery. *J. Polit. Econ.* **62**, 1924–1942

Graham M. (1948) *Rational Fishing of the Cod in the North Sea.* London: Edward Arnold.

Gulland J.A. (1971) *The Fish Resources of the Ocean.* 255 p. West Byfleet: Fishing News (Books).

Gulland J.A. (1974) *The Management of Marine Fisheries.* Bristol: Scientechnica.

Gulland J.A. (1977) The analysis of data and development of models. *In* J.A.Gulland (ed.) *Fish Population Dynamics*, pp.67–95. London: John Wiley.

Gulland J.A. and G.R.Williamson (1962) Transatlantic journey of a tagged cod. *Nature, Lond.* **195**(4844), 921

Harden-Jones F.R. (1968) *Fish Migration*. London: Edward Arnold.

ICNAF (1968) Report of the working group on joint biological and economic assessment of conservation action. *Ann. Proc. Int. Commn. N.W. Atlant. Fish*. **17**, 48–84

Innis H.A. (1954) *The Cod Fisheries*. Toronto: Univ. of Toronto Press.

Murphy G.I. (1966) Population biology of the Pacific sardine (*Sardinops caerulea*). *Proc. Cal. Acad. Sci*. **34**(1), 1–84

Murphy G.I. (1977) Clupeoids. *In* J.A.Gulland (ed.) *Fish Population Dynamics*, pp.283–308. London: Wiley-Interscience.

Nagasaki F. (1973) Long-term and short-term fluctuations in the catches of coastal pelagic fisheries around Japan. *J. Fish. Res. Board Can*. **30**(12), pt.2, 2361–2367

Parrish B.B. (ed.) (1973) Fish stocks and recruitment: Proceedings of a symposium held in Aarhus, 7–10 July 1970. *Rapp. Proc. Verb. Reun. Cons. Int. Explor. Mer*. **164**

Pella J.J. and P.K.Tomlinson (1969) A generalized stock production model. *Bull. I-ATTC*, **13**(3), 421–496

Russel E.S. (1942) *The Overfishing Problem*. Cambridge: Univ. Press.

Ryther J.H. (1969) Photosynthesis and fish production in the sea. *Science*, **166**, 72–76

Schaefer M.B. (1954) Some aspects of the dynamics of populations important to the management of marine fisheries. *Bull. I-ATTC*, **1**, 25–56

Schevill W.E. (ed.) (1974) *The Whale Problem: A Status Report*. Cambridge, Mass.: Harvard Univ. Press.

Townsend C.H. (1935) The distribution of certain whales as shown by logbook records of American whaleships. *Zoologica, N.Y*. **19**, 1–50

Chapter 16
Aquaculture

JAMES W. AVAULT, JR.

16.1 INTRODUCTION

Aquaculture is the deliberate culture of plants and animals in water, usually for commercial purposes. A species is stocked or confined in owned or leased waters, managed to produce maximum yield, harvested, and then utilized. Aquaculture is like agriculture where men produce predictable crops through management. Aquaculture is not, however, capture fisheries—the harvest of fishes from oceans and other natural waters.

Aquaculture includes the culture of species such as channel catfish (*Ictalurus punctatus*) and red swamp crawfish (*Procambarus clarkii*) for food, species such as largemouth bass (*Micropterus salmoides*) and bluegill (*Lepomis macrochirus*) for sportfishing, and species such as fathead minnow (*Pimephales promelas*) and golden shiner (*Notemigonus crysoleucas*) for fish bait. Species such as goldfish (*Carassius auratus*) are cultured for ornamental purpose. Various seaweed species are cultured for food additives, drugs, and other purposes.

Aquaculture is thought to have begun when man noticed fish that invaded inundated land. After waters receded, fish became stranded in potholes, and someone probably noted that some of these captive fish spawned and would eat food scraps. Another theory says that aquaculture began when wild fish were stored in floating cages to provide fresh fish later. In Cambodia, fish have been held in cages for centuries. It was only a short step before fish were deliberately stocked and fed in cages. This practice in Cambodia was probably the forerunner of today's cage culture and pen culture.

Aquaculture's earliest history began in the Indo-Pacific Region. Ling (1974) wrote that aquaculture began in China. Some emperors loved fish but it was difficult to get fish year-round in frigid northern China. An entrepreneur gained favor with the emperor when he stored fish and offered them to the emperor when fish were not normally available. Fish were no doubt also used as bribes, and eventually deliberate fish culture began.

The carp (*Cyprinus carpio*) was the first fish cultured. Artificial hatching of carp dates back to 2000 B.C. Aquaculture in China closely followed the development of silkworm culture, which dates back to 2698 B.C. The origin of aquaculture probably began during that period since silkworm pupae and feces were used to feed carp. Fan Lai wrote the first published work on carp culture in 460 B.C. Brown (1977) gave a detailed account of aquaculture in China and discussed the culture of other species. Today, aquaculture furnishes over 40 percent of China's fishery products.

Chinese who migrated to Taiwan, Thailand, Malaysia, Indonesia, and elsewhere in the Indo-Pacific Region brought with them knowledge of aquaculture, and culture of species other than carp began. Aquaculture is more important in the Indo-Pacific Region than in any other part of the world. More than 75 percent of the world's aquaculture production comes from this region. The milkfish (*Chanos chanos*) is cultured in brackish water ponds on a large scale in Taiwan, Indonesia, and the Philippines. In 1978, over 175,000 ha were devoted to milkfish culture in the Philippines alone. Indonesian aquaculture, mainly with milkfish, contributes significantly to the fishery, and in Java 60 percent of the fish consumed come from ponds. Around 1939, *Tilapia mossambica* was cultured in Java and the method spread throughout the Indo-Pacific Region. Other species of *Tilapia* were later cultured.

Japan is more noted for its aquaculture than perhaps any other country. Japan began commercial aquaculture approximately 150 years ago and produced 880,000 metric tons from shallow-sea culture and 67,000 metric tons from freshwater aquaculture in 1974 (Brown 1977). Japan has screened numerous species for aquaculture. A partial list of important species cultured in that country includes yellowtail (*Seriola quinqueradiata*), carp, eel (*Anguilla japonica*), shrimp (*Penaeus japonicus*), rainbow trout (*Salmo gairdneri*), oysters (several species), and the seaweed nori (*Porphyra tenera*).

Australian aquaculture is dominated by oyster production where about 10,000 metric tons are produced annually (Brown 1977). Hawaii cultures several species including milkfish. The freshwater prawn *Macrobrachium rosenbergii* shows potential as a culture species.

In Europe, aquaculture began over 2000 years ago when Romans cultured oysters. Carp culture developed during the Middle Ages and is widespread throughout much of Europe. Trout culture is well developed in Denmark and other Scandinavian countries. Several species of flatfishes are now cultured in England.

In Africa, although it dates back to 2500 B.C. aquaculture is not yet too important since capture fisheries have been emphasized. Some trout are cultured and *Tilapia* sp. culture developed on that continent during the 1950s.

In Latin America, carp are cultured in Mexico, and *Tilapia* sp. and the grass carp (*Ctenopharyngodon idella*) have also been introduced. Central America does not have a long history of aquaculture but has a potential for crustacean culture. In South America, aquaculture is best developed in Brazil where carp and *Tilapia* sp. are cultured.

In the United States, aquaculture of oysters began in the late 1850s, and mainly involved moving small oysters to better growing waters. Approximately 40 percent of the U.S. oyster production came from private aquaculture during 1976. The United States is one of the world's largest producers and consumers of oysters. Oyster production peaked in 1908 when 69,000 metric tons were produced, but production dropped to 9000 metric tons by 1973 due to pollution, diseases, predators, and low-cost imports (Glude 1977).

Commercial aquaculture of trout in the United States began during the 1870s. Private farms are now located in 38 states but trout production is concentrated in Idaho, Colorado, Wisconsin, Michigan and Pennsylvania. In 1977, approximately 200 farms produced 13,600 metric tons of trout. Culture of coho salmon (*Oncorhynchus kisutch*) and other salmon species began during the past decade. Approximately 60 metric tons of salmon were produced in 1972.

Commercial production of bait minnows began in the late 1920s and by 1971 over 32,780 ha were in production. Catfish farming began in the late 1950s and although less than 200 ha were in production by 1960, approximately 22,258 ha were in production during 1978. Crawfish farming began in south Louisiana during the mid 1950s and 24,000 ha were in production by 1980. This is the largest commercial crustacean aquaculture found in the United States.

Aquaculture of ornamental fish is practiced in several states but is most important in Florida. In 1972, approximately 150 tropical fish farms in Florida produced 97 million fish. Florida also imported an additional 53 million ornamental fish during 1972. The retail value of domestic and imported ornamental fish totaled $300 million in 1972.

16.2   CULTURE OF CHANNEL CATFISH

Several species of warmwater fish are cultured in the United States. The channel catfish is by far the most important in terms of number of states where catfish are cultured, ha in production, kg produced, and economic value of the industry. This species will serve as an example of the culture of warmwater fish. Management techniques for all species are similar since aquaculturists must be concerned with site selection, water quality, feeds and nutrition, parasites and diseases, and other management techniques.

16.2.1   Site selection

(*a*) *Soil requirements*. The soil should contain enough clay to hold water. Use a soil auger to take samples throughout the pond area. Soil with a history of pesticide use should be checked for pesticide residues.

(*b*) *Topography*. Land may be hilly or level. Pond construction may be simplified on hilly land since a dam can be placed across the narrow neck(s) of a valley, but fish farmers have less control over the pond's size, depth, or shape. Most fish farmers prefer building ponds on level ground.

(*c*) *Water.* An adequate supply of good quality water must be available. The volume of water needed to fill a pond is expressed in cubic meters. A one-ha pond with an average depth of 122 cm contains 12,200 cubic meters of water. A well producing 4000 liters per minute would fill this pond in about 50 hours. As a rule of thumb, a well producing 4000 liters per minute is adequate for a 16-ha fish farm. All ponds on a farm are not filled at the same time, and no pond should take more than 2 to 3 weeks to fill.

The water supply should be saturated with dissolved oxygen (DO), have a pH of 6.5 to 8.5, and have a total hardness and a total alkalinity of at least 50 ppm. Water temperature should range from 20 to 30°C, and water should be free of silt, pesticides, pollution, and wild fish. Well water is chronically low in DO, but splashing or aerating the water before it enters the pond will correct this. Excessive amounts of iron may be present but exposure to air will convert iron from the ferrous to the ferric state and it will then settle out.

(*d*) *Temperature.* Most catfish farms are located in the southern United States because of this region's longer growing season. The optimum temperature for growth is 30°C. Growth practically ceases when the water temperature drops to 12°C. Generally, warmwater species thrive where water temperature exceeds 21°C, whereas coldwater species grow best at water temperatures below 21°C.

(*e*) *Geographic location.* The farm should be located as close to the market as possible if live fish are hauled.

16.2.2   Pond construction

Although channel catfish are cultured in floating cages, tanks, and raceways, ponds are the mainstay of the industry.

(*a*) *Building the levee.* The entire area must be cleared of all brush, grass sod, and litter before construction begins. Such materials would cause leaks if put into the levee. Once cleared, the outline of the levee is staked off (Fig. 16.1). Two rows of stakes mark the top-width of the levee. If the levee will be 3.5 meters wide, the stakes are set in parallel rows 3.5 meters apart and extend the length of the levee. Toe stakes mark the base of the levee. They are parallel to and outside of the top-width stakes. The distance between top-width stakes and toe stakes depends on the slope of the levee and its height.

The levee has a certain minimum slope. The slope may be 2:1, 3:1, or greater. The toe stakes are set 2 meters from the top-width stakes for each one meter in levee height if a levee has a 2:1 slope. The slope of the levee depends on soil type, pond size, and potential wind and wave action. Soils containing clay hold together better, and a steeper slope can be used. Small ponds containing 2 ha or less require only a 2:1 slope, but larger ponds require a slope of 3:1 or greater.

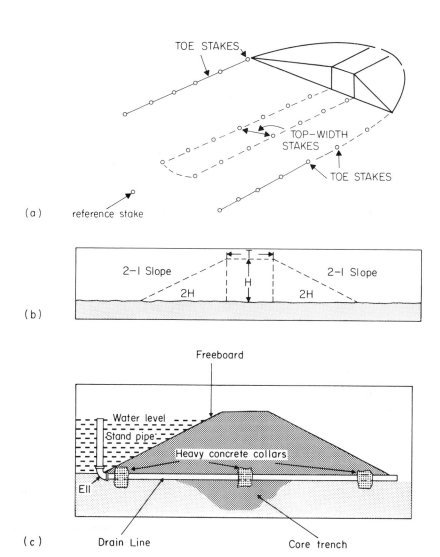

Fig. 16.1. (a) Portion of levee showing top-width, toe, and reference stakes; (b) Levee showing 2:1 slope, top-width (*T*), and height (*H*); (c) Cross section of levee showing drain line and stand pipe.

If a pond is 16 ha or larger, a 4:1 slope may be required because the levees are more likely to erode from wind and waves.

All levees must extend above the water level to prevent floods from washing levees. This extra height is known as freeboard.

(*b*) *Core trench.* A core trench dug the full length of the levee must be deep enough to hit clay subsoil. When completed, the core trench is filled with clay excavated from inside the proposed pond. Impervious clay must extend the full length of the levee and be well bound to the sub-soil to prevent leaks.

(*c*) *Drain line.* The drain line is installed at the lowest point in the pond and should fall 30 cm for each 30 meters of line. A stand pipe is joined to the drain pipe by a flexible ell. The pond can be drained by pushing the stand pipe into the water. The drain pipe should be 15 to 20 cm in diameter to drain ponds of one ha or smaller. Ponds larger than one ha require a 20 to 30 cm diameter drain pipe.

(*d*) *Shape and size of pond.* The pond bottom should be sloped from the shallow end to the deep end to facilitate draining. A fall of 6 cm per 30 meters is recommended. The pond bottom will erode at greater slopes. Ponds are usually 120 cm to 150 cm deep.

Pond shape affects construction costs and harvesting ease. Square ponds have less levee per ha and are cheaper to construct, but most fish farmers prefer ponds that are approximately $1\frac{1}{2}$ times longer than wide to facilitate harvest.

Larger ponds are less expensive to construct than smaller ponds because there is relatively less levee per ha of water. However, a farmer must also consider other factors such as harvesting and marketing. Experienced fish farmers find that ponds approximately 8 ha large are ideal for producing food fish. Ponds 0.5 to 1 ha in area are better suited for spawning and for rearing fingerlings.

16.2.3   Spawning and hatching

Good brood stock are essential (Fig. 16.2). They should be well fed but not overfed. They should not be overcrowded, and good water quality should be maintained. Parasite and disease problems must also be monitored, and brood stock should not be roughly handled at stocking.

Brood stock should be at least 3 years old and should weigh between 1.5 and 5 kg. Younger fish may spawn but spawning is not as reliable as with older brood stock. Fish heavier than 5 kg are difficult to handle. Moreover, egg production per female drops relative to body weight as fish increase in size. When the water temperature reaches a minimum of 22°C in the spring, brood stock are sorted according to sex. Males have a broad muscular head and are usually darkly pigmented under the jaw. Females have a narrow head and swell with eggs near spawning time.

Fifty pair are stocked per ha. Spawning containers, such as old milk cans, should be placed around the edge of the pond. The number of eggs spawned per female varies from about 6000 to 8000 eggs per kg of bodyweight. Eggs can be left

Fig. 16.2. Author with healthy brood channel catfish. Photo: James W. Avault, Jr.

in the spawning container for the male to fertilize and hatch or they can be brought into a hatchery, placed in a trough with a paddlewheel, and hatched artificially. The time required to hatch eggs varies with water temperature. Eggs require from 10 days to hatch at a water temperature of 21°C and 5 days to hatch at a temperature of 29°C. The optimum hatching temperature is about 27°C. Once eggs hatch, the sac fry rely on yolks for approximately 3 days before they swim up in search of food. Swim-up fry are usually fed in tanks for about a week or until they are feeding well. The best feed is a mash that contains about 35 percent protein.

16.2.4 Rearing fingerlings

Ponds for rearing fingerlings should be filled no more than a few days before fry are stocked to prevent a build-up of predators, especially predacious insects. Air-breathing insects can be controlled by spraying a mixture of diesel fuel and motor oil on the pond surface once a week. The spray should be applied when there is a slight breeze to help spread the oil mixture that clogs the insects' breathing tubes.

The number of fry stocked per ha governs their growth. A stocking rate of 50,000 fry per ha in summer will usually result in fingerlings 15 to 20 cm long by the fall. Rearing ponds can be stocked at higher or lower rates depending on the final size desired. Fish are fed about 3 to 4 percent of body weight daily.

16.2.5   Producing eating size fish

Fingerlings grow to eating size in production ponds. These ponds are usually stocked in early spring, and fingerlings grow to eating size by autumn. The number of fingerlings stocked per ha depends on the size desired at the end of growing season. Medium-sized fingerlings (10 to 15 cm) stocked at 3750 per ha usually weigh an average of 450 grams at the end of a 210-day growing season. The same sized fingerlings stocked at 5000 per ha will weigh an average of slightly less than 450 grams after 210 days. Larger fingerlings (25 cm long) stocked at 3000 per ha will weigh an average of about 900 grams at the end of a growing season.

(*a*) *Feeding Catfish.* Catfish are fed daily to produce an eating size fish. Good catfish feed contains 28 to 32 percent protein, no less than 5 percent fat, from 10 to 20 percent carbohydrate, and from 10 to 15 percent fiber. A minimum of 8 percent of the ration should be from fishmeal, and all feeds should contain necessary vitamins.

Feeds may either sink or float. Sinking pellets are cheaper but do not remain stable in water as long as floating pellets. Pellets vary in size, and many feed manufacturers produce several sizes of pellets to correspond with fish sizes.

Fish are fed a certain percentage of their bodyweight daily. When first stocked, fingerlings are fed approximately 4 percent of their bodyweight. They are fed 3 percent of their bodyweight daily when they weigh approximately 225 grams, and 2 percent over 300 grams. Feeding charts are available that consider the size of fish, water temperature, type feed, and other factors.

The cost of feed is the single largest variable cost incurred when growing fish. If, for example, a 1-ha pond is stocked with 5000 fish and fish weigh an average of 45 grams each, fingerlings will weigh a total of 225 kg. If each fish is grown to a half kg, standing crop will be 2500 kg, 2275 kg of which was produced in the pond. If the feed conversion rate (i.e., the ratio of feed to live weight produced) is 1.75, and feed costs 33 cents per kg, 3981 kg of feed would be required at a total cost of $1314.

(*b*) *Water management.* Oxygen depletion is the main water management problem. Generally, the level of dissolved oxygen should never be allowed to drop below 3 ppm. Most oxygen is added to water as a by-product from the photosynthetic activity of green plants, primarily phytoplankton. Wave action and agitation of water by aeration increases the amount of dissolved oxygen as water is exposed to the atmosphere. Dissolved oxygen can be reduced in pond water by lack of sunlight that limits photosynthesis, by a shortage of nutrients, especially phosphorous, which also curtails photosynthesis, by an increase in water temperature or salinity, by increased respiration of plants and animals particularly in overstocked ponds, and by oxidation of organic matter.

Dissolved oxygen levels can be maintained in pond water by avoiding overstocking or overfeeding. No more than 5000 fish per ha should be stocked and no more than 40 kg of feed per ha should be fed daily in ponds without water flow. When fish are overstocked, the organic matter in feces and the accumulation of uneaten feed may deplete oxygen. Low amounts of dissolved oxygen can be corrected by several means (see Chapter 12). Fresh water containing more dissolved oxygen can be pumped into ponds, and/or water can be recirculated or sprayed into the air. Phosphate fertilizer can be added to pond water to encourage regrowth of phytoplankton.

Off-flavor catfish is related to water quality. An earthy, musty or muddy taste may be imparted to catfish flesh from chemical compounds related to certain bacteria and blue-green algae. Growers should eat a baked fish without salt or other spices and delay harvesting if they detect a musty taste. The musty taste can be controlled by flushing the pond with fresh water or by holding fish for as long as a week in tanks with flowing water.

(*c*) *Fish health*. Fish health can be maintained by not overstocking, by proper feeding, and by ensuring that water contains sufficient dissolved oxygen. Wild fish may carry diseases and must be kept out of ponds. Water that contains wild fish should be filtered through a saran screen before it enters ponds.

Fish may still get parasites or diseases even with these precautions. Parasites and diseases may be controlled by correcting the problem, such as poor water quality, by changing the environment, or by treating fish.

The ciliated protozoan *Ichthyophthirius multifilis*, or Ich, may be controlled by changing the environment. This parasite is one of the most serious catfish parasites. It also affects other fish including ornamentals. Ich can be controlled by raising the water temperature above 28°C. It can also be controlled by holding infected fish in fast flowing water; the parasite cysts drop off the fish and flushed away.

Parasites and bacterial diseases can also be controlled by adding drugs to feed, by vaccinating fish, or by adding a chemical to water. Fish chemotherapy is regulated by the U.S. Department of Agriculture and the U.S. Food and Drug Administration, and all chemicals used to treat fish must be approved by these two agencies. Bacterial diseases may occur in the summer, especially if fish have been stressed by chronically low oxygen levels. Terramycin has been incorporated into fish feed for 7 to 10 days to combat these diseases. Fish with parasites can be treated with chemicals by dipping, salting, flushing, and prolonged treatment. Dipping involves dipping the infected fish or eggs in a solution of the desired chemical. Brood stock are usually salted when they are handled or transfered. Fish are dipped in a 3 percent salt solution for a minute or so. Most external parasites drop off, and the fish produces a new coat of protective mucous. In flushing, a chemical is added to the upper end of a trough and flows through and out of the trough. With prolonged treatment, a chemical is added to a container or pond and remains until it breaks down and becomes harmless.

Determining the correct amount of chemical required to treat for parasites is an important step in aquaculture. If, for example, fish in 40-liter aquaria are infected with a parasite and the recommended treatment requires adding 1.0 ppm of a particular chemical, then 40 mg of the chemical will give the desired concentration of 1 ppm in a 40-liter aquarium. Chemical treatment in ponds depends on the cubic meters of water in ponds. One gram of chemical per cubic meter is equivalent to a 1 ppm concentration.

(*d*) *How to increase production.* There are several ways to increase production and subsequent profits. Double cropping, i.e. alternating two crops in the same pond during the year, shows potential. In the South, catfish have been grown during warm months and rainbow trout have been grown during cooler months.

Production has been increased significantly in recent years by increased stocking rates, as many as 20,000 fish per ha, coupled with aeration or the lavish use of water to flush ponds. Flushing ponds helps remove excess fecal wastes and uneaten food to maintain desired oxygen levels. Aeration oxidizes nitrogenous wastes to less harmful by-products.

Polyculture, growing two or more species with differing food habits in the same pond, has been used for hundreds of years in China. A pond has several food niches—benthic, planktonic, higher plants, plus fish food organisms found in the water column. Ideally, various species are stocked to fully utilize all food niches. Researchers at Louisiana State University grew almost 6 metric tons of fish per ha in one growing season with a type of polyculture (Green *et al.* 1978). Channel catfish, confined in floating cages, were fed commercial pellets, and crawfish (*Procambarus clarkii*), bigmouth buffalo (*Ictiobus cyprinellus*), and paddlefish (*Polyodon spathula*) were stocked loose in the same pond. Crawfish and buffalo fed on waste feed that fell through the mesh of the baskets. The feces of catfish enriched the waters and encouraged growth of fish food organisms. Paddlefish fed on the resulting plankton.

16.2.6 Harvesting

Fish should not be fed several days before they are harvested so their digestive tracts are empty when fish are handled. Without food, fish can better withstand handling. Fewer feces are produced when fish are hauled in tanks; there will be fewer problems caused by bacteria. During harvest pond water is gradually removed, although growers must make sure that oxygen supplies are not depleted as the water level drops. Fish are usually harvested by pulling seines by hand or with tractors or winches through the water. It is best to harvest fish during cool weather since cool water holds more oxygen and fish are less active. Experienced farmers may harvest year round to meet market needs. In ponds with a smooth bottom, 70 to 90 percent of the fish can be seined out when the pond is half filled with water. The remaining fish are easily caught when they collect in a harvest pit at the drain.

16.2.7   Marketing and processing

Catfish farmers have several possible markets. Fingerlings and brood fish can be marketed to other farmers. Eating size fish can be sold to operators of pay lakes who charge sportsmen for fish they catch. Fish can be sold wholesale to cooperatives, jobbers, and processors. Some farmers sell directly to local customers and restaurants.

Approximately 50 percent of the eating size fish are sold to pay-lake operators and a large percentage is sold to processing plants. Live fish are delivered to both outlets. Processing plants usually stun fish with electricity, remove the head with a band saw, remove the entrails, and then remove the skin. The dressing percentage is about 60 percent, i.e., 60 percent meat and 40 percent waste.

Catfish, and finfish in general, are marketed in various forms or cuts. Most are sold as whole dressed fish. Catfish heads are removed, but the head is left on trout and salmon. Dressed fish may be filleted or steaked and sold fresh or frozen. Iced fish have a maximum shelf life of 11 or 12 days, while refrigerated fish have a shelf life of 8 days. Frozen fish packaged in polyethylene bags and frozen in a blast freezer have a shelf life as long as 12 months.

16.3   CULTURE OF TROUT AND SALMON

Trout and salmon are typical coldwater species. While catfish are grown in ponds at low densities, trout are typically grown in raceways with constantly flowing water at high densities. Salmon are cultured in high population densities in floating pens or cultured by ocean ranching.

16.3.1   Trout culture

(*a*) *Culture Facilities.* Though trout may be cultured in ponds, most commercial operators use concrete or earthen raceways (Fig. 16.3). Raceway dimensions vary, but most are 3 to 10 meters wide, 0.6 to 1.0 meters deep, and up to 2 km long. Raceways should not be deeper than 1 meter so all water is circulated and solids don't build up on the bottom. Raceways can be deeper than 1 meter if a vertically sliding gate is added at the water discharge end to flush out collected bottom solids.

(*b*) *Water supply.* The water should be saturated with dissolved oxygen, water temperature should range from 14 to 18°C, have a pH of 7.0 to 7.5, be relatively clear, and have a total alkalinity and total hardness of at least 50 ppm. An abundant supply of water should also be available at all times. Water should

Fig. 16.3. Raceway for growing trout. Photo courtesy U.S. Department of Agriculture, photo by Dick Gooby.

flow at 5 liters per second per ton of fish (Sedgwick 1976). This flow rate will maintain desired oxygen levels and will remove wastes. Too rapid a water flow, however, means that fish will expend more energy trying to maintain position, and will decrease feed conversion efficiency.

(*c*) *Fertilization and incubation of eggs.* Unlike catfish spawning where males choose their mates, rainbow trout are hand-stripped of their eggs and milt. The female's eggs are collected in a pan, and milt from a ripe male is then stripped into the pan. One male can provide enough milt to fertilize the eggs from several females. Eggs and sperm are gently mixed to ensure fertilization.

Hatching requires gently flowing water which contains at least 7 ppm of dissolved oxygen. There are several methods for incubating eggs, but most hatcherymen place eggs in incubating trays. Trays may be stacked like drawers in a dresser and they can be removed easily. The water supply allows eggs in all trays to receive a gentle flow of water. A flow of 1 liter per minute per 3000 eggs is needed, but it may be necessary to have a flow rate of 1 liter per minute per 1500 eggs (Sedgwick 1976). Dead eggs may have to be removed daily to prevent the spread of fungus, but some hatcheries use fungicides. Eggs hatch in 21 to 23 days at water temperature of 13 to 14°C.

(*d*) *Rearing fish*. Newly hatched sac fry are removed from trays and are stocked in troughs or tanks. After fry absorb yolks, they are fed several times a day. As fry grow, they may be thinned, sorted by size, and restocked in troughs or raceways. Fry in troughs or raceways are fed approximately 4 percent of their bodyweight daily. Depending on diet, water temperature, and other factors, market-size trout (20 to 35 cm long) can be produced in 7 to 14 months.

Parasites and diseases are common problems. Bacterial infections such as *Aeromonas salmonicida* and *A. hydrophilia* may be controlled with sulfamerazine doses of 200 mg per kg of live fish for 3 to 4 days. The material is incorporated into feed. Several common protozoan parasites include *Myxosoma cerebralis* or whirling disease, *Costia necatrix*, and *Hexamita truttae*. Whirling disease is difficult to control in fry fish. *Costia* is controlled by bathing fish in 250 ppm formalin for 1 hour. *Hexamita* may be controlled by adding 2 percent calomel to feed for 4 days. Viral diseases include infective pancreatic necrosis, infective haematopoitic necrosis, and viral haemorrahagic septicaemia. There are no cures for viral diseases and infected fish should be destroyed.

(*c*) *Marketing and processing*. Trout are marketed to fee-fishing establishments or are processed for the food market. There are an estimated 1000 to 2000 fee-fishing operations in the United States (Brown 1977). These operations may buy catchable fish or they may raise purchased fingerlings to catchable size. Processed trout are usually drawn and sold with the head left on. The restaurant trade prefers dressed fish weighing 225 to 335 grams.

16.3.2 Salmon culture

(*a*) *Pen culture*. The coho salmon and other salmonids are being raised in floating pens (Fig. 16.4). There are three phases in this method: 1) fresh water incubation of eggs in a hatchery, 2) rearing fry in fresh water tanks, and 3) growing fish to market size in salt water pens.

Salmon incubation procedures are similar to those used with trout. Incubation of coho salmon eggs requires about 60 days at water temperatures of 9 to 13°C. Silt deposition, low water temperatures, and flow rates can affect incubation. After hatching, fry feed on yolks for about 30 days and are then moved to tanks or raceways. Here they are fed commercial dry or moist pellets at rates of 3 to 7 percent of their bodyweight. Fish 5 or 6 months old are then stocked in salt water pens where they are fed daily until they reach market size.

Coho salmon may be sold when pan-sized. Fish are chill-killed, dressed with heads on, gutted, glazed, and placed in individual plastic bags. Frozen fish are marketed at various sizes ranging from 170 to 400 grams. Some culturists market fillets from larger fish.

As with any new type of aquaculture, growers who culture salmon in pens

Fig. 16.4. Floating pens used in culture of salmon. Photo: James W. Avault, Jr.

face several problems. Growers still depend on surplus eggs from state hatcheries, but they are now developing their own brood stock through ocean ranching. Diseases, especially the bacterium *Vibrio*, pose problems, and the U.S. FDA has approved only a few chemicals for aquaculture use. Site availability could be a problem because growers must use public waters. Legal restrictions and zoning ordinances must be considered. Feed must be available at a reasonable cost. National and state regulations will eventually control pollution from pen culture.

(*b*) *Ocean ranching.* Pacific salmon traditionally forage in the ocean and return to the stream from which they hatched when sexually mature. This homing instinct is the basis for ocean ranching. In ocean ranching, fish are spawned and hatched in hatcheries, grown in confinement, and released when they can make their way to the ocean. Fish return when they are ready to spawn. The best fish are used as brooders while surplus fish are dressed and marketed.

Ocean ranching is practiced in California, Washington, Oregon, and Alaska, using five species of salmon. The more important species are the pink salmon (*O. gorbusha*) and the chum salmon (*O. keta*).

The hatchery is the major facility. Spawning methods are similar to those used for trout. Salmon fry are confined in tanks where they grow for periods ranging from a few days or as long as a year before they are released. Growers feel that the survival rate of salmon increases when young are held in confinement for a relatively long time before release. The fish have a longer time to become imprinted when held for longer periods.

Although thousands of young may be released, often fewer than 1 percent survive and return. Hasler (1975) improved these odds. He noted that the homing instinct was based on the ability of salmon to recognize familiar smells. He added morpholine, an acrid-smelling liquid, to water where fish were confined and imprinted them to that smell. When fish eventually returned from the ocean, morpholine was again dripped into water where fish were released. In one study, 10 times as many salmon imprinted with morpholine returned as did non-imprinted fish.

16.4   CULTURE OF OTHER FISH SPECIES

Several other fish species are cultured throughout the world, mostly in ponds. Culture is usually extensive—fish feed mainly on pond organisms—as opposed to intensive—fish stocked at high rates and fed. Four important cultured fish species or groups are milkfish, *Tilapia* sp., various carps, and eels (Fig. 16.5).

The milkfish is well suited for culture. Milkfish are euryhaline, disease resistant, fast growing, and feed low on the food chain. Fry must, however, be captured along sandy beaches. Fry dealers purchase fry, and several middlemen may be involved before fingerlings are stocked into growing ponds. Fish feed on algae beds fertilized with chicken manure and inorganic fertilizers.

There are close to 100 known species of *Tilapia*, and about half a dozen species are now cultured. *Tilapia* are very hardy tropical fishes that can be fed agricultural wastes. *Tilapia* may spawn when only a few months old and produce numerous young, thus causing an overcrowded population. Stocking only males avoids this problem. Certain hybrid *Tilapia* crosses produce all males.

The Chinese culture several species of carp including the silver carp (*Hypophthalmichthys molitrix*), big head carp (*Aristichthys nobilis*), and the grass carp. Grass carp were introduced into the United States to control aquatic weeds. The catla (*Catla catla*), rohn (*Labeo rohita*), and mrigal (*Cirrhinus mrigala*) are important carp species in India.

Eel culture is widespread in Japan and is important in Europe. Wild young eels, elvers, are caught, confined in small ponds, and fattened with trash fish or a prepared ration containing primarily fish meal.

16.5   CULTURE OF CRAWFISH

Several crustaceans are cultured in the United States but crawfish are the only crustaceans cultured on a large scale. Crawfish culture is an example of extensive

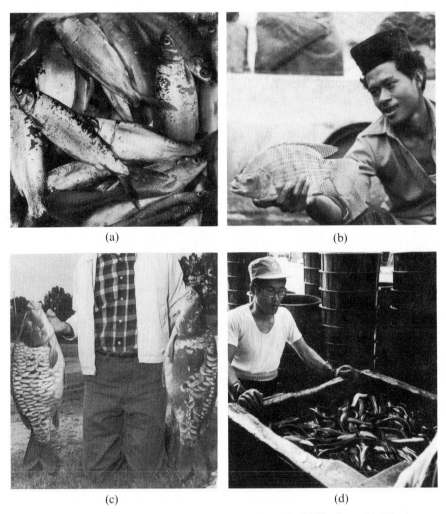

(a)   (b)

(c)   (d)

Fig. 16.5. (a) Eating-size milkfish, about 400 grams, from the Philippines; (b) *Tilapia* sp. at fish market in Indonesia; (c) A strain of carp (*Cyprinus carpio*); (d) Eels being readied for processing near Kyoto, Japan. Photos: James W. Avault, Jr.

aquaculture where the cultured species forage on aquatic vegetation. Crawfish farming, though practiced in several states, is primarily centered in Louisiana, where 18 million kg of wild and cultured crawfish were marketed in 1978. The red swamp crawfish (Fig. 16.6) and the white river crawfish (*P. acutus acutus*) are the species cultured, usually mixed together. The red swamp crawfish comprises approximately 90 percent of total production. A number of countries in Europe—Spain, Austria, Sweden, France, and West Germany, to name a few— have begun small scale crawfish farming. The red swamp crawfish has been successfully cultured in warm climates, as in southern Spain. The signal crawfish (*Pacifastacus leniusculus*) from California does well in colder areas.

Fig. 16.6. Red swamp crawfish. Photo: James W. Avault, Jr.

16.5.1 Culture ponds

Crawfish are grown in rice fields, open ponds, and in wooded ponds containing brush and trees left after construction of ring levees. Land used for wooded ponds is usually too boggy for agricultural or other commercial purposes. Ring levees, constructed with drag lines, are usually filled with water from bayous with a low lift pump. Open ponds are clear of brush and trees. Crawfish are cultured in rice field ponds after the rice is harvested.

16.5.2 Water supply

Most crawfish farmers use water from streams to fill ponds. Such water often contains low amounts of dissolved oxygen, high amounts of silt, and may also contain contaminants and trash fish. Aeration by splashing water on a wooden platform and screening water will help eliminate some of these problems. Ground water from wells is the best source of water and most rice farmers use water from this source. All water, regardless of source, should contain at least 3 ppm oxygen, have a total hardness of 50 to 100 ppm, and have a pH of 6.5 to 8.5. Water need be only 45 cm deep in wooded and open ponds. Most rice ponds contain only 15 cm of water, a level which is also suitable for crawfish.

16.5.3 Seeding ponds

The same seeding procedure is used for all three types of ponds. Brood crawfish are stocked in June at about 30 kg/ha. The stocking rate depends on the amount of protective vegetation and whether native crawfish are present. Once stocked, crawfish begin to burrow underground. Slowly removing water encourages further burrowing. Draining ponds also helps control fish and other predators and will help vegetation regrow.

Crawfish usually mate in the summer, although mating may take place at other times. The female stores sperm in a seminal recepticle until eggs are laid in the fall. Females produce about 300 eggs which are attached to the undersides of their tails by a sticky substance called glair. Females with eggs are said to be 'in berry'.

16.5.4 Forage for crawfish

Crawfish farmers do not feed a commercial ration but plant or encourage the growth of native aquatic plants that crawfish eat such as alligatorweed (*Alternanthera philoxeroides*), water primrose (*Ludwigia* sp), and smartweed (*Polygonum* sp). Millet may be planted also.

16.5.5 Reflooding ponds

Ponds are reflooded when young are found in burrows, usually in September and October. Brood crawfish and young emerge from burrows and begin feeding on tender green plants and on periphyton. Crawfish also feed on detritus and rice straw. Rice straw, or any organic matter, must contain 17 units of carbon or less to each unit of nitrogen before crawfish derive nutritive value from these substrates.

16.5.6 Harvesting

Funneled traps made of 19 mm mesh chicken wire are used to harvest crawfish (Fig. 16.7). Fish such as gizzard shad (*Dorosoma cepedianum*) are used as bait.

Fig. 16.7. Chicken wire trap with 19 mm mesh used for trapping crawfish. Photo: James W. Avault, Jr.

Approximately 25 traps per ha are baited, set, and checked daily. Trapping begins in November and continues until June, though the trapping season may be extended depending on the catch, market price, and other factors. Brood stock seeded in June are harvested first. Large, dark crawfish are easily distinguished from the light-colored young crawfish. First-caught crawfish should be examined to be sure that the tail is full of meat. If not, harvesting should be postponed. Brood crawfish with scanty tail meat are known as 'hollow tails'.

Harvesting will initially yield up to a 0.5 kg per trap or 12.5 kg per ha per trap day. The catch of crawfish drops later as brooders are removed and as water temperature drops during winter. Crawfish become lethargic when temperature drops below 12°C. Harvest resumes the following year around February to catch young.

Crawfish meat becomes tough during mid-June and the remaining crawfish begin to burrow. This is the signal for farmers to remove the water from the ponds. Since holdover crawfish are the brood stock for next year's crop, crawfish don't have to be restocked.

Harvesting is often done by professional trappers who may receive half the gross catch. Trappers may furnish traps, bait, boats and fuel. Though this method appears to be ineffectual, crawfish farming is profitable. Wooded ponds yield the lowest returns and may produce only an average of 200 kg/ha. Even so, very little management is required and costs are low. Open ponds or rice field ponds may produce more than 1000 kg/ha.

16.5.7   Marketing and processing

Crawfish are marketed live or as peeled tail meat. Most crawfish are marketed through one of approximately 40 processing plants in southern Louisiana. Early crawfish sold during winter months bring best prices. Prices drop as wild crawfish, which comprise 60 percent of the total crop, enter the market. At the processing plant, crawfish are held in porous sacks and stored in coolers at temperatures of about 3°C. The next day, live crawfish are either removed, washed, and sold to live markets or are readied for peeling.

Peeling crawfish are first blanched for about 5 minutes in boiling water that contains no spices or additives. Blanching kills the crawfish and bacteria and facilitates manual peeling. The par-boiled (blanched) crawfish are processed by manually separating the meat from the shell. A recently invented peeling machine should eventually replace much of the hand labor. Peeled tail meat is packaged in 0.45 kg bags and is iced or frozen. Refrigerated crawfish have a shelf life of about 5 days.

16.5.8   Double cropping rice/crawfish

Interest in double cropping rice and crawfish has increased recently. Rice is planted in March; crawfish are seeded in June and begin burrowing. Water is drained in August to facilitate rice harvest, and rice stubble is flooded in September. Crawfish feed on decaying rice stubble and on microorganisms associated with decaying organic matter. As much as 1500 kg/ha can be produced. Herbicides and most fungicides used in rice production appear to be relatively non-toxic to crawfish, but most insecticides are highly toxic. Approximately 240,000 ha of rice are grown in Louisiana and well over 1 million ha are grown in the United States, so this type of double cropping could be expanded.

16.6   CULTURE OF OTHER CRUSTACEANS

16.6.1   Freshwater prawns

Freshwater prawns, also known as river shrimp, belong to the genus

Macrobrachium, containing more than 100 species. The giant Malaysian prawn, *M. rosenbergii*, is the most widely cultured of these species (Fig. 16.8). When marketed, prawns weigh about 100 grams and are a gourmet food. Culture is complicated by the many life stages of the prawns, each of which requires special conditions.

Healthy mature prawns can breed year-round in captivity. Ripe ovaries of females are large, orange-colored masses in the dorsal and lateral parts of the cephalothorax. Mature females, weighing about 80 grams, produce about 60,000 eggs and can lay eggs twice within 5 months. Eggs, attached to the female's pleopods, hatch in about 19 days.

Fig. 16.8. Malaysian prawn. Photo: James W. Avault, Jr.

After eggs hatch, larvae go through 11 developmental stages during which they are planktonic and feed constantly. Living foods, mainly *Artemia* nauplii, are kept in culture tanks with the prawns. After approximately 17 to 18 days, larvae metamorphose to post larvae and settle to the bottom.

Juvenile prawns, 60 days old, may be stocked in ponds or raceways and fed commercially prepared crustacean rations, chicken feed, or other materials. Production in kg/ha has not been well established, but over 1000 kg/ha can be produced in ponds. Substrate or hiding places must be provided. Cannibalism is a problem because newly molted prawns are virtually defenseless.

Salinity and water temperature are critical parameters for survival of prawns.

Larval and post-larval stages require brackish water, containing an optimum 12 to 13 ppt salinity while adults grow best in fresh water. Temperature appears to be a more important factor than water salinity. The acceptable temperature range for adults and juveniles is about 28°C to 33°C, and for larvae is 24° to 30°C. Total water hardness should be 50 to 100 ppm.

Temperatures in the United States are generally too low for prawn culture. Intensive culture systems utilizing heated water will work but may not be economically feasible. Most commercial operations are located in Central America and in areas where temperatures are warm year-round.

Malaysian prawns are highly perishable. The best processing techniques chill-kill prawns in ice containers, blanch, and then re-ice. The entire process requires less than a minute per batch. Prawns blanched at 150°C for 15 seconds have a shelf life of about 6 days.

16.6.2   Penaeid shrimp

Shrimp in the genus *Penaeus* are farmed in two basic ways. In India, the Philippines, Taiwan, Ecuador and elsewhere, young ocean shrimp enter estuaries where they are netted on clusters of small plants. The shrimp are then moved to nurserys and grow-out ponds. Shrimp may also enter ponds directly with the tides. A more intensive method is used in Japan, where newly hatched larvae are cultured in tanks until they grow to eating size. Commercial crustacean rations are fed to juveniles. Shrimp can also be stocked and managed in ponds.

Although shrimp have been mated in captivity to obtain young, gravid females from ocean waters are usually brought into the laboratory to hatch young. Females may produce from 30,000 to more than 150,000 eggs, depending on the species and other factors. Eggs hatch to the first larval stage, nauplius, within 14 to 15 hours at a water temperature of about 28°C. The nauplius stage does not feed but lives off its yolk. The second stage, zoea, feeds on diatoms and other plankton cultured in the lab. The mysis stage feeds on *Artemia*. The next stage, post larvae, can then be stocked in grow-out facilities.

There is great market demand for shrimp, which are usually sold according to size or grades based on the number of shrimp per kg. A kg may contain 33 or fewer jumbo shrimp or 132 or more of the smallest grade. At present in the U.S., the cost of culture cannot compete with harvest of wild shrimp. Most shrimp culture operations are located in Central America and elsewhere.

16.7   CULTURE OF OYSTERS

Oyster species can be divided into two arbitrary groups. The flat oysters have both shells flat and include two commercially important species in the genus *Ostrea*. The cup-shaped oysters belong to the genus *Crassostrea*, with four commercial species. Members of this genus have flat uppershells away from the substrate and cup-shaped shells next to the substrate. The American oyster (*C. virginica*) has perhaps received the most attention. Galtsoff (1964) studied its life history and Bardach *et al.* (1972) and Iverson (1968) reported on its culture. American oysters are currently harvested from Canada to Mexico. Leading oyster growing states are Virginia, Maryland, Louisiana, and Florida. Two important species on the west coast of the U.S. are the native Olympia oyster (*Ostrea lurida*) and the Japanese oyster (*C. gigas*).

16.7.1   Water requirements
The American oyster can live in water temperatures as low as 1°C during winter in northern states and water temperatures as high as 36°C in Gulf states. Maximum rate of ciliary activity (feeding) occurs in water temperatures from 25 to 26°C (Galtsoff 1964). Ciliary movement declines at temperatures higher than 32°C. Successful mass spawning and setting occurs at temperatures of 20°C and above. The American oyster can tolerate salinity ranging from 5 to 30 ppt, but gonadal development and growth are inhibited at salinity levels outside this range.

16.7.2   Reproduction and life history
American oysters start spawning whenever the proper temperature is reached. One female may release as many as 100 million eggs. Fertilization by the male is external. Oysters in the genus *Ostrea*, however, brood eggs within the branchial chamber and release the larvae in the veliger stage (see Chapter 4). Fertilized eggs hatch in a few hours. The free swimming larvae (veligers) are weak swimmers and drift about in the currents for 2 to 3 weeks. Eventually larvae use their muscular feet and attach to a clean surface, called cultch. After attachment, larvae enter the 'spat' stage and begin shell development. Oysters reach sexual maturity in about a year, but require as many as 4 years to reach market size.

16.7.3   Methods for culturing oysters
(*a*) *Bottom Culture.* Bottom culture is an extensive process in which appropriate cultch is provided for the settling larvae. Some farmers use clean oyster shells on the bottom for cultch but posts, string, and other materials are also used. The oysters may remain in this settling area until they reach market size, but this area is usually too crowded, so spat oysters are moved to another growing location. Some farmers may periodically thin down the oyster population and redistribute them over a larger area.

Predation of young oysters by crabs and certain fish is often a problem when oysters are grown on the bottom. Oyster drills, such as *Urosalphinx cinerea* and *Eupleura caudata*, and the common starfish (*Asterias forbesi*) are major predators. Chlorinated hydrocarbon pesticides have been used to control these predators, but growers are concerned that filter-feeding oysters may concentrate these chemicals in their flesh.

There are other ways to control predators. An underwater plow may bury oyster drills, starfish, and others before seeding oysters. Suction devices have been used to remove predators and competitors. Oysters can also be dipped in a saturated solution of rock salt and then exposed to air for a while to kill most shellfish predators without harming oysters. Scuba divers may monitor oyster beds. Starfish mops, long iron beams with bundles of rope yarn, drug across the beds, will entangle starfish for removal. Quicklime applied through pumps at 300 kg/ha will control starfish.

Accumulated silt harms oysters, particularly young oysters. Proper site selection is important, but oysters may be moved periodically to another bed. Some growers routinely clean silt from oyster beds with suction dredges or high-pressure water jets.

Even though bottom culture of oysters appears to be a crude method, yields of 1000 kg/ha have been obtained by growers in Long Island Sound. Controlling siltation and predators could increase production in this area to 5000 kg/ha. Control of pollution from both industry and domestic wastes is equally important. Oyster grounds have been closed at an alarming rate because of pollution.

(*b*) *Off-bottom culture.* Off-bottom culture, in which oysters are suspended in the water column on hanging strings or sticks (Fig. 16.9), has several advantages. Growers have the option of moving rafts to better waters during pollution or predator problems or when plankton-rich waters are accessible. Bottom predators, competitors and siltation are less of a threat in off-bottom culture than in bottom culture. Oysters can utilize the entire water column when strings or sticks are used. Rafts place oysters in the upper strata of water where the most plankton (food) is found. Oysters can be grown in off-bottom cultures in about $2\frac{1}{2}$ years in northern states, approximately half the time needed to grow oysters on the bottom.

Off-bottom culture is not widely practiced in the United States because rafts or other cultural units may interfere with navigation and may cause other legal, social, and political problems. Off-bottom culture requires more labor which increases production costs. Some countries, however, such as New South Wales, France, the Netherlands, Japan, and Australia, culture oysters off-bottom on a large scale. In Japan, at least eight oyster species are cultured, with *C. gigas* the most important species.

Most Japanese growers purchase seed oysters from specialists in seed production. Oysters are cultured by raft method or long-line method. Bamboo rafts are lashed together in two layers at right angles to each other. Tarred ropes strung between two parallel pairs of floats are used in long line culture. Oysters can be harvested after 1 or 2 years old when they are cultured by these methods.

(a)

(b)

Fig. 16.9. (a) String culture of oysters Photo: James W. Avault, Jr. (b) Fisherman with seed mussels ready to transplant onto bouchot, Charron, France. Photo: Courtesy Hurlburt and Hurlburt (1975).

In New South Wales, spat of *C. commercialis* attach to tarred sticks which have been placed on rocks in estuarine areas. After spat fall, the sticks are moved to low salinity waters for approximately 2 years. Oysters are then scraped loose and placed in chicken-wire trays for another year before they are harvested.

16.7.4   Marketing and processing

Market oysters may be separated into three basic groups, though groups may vary among countries. In the United States, the largest and fattest oysters are usually sold to shucking houses where they are placed into containers. They may remain fresh for as long as 10 days when refrigerated or packed in ice.

Medium, round, and uniform oysters are used for the counter half-shell market. Oysters in the shell are usually sold by the dozen, and may be sold live. Small oysters are steam canned, a process in which salt is added and oysters are heat processed to avoid spoilage.

16.8   CULTURE OF OTHER MOLLUSKS

16.8.1   Mussels

The life cycle of mussels is similar to that of the American oyster. Females spawn 5 to 12 million eggs annually. A ciliated larva develops approximately 4 hours after fertilization and forms all organs after 10 weeks. The free-swimming veliger floats briefly before it attaches itself to a substrate with its byssus. Like oysters, mussels feed on plankton. Mussels are farmed by (1) raft or rope culture in Spain, (2) pole culture in France, and (3) bottom culture in the Netherlands.

(*a*) *Raft Culture.* Raft culture is practiced on the northwest Atlantic coast of Spain, where deep bays or rias extend far inland and are protected from ocean winds and waves (Hurlburt and Hurlburt 1975). Rafts, averaging 23 meters square, are anchored along the sides of bays in about 11 meters of water. Each raft supports up to 700 ropes, each of which is about 9 meters long.

Clumps of young seed mussels are gathered from shoreline rocks during the fall and wrapped onto the ropes. The seeds then attach themselves to the ropes with their byssus, and, in the spring, more seed mussels collect directly on the ropes. Mussels started in the fall mature to a length of approximately 6 cm after a year. Spring mussels mature in 18 months. The mussels have to be thinned down several times during the growing period; they are stripped off the ropes and are wound onto new ropes with netting or string.

A single rope produces over 113 kg of live mussels annually, and a 700-rope raft produces as much as 40,816 kg of drained meat annually. Since one ha of water can support between 7 and 12 rafts, an intensively cultivated area can produce more than 280,000 kg of pure meat per ha per year.

At harvest, the mussel-laden ropes are hoisted aboard a work-boat with a winch. The rope is shaken, and the mussels collect in a mesh basket. Small mussels are rewrapped back onto ropes. Mussels to be cooked and canned go directly to factories. Mussels sold fresh are depurated for 48 hours by pumping sea water in tanks that contain mussels on racks. Chlorine is added and is allowed to evaporate. Clean mussels are placed in 15-kg mesh bags, rinsed, and drained for 3 hours before shipment.

(b) *Pole culture.* The pole or 'Bouchot' method of culturing mussels is used in France where large tides may be 15 meters higher than low water tides. Oak poles are driven into the ocean floor in rows so that the tops of the poles are exposed at low tide. The bottom of the poles are wrapped with a smooth plastic to discourage starfish and crabs from attaching themselves. Seed mussels gather on ropes wrapped around the poles (Fig. 16.9). Mussels grow, are thinned down, and are rewrapped on poles.

Mussels reach harvestable size in 12 to 18 months. After harvest by boat or by workers on foot, the mussel clumps are broken apart, washed, and sized. Marketable mussels are placed in 20-kg sacks and sold fresh.

Large farms may have as many as 75,000 poles, and each pole will yield 9 to 11 kg of live mussel, or about 4.5 kg of meat per year. One ha will yield over 4480 kg of meat annually.

(c) *Bottom culture.* In the Netherlands, mussel seed is dredged up by boat and transplanted to culture plots in water 3 to 6 meters deep. The mussels mature to marketable size in about 20 months. About 80 percent of the mussels produced in the Netherlands are sold fresh in France and Belgium. The remaining 20 percent is canned for wider distribution.

16.8.2 Clams

The hard clam or quahog (*Mercenaria mercenari*) can be found along the Atlantic coast of the United States from Maine to Florida in the intertidal zone to water as deep as 15 meters. The clams, as oysters and mussels, release eggs in the water where they are fertilized. The planktonic veliger drifts about for a week or so. The clam develops a strong muscular foot, siphon, and gills. At this stage, the young clam uses its byssus to secure itself to sand grains or other substrate. The clam eventually uses its foot to dig into the bottom where it remains buried for the rest of its life. Clams may live for 25 years and grow as long as 14 cm. The soft clam's (*Mya arenaria*) appearance and life cycle is similar to that of the hard clam. It is most abundant along the Atlantic coast from Labrador to North Carolina.

The hard clam and soft clam can be farmed using procedures similar to those required for bottom culture of oysters. Clams do not, however, require cultch on which to settle since they settle directly on the bottom. Some farms located in or near natural spawning areas allow larvae to drift naturally into fenced-in areas to settle. Hatchery-grown stock can be purchased if natural seed is not available.

Crushed stone or gravel can be placed on the bottom of the growing area. Clams that burrow under this cover are safe from predatory crabs and fish. A dredger harvests clams by first loosening soil with water pressure and then sucking up clams. Clams are also harvested with hoes and hand labor. Clam culture is not very widespread, but is most prevalent in Japan where some several species are cultured.

16.9   POTENTIAL OF AQUACULTURE

The Food and Agriculture Organization (FAO) of the United Nations stated that world aquaculture production totaled 5 million metric tons in 1973 and increased to 6 million metric tons only two years later (Pillay 1976). Table 16.1 lists production of fisheries products from aquaculture in selected countries. The President's Advisory Committee (1967) projected that pond culture alone will yield 15 million tons by the year 2000. FAO's Indicative World Plan for Agricultural Development (1969) predicted that production would increase to 25 million metric tons by 1985. Other more optimistic predictions note that 40 to 50 million metric tons annually can eventually be produced. By comparison, the world's annual catch of fisheries products from oceans has leveled off at approximately 65 to 70 million metric tons. Of this total, roughly 70 percent is edible and 30 percent is reduced to fish meal.

Fisheries products will play an important role in providing food as world population increases. The world population was estimated at 250 million 2000 years ago. This population doubled by the year 1600 and reached 1 billion around 1830. There were 2 billion people in 1930, and the world population exceeded 4 billion people in 1979. The world population is expected to double again in less than 40 years.

Fisheries products are quality foods that are 90 to 100 percent digestible. They have a much greater ratio of easily digestible muscle protein to connective tissue protein than mammals. Fish contain more polyunsaturated fatty acids and relatively less cholesterol than beef or poultry. Shellfish, such as lobster, contain slightly more cholesterol than other fisheries products. Vitamin-rich fish also contain needed minerals and many trace minerals. Fish are high in protein and contain much less fat than beef and pork.

The governments of several countries have made aquaculture a priority. The Philippines, for example, has made a massive effort to expand fish production through aquaculture. They recognize that highly nutritious fish can be produced more cheaply than beef or pork. Increasing interest in aquaculture caused the FAO to hold a global meeting in 1976 to discuss aquaculture's status and potential. Interest in aquaculture is increasing in the United States also. Several aquaculture bills have been introduced in the House of Representatives and Senate to develop a National Aquaculture Plan.

Table 16.1. Estimated world production through aquaculture in some leading countries in 1975.

	Metric tons
FINFISH	
China—all provinces excluding Taiwan Province	2,200,000
Taiwan—Province of China	81,236
India	490,000
U.S.S.R.	210,000
Japan	147,291
Indonesia	139,840
The Philippines	124,000
Thailand	80,000
Bangladesh	76,485
Nigeria	75,000
Poland	38,400
Vietnam, Republic of	30,000
Yugoslavia	27,000
Romania	25,000
Hungary	23,515
U.S.A.	22,333
SHRIMPS AND PRAWNS	
India	4,000
Indonesia	4,000
Thailand	3,300
Japan	2,779
OYSTERS	
Japan	229,899
U.S.A.	129,060
France	71,448
Korea, Republic of	56,008
Mexico	45,000
MUSSELS	
Spain	160,000
Italy	30,000
France	17,000
Germany, Federal Republic of	14,000
The Netherlands	10,000
CLAMS	
Korea, Republic of	24,920
Taiwan, Province of China	13,898
SCALLOPS	
Japan	62,600
COCKLES AND OTHER MOLLUSCS	
Malaysia (cockles)	28,000
Taiwain, Province of China	1,243
Korea, Republic of	733

Adapted from T.V.R. Pillay, (1976).

Fish are the ideal animals to culture for food. They can be grown on many lands that are unfit for agricultural crops. This is a particularly important advantage in several countries where there is little, if any, potential arable land available for agricultural crops. Fish can also be rotated with land crops and can convert agricultural wastes to fish flesh. The feed conversion efficiency of fish is considerably better than that of other animals. Cold-blooded fish do not waste energy maintaining body temperature as do mammals and poultry and conserve more energy through their natural buoyancy in water.

Aquaculture production can be increased dramatically on existing operations and by expanding aquaculture to new areas. Yields can be increased on existing operations by carefully selecting cultured species with emphasis on species low on the food chain. Production from species close to the first link in the food chain, such as filter-feeding mollusks, can increase more than production from species that feed further up on the food chain. Polyculture utilizes all food niches in ponds. Yields of 10 tons per ha per year have been achieved through polyculture in India and Israel (Avault and Smitherman 1976). Intensive culture involving high density stocking, feeding, and aeration has produced as much as 25 tons per ha annually in Israel.

Double cropping increases food production. Catfish may be cultured during warm months and harvested in the fall. Instead of leaving ponds idle, a cold water species such as the rainbow trout can be cultured during winter months. Some fish can be double cropped with an agricultural crop. An example is rice/crawfish double cropping.

A fish crop and agricultural crops can be grown simultaneously. In the Philippines, for example, tilapia or carp are grown together in rice fields (Fig. 16.10). Fish are not fed but production may reach 200 kg/ha. Rice production often increases when fish eat harmful insects, fertilize waters with their feces, and tiller the soil with their feeding activities.

Likewise, animal crops and fish crops can be grown together. In some countries hog wastes are washed into ponds where certain *Tilapia* species feed directly on wastes. These wastes also fertilize pond water. Ducks are also grown with fish in several countries, with duck feces providing feed and fertilizer. As much as 3500 kg of fish per ha are produced from duck feces in Taiwan.

In the United States, the thermal effluents from power plants could result in winter aquaculture in some areas. Catfish production has increased dramatically when ponds are aerated continuously. Previous pond management techniques

---

Fig. 16.10. Examples of mixed agriculture and aquaculture. (a) Fish and rice being grown together in Philippines; (b) Hogs being grown on pond banks of fish ponds in Philippines. Fresh manure of up to 40 hogs can be used as feed/fertilizer for each ha of water; (c) Ducks grown in fish ponds in Philippines; up to 500 ducks can be grown per ha of water. Photos: James W. Avault, Jr.

(a)

(b)

(c)

yielded 1 metric ton of catfish per ha. Farmers in the Mississippi Delta are now producing as much as 6 metric tons per ha with aeration.

Aquaculture could be expanded to millions of ha of land. In Asia, except for rice land, none of this land can now be used to produce traditional land crops because of various problems such as acid soils. There are 11 million ha of swamps and floodlands, 963,000 ha of freshwater reservoirs, 85 million ha of irrigated rice fields and 1 million ha of brackish and saltwater swamps in Asia that could be used for aquaculture. There are extensive saline swamps and estuaries in Africa and Latin America that could increase aquacultural production by at least 10-fold.

In summary, aquaculture will play a significant role in future food production. Many countries now rely on aquaculture to provide much of their needed high quality food. Fish is cheaper to produce than beef or pork, can convert wastes to fish flesh, grow well on poor lands, and can be integrated with both field and other animal crops. If present trends continue, fish production from aquaculture will play an ever more important role in world food production.

Food production must increase dramatically in the decades ahead to adequately feed an increasing population. We now also confront problems resulting from the depletion of non-renewable resources. The actions of governments, researchers, and growers will determine whether the potential production from aquaculture will be fully utilized to meet these challenges.

16.10   REFERENCES

Avault James W.Jr. and R.O.Smitherman (1976) Pond culture of fin fish. *In* T.V.R.Pillay (ed.), *FAO Tech.Conf. on Aqua. FAO Fish. Rep. No.* 188. pp.6–8.

Bardach John E., John H.Ryther, and William O.McLarney (1972) *Aquaculture, the Farming and Husbandry of Freshwater and Marine Organisms.* New York: Wiley-Interscience.

Brown E. Evan (1977) *World Fish Farming: Cultivation and Economics.* Westport, Connecticut: The Avi Publishing Company.

FAO (1969) *Provisional Indicative World Plan for Agricultural Development.* vols. 1 and 2.

Galtsoff P.S. (1964) The American oyster. U.S. Fish and Wildlife Service, *Fish. Bull.* **64**.

Glude John B., (ed.) (1977) NOAA aquaculture plan. Prepared by *NOAA, NMFS, Office of Sea Grant.*

Green Lee M., James S. Tuten, and James W.Avault, Jr. (1978) Polyculture of red swamp crawfish (*Procambarus clarkii*) and several North American fish species. *In Proc. Fourth Inter. Symp.* of the *Inter. Assoc. of Astacology*, **5**, 287–298.

Hasler Arthur D. (1975) How the salmon comes home. *University of Wisconsin Sea Grant College Program* WIS-S G-76-361.

Hurlburt C. Graham and Sarah W.Hurlburt (1975) Blue gold: mariculture of the edible blue mussel (*Mytilus edulis*). *Marine Fish. Rev.* **37**, 10–18.

Iverson E.S. (1968) *Farming the Edge of the Sea.* London: Fishing News (Books).

Lawrence J.M. (1949) Construction of farm fish ponds. *Agri. Exper. Sta. of the Al. Polytechnic Inst. Cir. No. 95.*

Ling S.W. (1974) Keynote address. *In* J.W.Avault, Jr. (ed.), *Proc. Fifth Annu. Work. World Mariculture Soc.,* vol. 5, pp.19–25. Louisiana: Louisiana State Univ., Division of Continuing Education.

Pillay T.V.R. (1976) The state of aquaculture, 1975. *In* T.V.R.Pillay (ed.), *FAO Tech. Conf. on Aqua. FAO Fish. Rep.*

President's Science Advisory Committee on the World Food Supply (1967) *The World Food Problem* (3 vols.). Wash., D.C. Superintendent of Documents.

Sedgwick S. Drummond (1976) *Trout Farming Handbook* 2nd ed. Great Britain: Hollen Street Press.

Index